Two Manuals Combined into One Book

Physical Security

And

Military Working Dogs

Department of Defense

Pentagon
Publishing

***FM 3-19.30**

Field Manual
No. 3-19.30

Headquarters
Department of the Army
Washington, DC, 8 January 2001

PHYSICAL SECURITY

Contents

Distribution Restriction: Approved for public release; distribution is unlimited.

*This publication supersedes FM 19-30, 1 March 1979.

Page

Page

Preface

This field manual (FM) sets forth guidance for all personnel responsible for physical security. It is the basic reference for training security personnel. It is intended to be a "one-stop" physical-security source for the Department of Defense (DOD), the Department of the Army (DA), and other proponents and agencies of physical security.

Prevention and protection are the two primary concerns of physical security. Both serve the security interests of people, equipment, and property. These interests must be supported at all staff and command levels; and this support must be unified in joint, multinational, and interagency operations.

Support to joint, multinational, and interagency operations relies on the fact that the Army will not conduct operations alone. Additionally, force-projection operations conducted by the military will involve the integration of war-fighting capabilities with stability and support operations. This manual's primary focus is the articulation of a balanced understanding of physical security for joint, multinational, and interagency operations throughout the environments of peacetime, conflict, and war (whether in the continental United States [CONUS] or outside the continental United States [OCONUS]).

Physical security must integrate the various capabilities of joint, multinational, and interagency operations in pursuit of a seamless connection between the strategic, operational, and tactical levels of war. Physical security must also address an expanded range of threats that embraces not only traditional threat components of war, but also nontraditional threats generated by guerrillas, terrorists, criminals, and natural or man-made disasters. In addition, physical security must address the concept of Homeland Defense due to the aforementioned threats.

Homeland Defense is the military's role in the United States (US) government's principal task of protecting its territory and citizens. This is accomplished by joint, interagency, and multijurisdictional organizations. Homeland Defense includes—

- Supporting domestic authorities for crisis and consequence management with regard to weapons of mass destruction (WMD).
- Protecting national-security assets (such as installations) and deploying forces and ensuring the availability, integrity, and adequacy of other critical assets.
- Deterring and defending against strategic attacks while maintaining freedom of action through antiterrorism and force-protection operations.

With this in mind, it is essential to address the five pillars of force protection—combating terrorism, physical security, personal security, law enforcement, and operations security (OPSEC). Physical security is a central component of force protection and provides an integrated venue to express support for operations. Physical security is a primary-leader task and an inherent part of all operations to protect soldiers, family members, civilians, and resources. This function directly supports the Army's universal task list.

While the effects of these changes (when viewed individually) appear revolutionary, the basic activities remain relatively unchanged, though executed under different conditions and standards. Another component that remains unchanged is our reliance upon quality soldiers and leaders well versed in physical-security fundamentals. Leaders will be challenged to ensure that they are functionally proficient; possess an understanding of physical-security operations; are educated in joint, multinational, and interagency operations; and have the ability to perform physical-security functions in support of full-dimension operations.

Appendix A contains an English-to-metric measurement conversion chart. Appendix B is a sample installation crime-prevention handbook. This handbook is designed to assist commanders in developing crime-prevention programs for their installation and units.

The proponent of this publication is HQ TRADOC. Send comments and recommendations on DA Form 2028 directly to Commandant, US Army Military Police School (USAMPS), ATTN: ATSJ-MP-TD, Directorate of Training, 401 Engineer Loop, Suite 2060, Fort Leonard Wood, Missouri 65473-8926.

Unless this publication states otherwise, masculine nouns and pronouns do not refer exclusively to men.

Chapter 1

Physical-Security Challenges

Physical security is defined as that part of security concerned with physical measures designed to safeguard personnel; to prevent unauthorized access to equipment, installations, material, and documents; and to safeguard against espionage, sabotage, damage, and theft. As such, all military operations face new and complex physical-security challenges across the full spectrum of operations. Challenges relative to physical security include the control of populations, information dominance, multinational and interagency connectivity, antiterrorism, and the use of physical-security assets as a versatile force multiplier.

OVERVIEW

1-1. Reductions in manpower and funding are critical challenges to physical security. Manpower for supporting physical-security activities is reduced through deployments and cutbacks. The rapid evolution of physical-security-equipment technology also lends to physical-security challenges, which are exponentially multiplied by the introduction of the information age.

1-2. Physical-security challenges must be understood, and measures must be taken to minimize them to enhance force protection. Leaders must create order when coming upon a situation; and when they depart, some semblance of that order must remain. They must be aware of the human-dimension factors and ensure that their soldiers do not become complacent. It was human error rather than modern technology that took lives in the bombings of the African embassy. Warning was given, but not heeded. Complacency became a physical-security challenge.

AUTOMATED INFORMATION SYSTEMS

1-3. Success on past battlefields has resulted not so much from technological advances, but from innovative ways of considering and combining available and new technologies as they apply to war fighting. Some of these technologies dealt with disseminating and processing information. For example, the telegraph, the telephone, the radio, and now the computer have redefined the fire-support paradigm.

1-4. As the armed forces move into the technological age, a greater need for physical-security measures is required. The risks associated with automated information systems (AISs) are widespread because computers are used for everything. Army Regulation (AR) 380-19 outlines the requirements that commanders and managers need for processing unclassified and classified information and for securing media, software, hardware, and different systems.

1-5. The threat to AISs and information systems security (ISS) involves deliberate, overt, and covert acts. This includes the physical threat to tangible property, such as the theft or destruction of computer hardware. Also included is the threat of electronic, electromagnetic-pulse, radio-frequency (RF), or computer-based attacks on the information or communications components that control or make up critical Army command and control (C^2) infrastructures. In most cases, the threat's target is the information itself rather than the system that transmits it. The threat comes from a range of sources, including the following:

- Unauthorized users (such as hackers) are the main source of today's attacks, primarily against computer-based systems. The threat they pose to AIS networks and mainframe computers is growing.
- Insiders are those individuals with legitimate access to an AIS. They pose the most difficult threat to defend against. Whether recruited or self-motivated, the AIS insider has access to systems normally protected by ISS against an attack.
- Terrorists once had to operate in the immediate vicinity of a target to gain access to or collect intelligence on that target. The proximity to the target risked exposure and detection. Today, a terrorist can accomplish most target selection, intelligence collection, and preoperational planning by gaining access through a computer network. He can increase his probability of success by using computer systems to reduce his "time on target." Terrorist access to an AIS also increases the threat of critical-data destruction or manipulation. Although his presence would be virtual, the potential for damage to Army C^2 systems could be equal to or greater than that achieved by physical intrusion, especially when used as a force multiplier in conjunction with a traditional terrorist attack. Therefore, while traditional preventive measures are still needed to protect unwanted access to information, the information age has added additional concerns for the commander and new opportunities for those with hostile intent.
- Non-state- and state-sponsored groups provide additional challenges. In many cases, it is difficult to confirm state sponsorship of threat activity against an AIS, no matter how apparent the affiliation might seem. Activists of all persuasions are increasingly taking advantage of information-age technology. Neither AISs nor ISS are immune from an adversary's interest in exploiting US military information systems or disrupting communication infrastructures. The availability of low-cost technology and the proliferation of an AIS increase the risk to the Army by potential adversaries.
- Foreign-intelligence services (FIS), both civil and military, are continually active and are another source of contention concerning information systems. In peacetime, they are increasingly targeted against US commercial and scientific interests, rather than military information. With little effort, this peacetime intrusiveness could easily be refocused on AISs and ISS using a wide range of information operations tactics.

- Political and religious groups are other potential adversaries to AISs and ISS. The world's political climate is diverse and complicated. It embraces traditional mainstream political values, as well as radical religious fundamentalism and political extremism. When political or religious viewpoints also incorporate anti-US sentiment, US information infrastructures (including AISs) are increasingly at risk of penetration or exploitation by these potential adversaries.

1-6. When considering an AIS, physical security is more than just safeguarding the equipment. It includes the following elements:

- Software is marked for each system and secured when not in use.
- Initial logon is password-protected (at a minimum).
- Passwords are a minimum of eight characters, using a mixture of letters and numerals.
- Access to an AIS is allowed only to authorized and cleared personnel (per AR 380-19).

1-7. Classified material is entered and transmitted only on approved devices with the following considerations:

- Approved classified devices are operated in a secured environment.
- Classified devices are secured in appropriate containers when not in use.
- Secure telephone unit–III (STU-III) keys are secured in an appropriate safe when not in use (as outlined in AR 380-19).

1-8. Additional information regarding AISs can be found in ARs 380-5 and 380-19. Required training of personnel working with an AIS is located in AR 380-19.

OPSEC AND THE THREAT

1-9. OPSEC is a process of identifying critical information and subsequently analyzing friendly actions attendant to military operations and other activities. The threat is identified using the factors of mission, enemy, terrain, troops, time available, and civilian considerations (METT-TC). The threat defines the physical-security challenges. Implementing physical-security measures supports OPSEC. Providing soundproof rooms for conducting briefings is a simple but invaluable measure.

1-10. Another issue to consider when evaluating physical-security challenges is what actions to take in case of political implications interfering with physical-security measures. In the devastating event at Khobar Towers, a warning was given but not everyone received it. It took too long to evacuate the building after the warning was issued because a cohesive plan was not in place.

1-11. Commanders can minimize the challenges to physical security through proactive measures. They should periodically change the physical-security posture of their area of responsibility to throw off perpetrators.

Chapter 2

The Systems Approach

Commanders must ensure that appropriate physical-security measures are taken to minimize the loss of personnel, supplies, equipment, and material through both human and natural threats. Commanders commonly exercise those protective responsibilities through the provost marshal (PM) and/or physical-security officer and the force-protection officer. The force-protection officer must coordinate with several different agencies to complete his mission. For example, the Army's Intelligence and Counterintelligence Program (see Appendix C) provides information that will be used to complete the unit's crisis-management plan (see Appendix D).

PROTECTIVE SYSTEMS

2-1. The approach to developing protective measures for assets should be based on a systematic process resulting in an integrated protective system. The protective system focuses on protecting specific assets against well-defined threats to acceptable levels of protection. The system is organized in-depth and contains mutually supporting elements coordinated to prevent gaps or overlaps in responsibilities and performance.

2-2. Effective protective systems integrate the following mutually supporting elements:

- Physical protective measures, including barriers, lighting, and electronic security systems (ESSs).
- Procedural security measures, including procedures in place before an incident and those employed in response to an incident. (These include procedures employed by asset owners and those applied by and governing the actions of guards.)
- Terrorism counteraction measures that protect assets against terrorist attacks.

2-3. The following determinations are made when considering system-development procedures:

- The resources available.
- The assets to be protected.
- The threat to those assets.
- The risk levels applicable to those assets.
- The applicable regulatory requirements for protecting the assets.
- The applicable level of protection for those assets against the threat.
- Additional vulnerabilities to the assets (based on the threat).

SYSTEMS DEVELOPMENT

2-4. AR 190-51, DA Pamphlet (Pam) 190-51, and Technical Manual (TM) 5-853-1 are useful tools for developing protective systems using the systems approach. The key to applying these tools successfully is to use a team approach. A team may include physical-security, intelligence, and operations personnel; the installation engineers; and the user of the assets. It may also include representatives from the multinational, host-nation (HN), and local police as well as the regional security office from the embassy.

ASSETS

2-5. Protective systems should always be developed for specific assets. The goal of security is to protect facilities and buildings and the assets contained inside. The risk-analysis procedure in DA Pam 190-51 is used to identify assets. This procedure is applied to all mission-essential or vulnerable areas (MEVAs) according to AR 190-13. It represents the majority of assets with which DOD is commonly concerned. These assets include—

- Aircraft and components at aviation facilities.
- Vehicle and carriage-mounted or -towed weapons systems and components at motor pools.
- Petroleum, oil, and lubricants (POL).
- Controlled medical substances and other medically sensitive items.
- Communication and electronics equipment; test, measurement, and diagnostic equipment (TMDE); night-vision devices (NVDs); and other high-value precision equipment and tool kits.
- Organizational clothing and individual equipment stored at central-issue facilities.
- Subsistence items at commissaries, commissary warehouses, and troop-issue facilities.
- Repair parts at installation-level supply activities and direct-support (DS) units with authorized stockage lists.
- Facilities-engineering supplies and construction materials.
- Audiovisual equipment, training devices, and subcaliber devices.
- Miscellaneous pilferable assets (not included above) and money.
- Mission-critical or high-risk personnel.
- General military and civilian populations.
- Industrial and utility equipment.
- Controlled cryptographic items.
- Sensitive information (included in TM 5-853-1, but not included in DA Pam 190-51).
- Arms, ammunition, and explosives (AA&E).
- Installation banks and finance offices.

RISK LEVELS

2-6. DA Pam 190-51 provides a procedure for determining risk levels—assessing the value of the assets to their users and the likelihood of compromise. These factors are assessed by answering a series of questions leading to value and likelihood ratings.

2-7. Asset value is determined by considering the following three elements:

- The criticality of the asset for its user and the Army as a whole.
- How easily the asset can be replaced.
- Some measure of the asset's relative value.

2-8. The relative value differs for each asset. For some assets, the relative value is measured in terms of monetary cost.

2-9. The likelihood of the threat is assessed for each applicable aggressor category by considering the asset's value to the aggressor, the history of or potential for aggressors attempting to compromise the asset, and the vulnerability of the asset based on existing or planned protective measures.

REGULATORY REQUIREMENTS

2-10. The risk level is the basis for determining the required protective measures for assets covered in AR 190-51. For each asset type, there may be physical protective measures, procedural security measures, and terrorism counteraction measures. These measures are specified by risk level. The measures identified in AR 190-51 are the minimum regulatory measures that must be applied for the identified threat level. The minimum regulatory measures for AA&E are based on the risk category established in AR 190-11.

ANTITERRORISM/FORCE-PROTECTION CONSTRUCTION STANDARDS

2-11. In accordance with DOD Instruction 2000.16, the commanders in chief (CINCs) have developed standards for new construction and existing facilities to counter terrorism threat capabilities within the area of responsibility. These construction standards have specific requirements for such measures as standoff distance, perimeter barriers, building construction, and parking. The DOD construction standard provides for minimum standards that must be incorporated into all inhabited DOD structures regardless of the identified threat. These standards provide a degree of protection that will not preclude the direct effects of blast but will minimize collateral damage for buildings and people and will limit the progressive collapse of structures. These standards add relatively little cost, may facilitate future upgrades, and may deter acts of aggression. (All services have adopted common criteria and minimum standards to counter antiterrorism/force-protection [AT/FP] vulnerabilities and terrorism threats.) Protection to identified threat levels is described in the following paragraphs. Physical-security personnel must be familiar with the CINC and DOD AT/FP construction standards because these standards may affect elements of physical-security plans and how individual facilities are secured.

THREAT IDENTIFICATION

2-12. The threat must be described in specific terms to help determine the assets' vulnerabilities or to establish protective measures. This description should include the tactics that aggressors will use to compromise the asset (weapons, tools, and explosives are likely to be used in an attempt). For example, the threat might be described as a moving vehicle bomb consisting of a 4,000-pound vehicle containing a 500-pound explosive. Another example would be a forced-entry threat using specific hand, power, or thermal tools. These types of threat descriptions (called the design-basis threat) can be used to design detailed protective systems to mitigate the attacks. TM 5-853-1 and DA Pam 190-51 contain procedures for establishing design-basis threat descriptions in the format described above. These procedures can be used together or separately. Threats listed in the TM will be summarized later in this chapter. When using the TM as a lone source or in conjunction with DA Pam 190-51, the following actions occur:

- When the TM process is used alone, the user goes through an identical process to that in DA Pam 190-51 up to the point where the risk level would be determined. In TM 5-853-1, the value and likelihood ratings are used differently than in DA Pam 190-51. The likelihood rating is used to determine the weapons, tools, and explosives that will be used by a particular aggressor in carrying out a specific tactic. In this procedure, higher likelihood ratings result in more severe mixes of weapons, tools, and explosives. The assumption is that the more likely the attack, the more resources the aggressor is likely to use in carrying out the attack.
- When the procedure in TM 5-853-1 is used in conjunction with the results of the DA Pam 190-51 risk analysis, the likelihood rating is taken directly from the risk analysis and applied as described above.

LEVEL OF PROTECTION

2-13. The level of protection applies to the design of a protective system against a specified threat (for example, a bomb, breaking and entering, pilfering, and so forth). The level of protection is based on the asset's value rating from either DA Pam 190-51 or TM 5-853-1. The level increases as the asset's value rating increases. There are separate levels of protection for each tactic. TM 5-853-1 provides detailed guidance on how to achieve the levels of protection, and Chapter 3 of this manual provides a summary of the levels of protection as they apply to various tactics.

VULNERABILITIES

2-14. Vulnerabilities are gaps in the assets' protection. They are identified by considering the tactics associated with the threat and the levels of protection that are associated with those tactics. Some vulnerabilities can be identified by considering the general design strategies for each tactic described in TM 5-853-1 and as summarized in Chapter 3 of this manual. The general design strategies identify the basic approach to protecting assets against specific tactics. For example, the general design strategy for forced entry is to provide a way to detect attempted intrusion and to provide barriers to delay the aggressors until a response force arrives. Vulnerabilities may involve inadequacies in intrusion-detection systems (IDSs) and barriers. Similarly, the general design strategy for a moving vehicle bomb is to keep the vehicle as far from the facility as possible and to harden the facility to resist the explosive at that distance. Vulnerabilities may involve limited standoff distances, inadequate barriers, and building construction that cannot resist explosive effects at the applicable standoff distance.

PROTECTIVE MEASURES

2-15. Where vulnerabilities have been identified, protective measures must be identified to mitigate them. AR 190-13, AR 190-51, DA Pam 190-51, and TM 5-853-1 are effective tools for developing protective measures. The key to effective development of protective systems is a partnership between physical-security personnel and the installation engineers. Appendix E of this manual discusses information for office security, which should be listed in the physical-security plan (see Appendix F). Appendix G discusses personal-protection measures.

THE INTEGRATED PROTECTIVE SYSTEM

2-16. Protective systems integrate physical protective measures and security procedures to protect assets against a design-basis threat. The characteristics of integrated systems include deterrence, detection, defense, and defeat.

DETERRENCE

2-17. A potential aggressor who perceives a risk of being caught may be deterred from attacking an asset. The effectiveness of deterrence varies with the aggressor's sophistication, the asset's attractiveness, and the aggressor's objective. Although deterrence is not considered a direct design objective, it may be a result of the design.

DETECTION

2-18. A detection measure senses an act of aggression, assesses the validity of the detection, and communicates the appropriate information to a response force. A detection system must provide all three of these capabilities to be effective.

2-19. Detection measures may detect an aggressor's movement via an IDS, or they may detect weapons and tools via X-ray machines or metal and explosive detectors. Detection measures may also include access-control elements that assess the validity of identification (ID) credentials. These control elements may provide a programmed response (admission or denial), or they may relay information to a response force. Guards serve as detection elements, detecting intrusions and controlling access.

2-20. Nuclear, biological, and chemical (NBC) detection systems must be used to measure and validate acts of aggression involving WMD. NBC detection systems should also be used to communicate a warning.

DEFENSE

2-21. Defensive measures protect an asset from aggression by delaying or preventing an aggressor's movement toward the asset or by shielding the asset from weapons and explosives. Defensive measures—

- Delay aggressors from gaining access by using tools in a forced entry. These measures include barriers along with a response force.
- Prevent an aggressor's movement toward an asset. These measures provide barriers to movement and obscure lines of sight (LOSs) to assets.
- Protect the asset from the effects of tools, weapons, and explosives.

2-22. Defensive measures may be active or passive. Active defensive measures are manually or automatically activated in response to acts of aggression. Passive defensive measures do not depend on detection or a response. They include such measures as blast-resistant building components and fences. Guards may also be considered as a defensive measure.

DEFEAT

2-23. Most protective systems depend on response personnel to defeat an aggressor. Although defeat is not a design objective, defensive and detection systems must be designed to accommodate (or at least not interfere with) response-force activities.

SECURITY THREATS

2-24. Security threats are acts or conditions that may result in the compromise of sensitive information; loss of life; damage, loss, or destruction of property; or disruption of mission. Physical-security personnel and design teams must understand the threat to the assets they are to protect in order to develop effective security programs or design security systems. Historical patterns and trends in aggressor activity indicate general categories of aggressors and the common tactics they use against military assets. Aggressor tactics and their associated tools, weapons, and explosives are the basis for the threat to assets.

THREAT SOURCES

2-25. There are many potential sources of threat information. Threat assessment is normally a military-intelligence (MI) responsibility. MI personnel commonly focus on such security threats as terrorists and military forces. Within the US and its territories, the Federal Bureau of Investigation (FBI) has primary responsibility for both foreign and domestic terrorists. The FBI, the US Army Criminal Investigation Command (USACIDC [CID]), and local law-enforcement agencies are good sources for physical-security personnel to obtain criminal threat information. Coordinating with these elements on a regular basis is essential to maintaining an effective security program.

THREAT CATEGORIES

2-26. Security threats are classified as either human or natural. Human threats are carried out by a wide range of aggressors who may have one or more objectives toward assets such as equipment, personnel, and operations. Aggressors can be categorized and their objectives can be generalized as described below. (See DA Pam 190-51 and TM 5-853-1 for more information.)

Aggressor Objectives

2-27. Four major objectives describe an aggressor's behavior. Any one of the first three objectives can be used to realize the fourth. These objectives include—

- Inflicting injury or death on people.
- Destroying or damaging facilities, property, equipment, or resources.
- Stealing equipment, materiel, or information.
- Creating adverse publicity.

Aggressor Categories

2-28. Aggressors are grouped into five broad categories—criminals, vandals and activists, extremists, protest groups, and terrorists. Hostile acts performed by these aggressors range from crimes (such as burglary) to low-intensity conflict threats (such as unconventional warfare). Each of these categories describes predictable aggressors who pose threats to military assets and who share common objectives and tactics.

- Criminals can be characterized based on their degree of sophistication. They are classified as unsophisticated criminals, sophisticated criminals, and organized criminal groups. Their common objective is the theft of assets; however, the assets they target, the quantities they seek, their relative efficiency, and the sophistication of their actions vary significantly. Vandals and activists may also be included under this category.
- Vandals and activists are groups of protesters who are politically or issue oriented. They act out of frustration, discontent, or anger against the actions of other social or political groups. Their primary objectives commonly include destruction and publicity. Their selection of targets will vary based on the risk associated with attacking them. The degree of damage they seek to cause will vary with their sophistication.
- Extremists are radical in their political beliefs and may take extreme, violent actions to gain support for their beliefs or cause.
- Protesters are considered a threat only if they are violent. Lawful protesters have to be considered, but significant protective measures and procedures are not normally needed to control their actions. The presence of extremists or vandals/activists at a peaceful protest increases the chance of the protest becoming violent.
- Terrorists are ideologically, politically, or issue oriented. They commonly work in small, well-organized groups or cells. They are sophisticated, are skilled with tools and weapons, and possess an efficient planning capability. There are three types of terrorists—CONUS, OCONUS, and paramilitary OCONUS.
  - CONUS terrorists are typically right- or left-wing extremists operating in distinct areas of the US.
  - OCONUS terrorists generally are more organized than CONUS terrorists. They usually include ethnically or religiously oriented groups.

- Paramilitary OCONUS terrorist groups show some military capability with a broad range of military and improvised weapons. Attacks by OCONUS terrorists are typically more severe.

2-29. Natural threats are usually the consequence of natural phenomena. They are not preventable by physical-security measures, but they are likely to have significant effects on security systems and operations. They may require an increase in protective measures either to address new situations or to compensate for the loss of existing security measures. They may reduce the effectiveness of existing security measures by such occurrences as collapsed perimeter fences and barriers, inoperable protective lighting, damaged patrol vehicles, and poor visibility. Natural threats and their effects relative to security include the following:

- Floods may result in property damage, destruction of perimeter fences, and damage to IDSs. Heavy rains or snowfalls may have similar effects even if they do not result in flooding.
- Storms, tornadoes, high winds, or rain may cause nuisance alarms to activate and cause damage to IDSs. They may limit the visibility of security personnel and may affect closed-circuit television (CCTV) systems. Winds may also disrupt power or communication lines and cause safety hazards from flying debris.
- Earthquakes may cause nuisance alarms to activate or may disrupt IDSs. They may also cause broken water or gas mains, fallen electrical or communication lines, and weakened or collapsed buildings.
- Snow and ice can make travel on patrol roads difficult, may delay responses to alarms, may impede the performance of IDSs, and may freeze locks and alarm mechanisms. Heavy ice may also damage power and communication lines.
- Fires may damage or destroy perimeter barriers and buildings, possibly leaving assets susceptible to damage or theft.
- Fog can reduce the visibility of security forces, thereby requiring additional security personnel. It may also increase the response time to alarms and reduce the effectiveness of security equipment such as CCTV systems.

Aggressor Tactics

2-30. Aggressors have historically used a wide range of offensive strategies reflecting their capabilities and objectives. These offensive strategies are categorized into 15 tactics that are specific methods of achieving aggressor goals (see TM 5-853-1). Separating these tactics into categories allows facility planners and physical-security personnel to define threats in standardized terms usable as a basis for facility and security-system design. Common aggressor tactics include—

- **Moving vehicle bomb.** An aggressor drives an explosive-laden car or truck into a facility and detonates the explosives. His goal is to damage or destroy the facility or to kill people. This is a suicide attack.
- **Stationary vehicle bomb.** An aggressor covertly parks an explosive-laden car or truck near a facility. He then detonates the explosives either by time delay or remote control. His goal in this tactic is the

same as for the moving vehicle bomb with the additional goal of destroying assets within the blast area. This is commonly not a suicide attack. It is the most frequent application of vehicle bombings.

- **Exterior attack.** An aggressor attacks a facility's exterior or an exposed asset at close range. He uses weapons such as rocks, clubs, improvised incendiary or explosive devices, and hand grenades. Weapons (such as small arms) are not included in this tactic, but are considered in subsequent tactics. His goal is to damage the facility, to injure or kill its occupants, or to damage or destroy assets.
- **Standoff weapons.** An aggressor fires military weapons or improvised versions of military weapons at a facility from a significant distance. These weapons include direct (such as antitank [AT] weapons) and indirect LOS weapons (such as mortars). His goal is to damage the facility, to injure or kill its occupants, or to damage or destroy assets.
- **Ballistics.** The aggressor fires various small arms (such as pistols, submachine guns, shotguns, and rifles) from a distance. His goal is to injure or kill facility occupants or to damage or destroy assets.
- **Forced entry.** The aggressor forcibly enters a facility using forced-entry tools (such as hand, power, and thermal tools) and explosives. He uses the tools to create a man-passable opening or to operate a device in the facility's walls, doors, roof, windows, or utility openings. He may also use small arms to overpower guards. His goal is to steal or destroy assets, compromise information, injure or kill facility occupants, or disrupt operations.
- **Covert entry.** The aggressor attempts to enter a facility or a portion of a facility by using false credentials or stealth. He may try to carry weapons or explosives into the facility. His goals include those listed for forced entry.
- **Insider compromise.** A person authorized access to a facility (an insider) attempts to compromise assets by taking advantage of that accessibility. The aggressor may also try to carry weapons or explosives into the facility in this tactic. His goals are the same as those listed for forced entry.
- **Visual surveillance.** The aggressor uses ocular and photographic devices (such as binoculars and cameras with telephoto lenses) to monitor facility or installation operations or to see assets. His goal is to compromise information. As a precursor, he uses this tactic to determine information about the asset of interest.
- **Acoustic eavesdropping.** The aggressor uses listening devices to monitor voice communications or other audibly transmitted information. His goal is to compromise information.
- **Electronic-emanations eavesdropping.** The aggressor uses electronic-emanation surveillance equipment from outside a facility or its restricted area to monitor electronic emanations from computers, communications, and related equipment. His goal is to compromise information.

- **Mail-bomb delivery.** The aggressor delivers bombs or incendiary devices to the target in letters or packages. The bomb sizes involved are relatively small. His goal is to kill or injure people.
- **Supplies-bomb delivery.** The aggressor conceals bombs in various containers and delivers them to supply- and material-handling points such as loading docks. The bomb sizes in this tactic can be significantly larger that those in mail bombs. His goal is to damage the facility, kill or injure its occupants, or damage or destroy assets. Appendix H addresses the actions to take when a bomb is suspected.
- **Airborne contamination.** An aggressor contaminates a facility's air supply by introducing chemical or biological agents into it. His goal is to kill or injure people.
- **Waterborne contamination.** An aggressor contaminates a facility's water supply by introducing chemical, biological, or radiological agents into it. These agents can be introduced into the system at any location with varying effects, depending on the quantity of water and the contaminant involved. His goal is to kill or injure people.

2-31. The aforementioned tactics are typical threats to fixed facilities for which designers and physical-security personnel can provide protective measures. However, some common terrorist acts are beyond the protection that facility designers can provide. They cannot control kidnappings, hijackings, and assassinations that take place away from facilities or during travel between facilities. Protection against these threats is provided through operational security and personal measures (see Appendices G and I), which are covered in doctrine relative to those activities and are under the general responsibility of the CID.

TACTICAL ENVIRONMENT CONSIDERATIONS

2-32. When determining the assets and threats, the same considerations should be given to the systems approach in the tactical environment as when in the cantonment area. The same process of determining the assets, their risk level, and any regulatory guidance apply. Identifying potential threats and the level of protection required for the assets are necessary. Commanders and leaders must also identify additional vulnerabilities and other required protective measures. Commanders are not expected to have the same physical protective measures due to the impact of resources, budget, location, and situations.

2-33. Commanders must consider the various tactics used by aggressors and use their soldiers' abilities to counteract these tactics. Considerations for specific assets (such as military-working-dog [MWD] and explosive-ordnance-disposal [EOD] teams and their abilities to detect and disassemble a bomb) must be identified. Units must have the ability to improvise in a tactical environment. Their training and resourcefulness will compensate for shortcomings in the field.

2-34. The systems approach to security provides focus and integration of resources. Protective systems are mutually supporting and systematically developed to negate the threat. Commanders conduct an intelligence preparation of the battlefield (IPB) and vulnerability assessments (VAs) to determine risks. Security resources and measures are applied to mitigate risks and to deter, detect, defend, and defeat the threat.

Chapter 3

Design Approach

Developing protective systems to protect assets depends on an effective partnership between engineers and physical-security personnel. Physical-security personnel need to understand the basic approaches the engineers will take in laying out protective systems. Engineers must understand the issues involved with ensuring that anything they lay out is compatible with security operations and the operations of the asset users. The best way to ensure a viable design is through teamwork. This chapter provides a summary of the basic approaches to protecting assets against threats (the design strategies). Understanding these strategies is critical to being an effective team member in developing protective systems.

DESIGN STRATEGIES

3-1. There are separate design strategies for protecting assets from each tactic described in Chapter 2. There are two types of strategies associated with each tactic—the general-design and specific-design strategies. The general-design strategy is the general approach to protecting assets against tactics. The specific-design strategy refines the general-design strategy to focus the performance of the protective system on a particular level of protection. (See TM 5-853-1 for more information.)

PROTECTIVE MEASURES

3-2. Protective measures are developed as a result of the general- and specific-design strategies. These protective measures commonly take the form of site-work, building, detection, and procedural elements.

- Site-work elements include the area surrounding a facility or an asset. Technically, they are associated with everything beyond 5 feet from a building. They can include perimeter barriers, landforms, and standoff distances.
- Building elements are protective measures directly associated with buildings. These elements include walls, doors, windows, and roofs.
- Detection elements detect such things as intruders, weapons, or explosives. They include IDSs, CCTV systems used to assess intrusion alarms, and weapon and explosive detectors. These elements can also include the guards used to support this equipment or to perform similar functions.
- Procedural elements are the protective measures required by regulations, TMs, and standing operating procedures (SOPs). These elements provide the foundation for developing the other three elements.

VEHICLE BOMBS

3-3. Vehicle-bomb tactics include both moving and stationary vehicle bombs. In the case of a moving vehicle bomb, the aggressor drives the vehicle into the target. This is commonly known as a suicide attack. In a stationary vehicle bomb, he parks the vehicle and detonates the bomb remotely or on a timed delay.

GENERAL-DESIGN STRATEGY

3-4. Blast pressures near an exploding vehicle bomb are very high, but they decrease rapidly with distance from the explosion. The design strategy for these tactics is to maintain as much standoff distance as possible between the vehicle bomb and the facility and then, if necessary, to harden the facility for the resulting blast pressures. Barriers on the perimeter of the resulting standoff zone maintain the required standoff distance. The difference between moving and stationary vehicle-bomb tactics is that the aggressor using the moving vehicle bomb will attempt to crash through the vehicle barriers; the aggressor using the stationary vehicle bomb will not. Therefore, vehicle barriers for the moving vehicle bomb must be capable of stopping a moving vehicle at the perimeter of the standoff zone. For a stationary vehicle bomb, vehicle barriers must mark the perimeter of the standoff zone, but they are not required to stop the moving vehicle. They only need to make it obvious if an aggressor attempts to breach the perimeter.

LEVELS OF PROTECTION

3-5. There are three levels of protection for vehicle bombs—low, medium, and high. The primary differences between the levels are the degree of damage allowed to the facility protecting the assets and the resulting degree of damage or injury to the assets.

- **Low.** The facility or the protected space will sustain a high degree of damage but will not collapse. It may not be economically repairable. Although collapse is prevented, injuries may occur and assets may be damaged.
- **Medium.** The facility or the protected space will sustain a significant degree of damage, but the structure will be reusable. Occupants and other assets may sustain minor injuries or damage.
- **High.** The facility or the protected space will sustain only superficial damage. Occupants and other assets will also incur only superficial injury or damage.

SITE-WORK ELEMENTS

3-6. The two primary types of site-work elements for vehicle bombs are the standoff distance and vehicle barriers. The vehicle's speed must also be taken into consideration.

Standoff Distance

3-7. The standoff distance is the maintained distance between where a vehicle bomb is allowed and the target. The initial goal should be to make that distance as far from the target facility as practical. Figure 3-1 shows the distances required to limit building damage to particular levels (including the levels of protection described above) for a range of bomb weights. All bomb weights are given in terms of equivalent pounds of trinitrotoluene (TNT), which is a standard way of identifying all explosives regardless of their composition. The example in Figure 3-1 is a building of conventional construction (common, unhardened construction). Buildings built without any special construction at these standoff distances will probably withstand the explosive effects. Conventionally constructed buildings at standoff distances of less than those shown in Figure 3-1 will not adequately withstand blast effects. (Refer to TM 5-853-1 for information on hardening buildings to resist a blast.) Do not allow vehicles to park within the established standoff distances. Recognize that this restriction can result in significant operational and land-use problems.

3-8. **Exclusive Standoff Zone.** When an exclusive standoff zone is established, do not allow vehicles within the perimeter unless they have been searched or cleared for access. The zone's perimeter is established at the distance necessary to protect the facility against the highest threat explosive. All vehicles should be parked outside the exclusive standoff zone; only

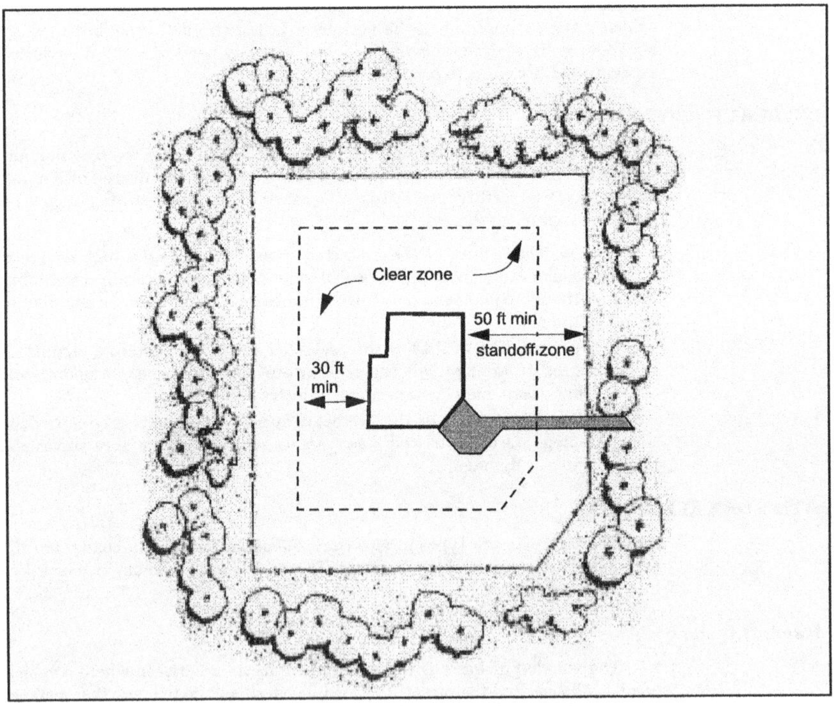

Figure 3-1. Standoff Distance

maintenance, emergency, and delivery vehicles should be allowed within the zone after being searched. Figure 3-2 shows an exclusive standoff zone.

3-9. **Nonexclusive Standoff Zone.** A nonexclusive standoff zone is established in a location having a mixture of cars and trucks (with relatively few trucks). A nonexclusive standoff zone takes advantage of aggressors being able to conceal a smaller quantity of explosives in a car than they can in a truck. Therefore, a nonexclusive standoff zone includes inner and outer perimeters. The inner perimeter is set at a distance corresponding to the weight of explosives that can be concealed in cars. The outer perimeter is set at a distance associated with the weight that can be placed in trucks (refer to TM 5-853-1). With these two perimeters, cars can enter the outer perimeter without being searched but they cannot enter the inner perimeter. Trucks cannot enter the outer perimeter, since it is established based on what they can carry. Figure 3-3 shows a nonexclusive standoff zone. The nonexclusive standoff zone provides the advantages of allowing better use of the parking areas and limiting the number of vehicles that need to be searched at the outer perimeter.

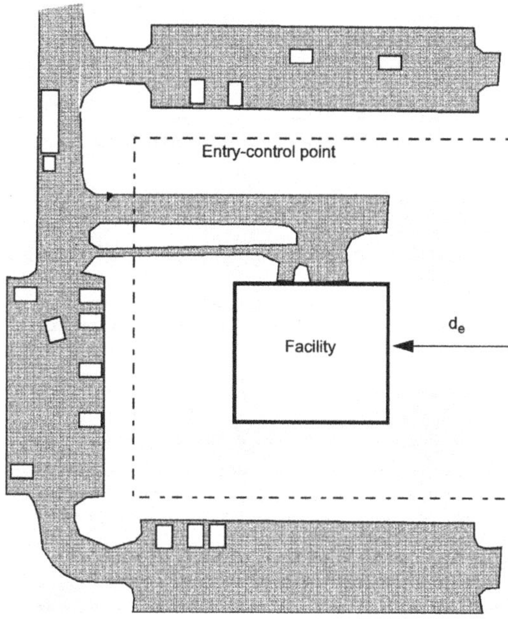

d_e = exclusive standoff-zone distance

Figure 3-2. Exclusive Standoff Zone

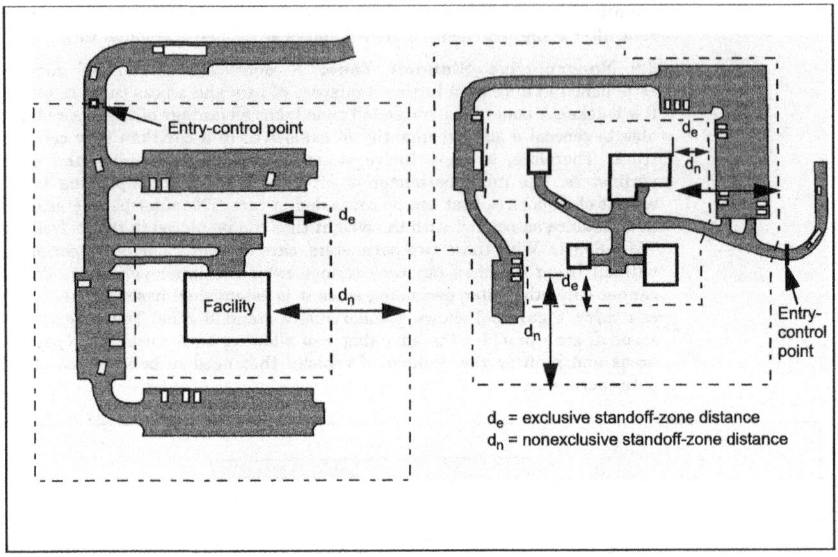

Figure 3-3. Nonexclusive Standoff Zone

Vehicle Barriers

3-10. Two types of vehicle barriers are used for vehicle bombs—perimeter and active barriers. The type of barrier used for a moving vehicle bomb differs from the barrier used for a stationary vehicle bomb. The barrier used for a stationary vehicle bomb does not have to stop a vehicle's motion. The goal for that barrier is to make anybody driving through the barrier noticeable. The assumption is that the aggressor's goal in the stationary vehicle bomb is to park the vehicle and sneak away without being noticed. Crashing through a barrier would be noticeable. Barriers for the moving vehicle bomb need to stop the vehicle's motion; they must be much more substantial.

3-11. **Perimeter Barriers.** Perimeter barriers are fixed barriers placed around the entire perimeter of a standoff zone. Anything that presents a fixed obstacle will work for the stationary vehicle bomb. Common applications include chain-link fences, hedges made of low bushes, and high (over 8 inches) curbs. Aggressors driving through such barriers are likely to be noticed. Barriers capable of stopping moving vehicles include chain-link fences reinforced with cable, reinforced concrete "Jersey barriers", pipe bollards, planters, ditches, and berms. When barriers such as the Jersey barriers and planters are used to stop moving vehicles, they must be anchored into the ground to be effective. The cables in the reinforced fence also have to be anchored into the ground or partially buried. Spaces between barriers should be no greater than 4 feet. Figure 3-4 shows common perimeter barriers for stationary or moving vehicle bombs. Refer also to TM 5-853-1.

3-12. **Active Barriers.** Active barriers are placed at openings in perimeters where vehicles need to enter or exit. These barriers must be able to be raised and lowered or moved aside. For the stationary vehicle bomb, barriers can be as simple as chain-link, pipe, or wooden gates that can be raised and lowered. Aggressors crashing through any of these or similar obstructions will likely draw attention. For the moving vehicle bomb, the barriers are heavy structures and have many construction and operations considerations

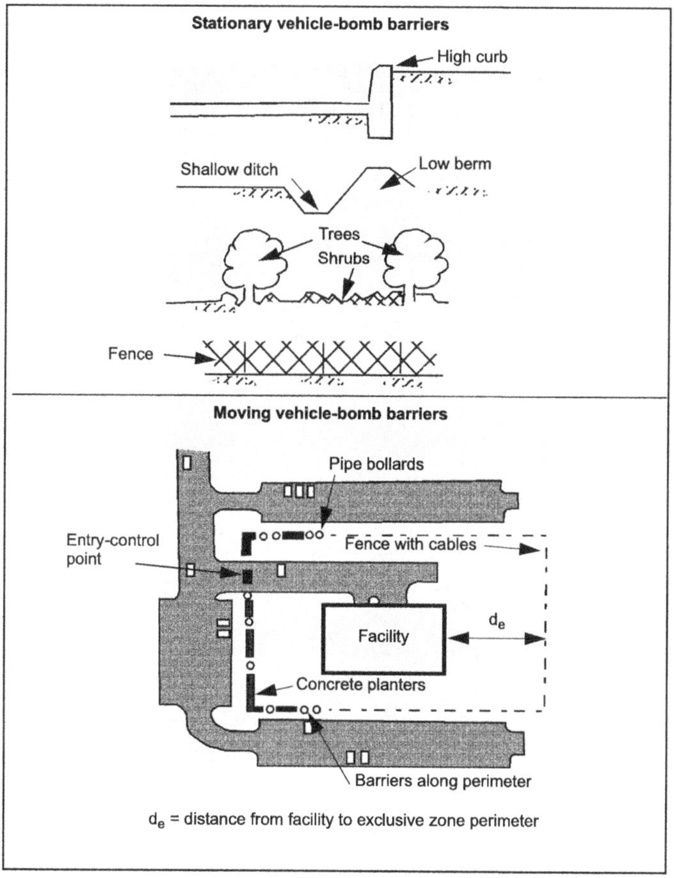

Figure 3-4. Perimeter-Barrier Application

associated with them. These barriers may stop vehicles weighing up to 15,000 pounds and travelling 50 miles per hour. They commonly cost tens of thousands of dollars (refer to TM 5-853-1). Some common active vehicle barriers are shown in Figure 3-5. For temporary or deployed conditions, park a vehicle across an opening and move it aside to grant access.

Speed Control

3-13. It is important to control the speed of a vehicle approaching a barrier used for a moving vehicle bomb. The energy from a vehicle that a barrier must stop increases as its speed increases. The energy also increases with more weight, but the effect of speed is much greater. Therefore, decreasing the vehicle's speed results in smaller and less costly barriers. The best way to limit a vehicle's approach speed to perimeter barriers is to place or retain obstacles in potential approach paths. The vehicles are forced to reduce speed when going around these obstacles. The same principle applies for road approaches. Placing obstacles in a serpentine pattern on the road forces a vehicle to reduce its speed (see Figure 3-6, page 3-8). If the vehicle hits the obstacles instead of going around them, they are still slowed down. Other means to slow vehicles include forcing them to make sharp turns and installing traffic circles.

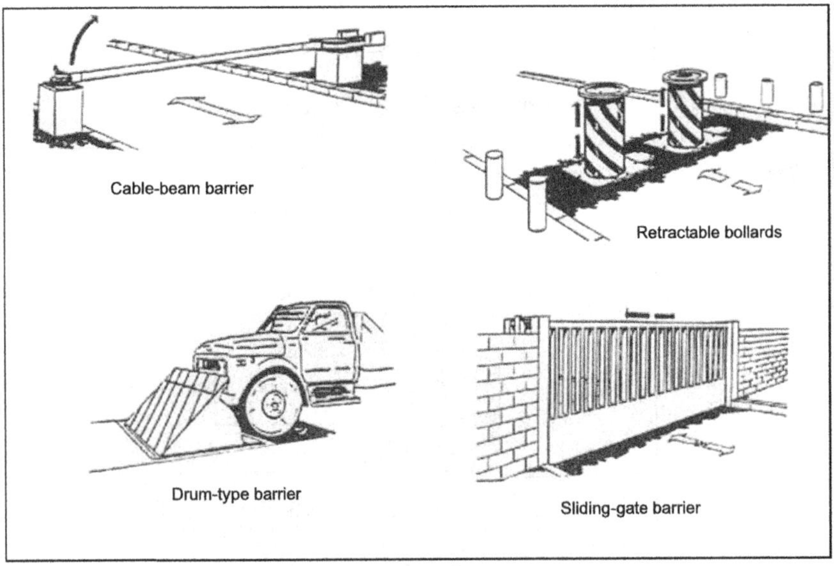

Figure 3-5. Active Vehicle Barriers

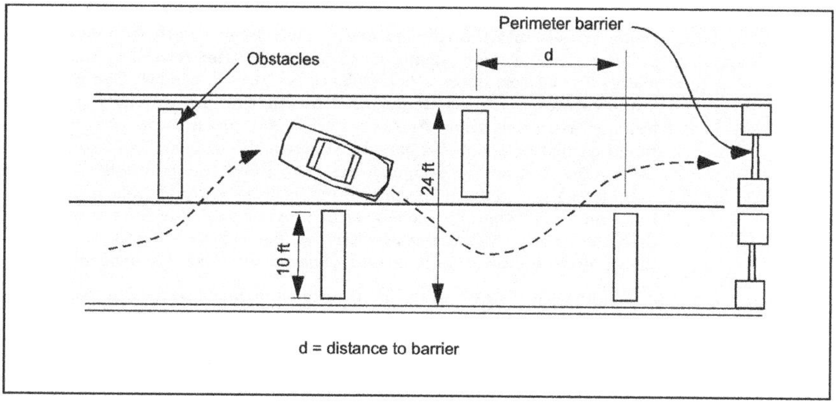

Figure 3-6. Serpentine Pattern

BUILDING ELEMENTS

3-14. Once the standoff distance is established and the site has been laid out, the designers can select the building components necessary to protect the assets against the threat explosives at the standoff distance. The building components include the walls, roofs, doors, and windows. Detailed design issues related to these building elements are covered in TM 5-853-1.

Walls and Roofs

3-15. If the distances shown for the desired damage levels in Figure 3-1, page 3-3, cannot be enforced, the building's walls and roofs will need to be strengthened. This can be achieved in new construction by using reinforced masonry or reinforced concrete in the walls and reinforced concrete in the roof. When the standoff distance is not available for existing construction, a more detailed analysis may be required to determine what the explosion's impact will be on the structure. When the construction is inadequate, more standoff distance should be investigated or the engineers should apply specialized techniques for retrofitting the construction to increase its strength.

Windows

3-16. Historically, glass fragments have caused about 85 percent of injuries and deaths in bomb blasts. There are two basic approaches to mitigating the effects of bomb blasts on glass—retrofitting the windows with film or curtains and using blast-resistant glazing.

3-17. **Retrofitting Windows**. One of the most common means of decreasing the hazards from broken glass is to install fragment-retention film on the glass. The film is a plastic (polyester) sheet that adheres to the window glass with a special adhesive. The film does not strengthen the glass; but when the glass breaks, it keeps the fragments from spreading throughout the room. The glass fragments stick to the film, and the film either stays in the window frame or falls into the room in one or more large, relatively nonhazardous pieces instead of many small, lethal pieces. Another retrofit approach is to install a blast curtain or a heavy drape behind the window. The curtain or drape catches the glass fragments. The curtains are generally used with fragment-retention film. Another retrofit technique is to use fragment-retention film with a metal bar placed across the window. This "catcher bar" catches the window. The designs for this and other types of retrofit devices are complicated and require specialized engineering-analysis tools. The retrofit techniques are generally thought of as providing a lower level of protection than the glazing replacement techniques. For deployed locations, removing the windows and covering them with plywood minimizes the danger.

3-18. **Blast-Resistant Glazing.** To achieve higher levels of protection, the window glass must be replaced and the window frame should be reinforced. Because of its expense, this procedure is generally limited to new construction and major renovations. Special blast-resistant glazing and frames are available that use either tempered glass or a plastic glazing (such as polycarbonate). Another promising type of blast-resistant glazing is laminated glass, in which several layers of common glass are adhered together with a special interlayer. The resulting laminated construction is usually stronger than common glass while retaining the same thickness. The interlayer acts similarly to fragment-retention film. For deployed locations, a means of minimizing the danger of windows is to remove them and replace them with plywood.

Doors

3-19. Doors are another building component particularly vulnerable to an explosive blast. Common metal and wood doors provide little resistance to a blast. The two ways to address the problem of doors is to install them in foyers or to replace them. Glass doors or doors containing windows should be avoided.

Foyers

3-20. Door hazards can be reduced by installing doors in foyers during construction or by adding foyers to existing buildings. When a door is located in a foyer and the outer door fails, the outer door flies into a wall instead of the building's interior (see Figure 3-7, page 3-10). The inner door then has a greater chance of remaining intact. This option generally provides a low level of protection.

3-21. Another option is to replace the doors with specially constructed blast-resistant doors and frames. These doors are commercially available and can provide a high level of protection, but they are very expensive and heavy. The doorframe must be made of the same type of material and provide the same level of protection as the door.

DETECTION ELEMENTS

3-22. Detection elements for vehicle bombs are limited to the use of guards to control access into standoff zones. The guards search vehicles seeking entry

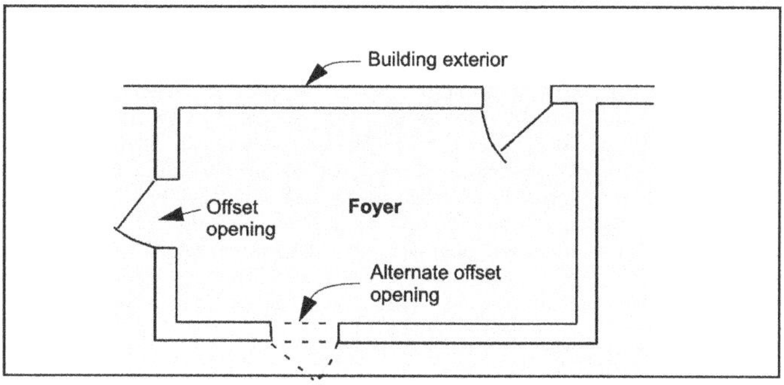

Figure 3-7. Door in Foyer

into the perimeter through an entry-control point. The recommended levels of searches depend on the required level of protection (see TM 5-853-1). Guards can be stationed at entry-control points continuously, or they can be summoned to an entry-control point when access is needed. The latter is commonly the case for the inner perimeter of exclusive standoff zones where only delivery and maintenance vehicles need access.

EXTERIOR ATTACK

3-23. An exterior attack is a physical attack using weapons such as rocks, clubs, improvised incendiary devices (IIDs) such as Molotov cocktails, explosives such as improvised explosive devices (IEDs), and hand grenades. The explosives can be thrown at or placed near a facility's exterior. Examples of IEDs for this tactic range from pipe bombs and hand grenades to briefcase-sized explosives.

GENERAL-DESIGN STRATEGY

3-24. Because the exterior attack is directed at a facility's exterior surfaces, the general-design strategy is to keep aggressors away from the facility (at a standoff distance) and, if necessary, to harden the facility's exterior components to resist the effects of weapons and explosives. A standoff distance from the facility reduces the degree of hardening required to resist weapons effects. When briefcase-sized bombs are a threat, an obstacle-free zone should be established around the facility and the explosives placed within should be detected and disarmed.

LEVELS OF PROTECTION

3-25. The levels of protection for exterior attacks are similar to those for vehicle bombs. Levels of protection vary based on the level of building damage and asset injury or damage allowed. However, due to the limited sizes of explosives involved in this tactic, the damage to the building will be much more localized and injuries or damage to assets will be confined to smaller areas.

SITE-WORK ELEMENTS

3-26. Site-work elements for exterior attacks are relatively limited because the explosive weights are more limited. Large standoff distances are not a consideration. The common approach to site-work elements is to lay out a standoff zone of about 50 feet and to provide a fence or perimeter barrier about 7 feet high. The purpose of the standoff is to make it harder for aggressors to throw pipe bombs and hand grenades at targets inside the perimeter. Trees can be left around the perimeter to make it harder for aggressors to throw explosives over the fence. The remaining component of site-work elements is a clear zone around the facility. A clear zone is applied so that anything placed in that area can be detected visually. This limits the aggressor's ability to place explosives near the target facility.

BUILDING ELEMENTS

3-27. Building elements for exterior attacks are similar to those for vehicle bombs. For small IEDs and IIDs, the building-element requirements do not increase the cost of the building significantly. For larger, briefcase-sized bombs, the measures are more significant than for incendiary devices but less than for vehicle bombs.

Walls and Roofs

3-28. Walls and roofs are not a problem with small explosives. Conventional construction normally provides adequate protection. Walls with 6-inch reinforced concrete or 8-inch, grout-filled, reinforced masonry will withstand the effects of typical pipe bombs or hand grenades. The corresponding roof construction is 6-inch reinforced concrete. In the case of briefcase-sized bombs, considerations similar to those discussed for vehicle bombs need to be employed.

Windows

3-29. A significant goal when constructing windows is to make them difficult to throw an explosive or incendiary device through, especially when considering smaller explosives. This is accomplished by constructing smaller windows or making narrow windows (see Figure 3-8, page 3-12). For existing windows, parts of the windows can be covered to achieve a narrow effect. These windows still may be susceptible to breakage due to explosive effects, even from the smaller explosives. This problem is solved by installing 3/4-inch-thick plastic (polycarbonate) glazing or by raising the windows over 6 feet high to develop a small standoff distance (as shown in Figure 3-9, page 3-12). A 3/4-inch glazing will also stop grenade fragments. Fragment-retention film, a blast curtain, or a heavy drape as described in vehicle-bomb tactics are also good applications for small bombs.

Conventionally Constructed Doors

3-30. Doors are not a significant problem with small bombs and incendiary devices. Generally, metal doors are adequate for incendiary devices, and doors placed in foyers (as shown in Figure 3-7) are adequate for pipe bombs and hand grenades. A similar application for briefcase-sized bombs would provide

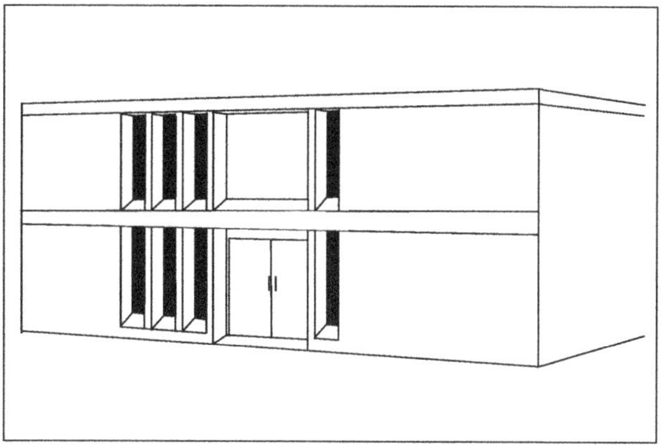

Figure 3-8. Narrow Recessed Windows

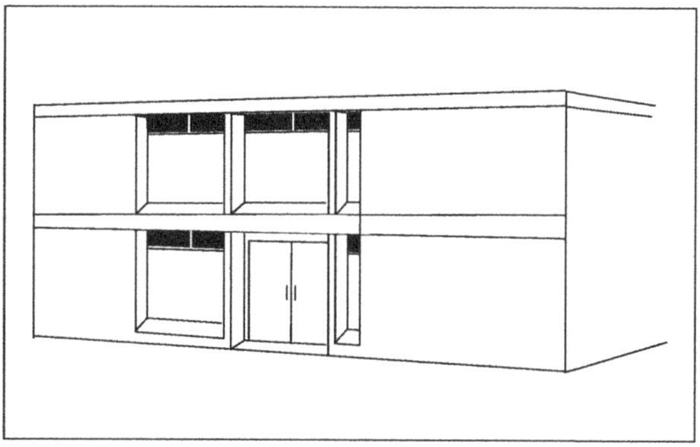

Figure 3-9. Narrow Raised Windows

only a low level of protection. To achieve higher levels of protection for briefcase-sized bombs, blast-resistant doors must be installed.

3-31. The requirements to meet the levels of protection for larger explosives are similar to those described for vehicle bombs, but they will not stop grenade fragments. Fragment-retention film and drapes or curtains can provide a low level of protection, but blast-resistant glazing is required to achieve a higher level of protection.

DETECTION ELEMENTS

3-32. Other than awareness of aggressor activity on or outside the site, detection is only a specific design goal where briefcase-sized bombs are anticipated. When that is the case, the clear zone around the building must be visually monitored so that any objects placed in it are detected. At higher levels of protection, visual surveillance is augmented by IDSs.

STANDOFF WEAPONS

3-33. The standoff-weapons tactic includes the use of AT weapons and mortars. In both of these tactics, the aggressor fires weapons at assets located in the protected facility from a distance. An AT-weapon attack requires a clear LOS to the target, while mortars can fire over obstacles and only need a clear line of flight.

GENERAL-DESIGN STRATEGY

3-34. Standoff-weapons attacks cannot be detected reliably before they occur. Protective design to resist these tactics relies on blocking LOSs to protected areas of a facility or hardening the facility to resist the particular weapon's effects. The approaches to protection against mortars and AT weapons differ from each other and will be discussed separately. Detection measures are not applicable for these tactics.

LEVELS OF PROTECTION

3-35. There are two levels of protection against both mortars and AT weapons. For AT weapons, the low level of protection depends on detonating the AT round before it hits the target facility. The high level of protection avoids the risk associated with that and hardens the building to resist the direct impact of the AT round.

3-36. For mortars, the low level of protection involves allowing some areas of the facility to be sacrificed. Those spaces provide a buffer to the assets to be protected. The assets within the sacrificial areas and the areas themselves may be destroyed. At the high level of protection, the building's exterior fully resists the mortar rounds and there are no sacrificial areas.

SITE-WORK ELEMENTS

3-37. The primary site-work element for standoff weapons is to obstruct LOSs from vantage points outside of the site. With AT weapons, the aggressor cannot hit what he cannot see. This is not true with mortars, but blocking LOSs from mortar firing points helps to make targeting more difficult. The LOSs are blocked by using trees, other buildings, vehicle parking areas, or fences. Another site-work element, a predetonation screen, applies only to an AT weapon. When using a predetonation screen, the AT round is detonated on the screen and its effects are dissipated in the distance between the screen and the target (see Figure 3-10, page 3-14). Any screen material (such as a

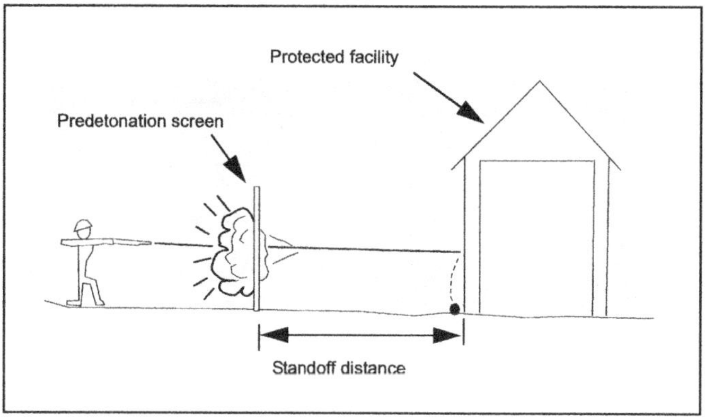

Figure 3-10. Predetonation Screen

wooden fence) will detonate the round unless it has spaces in it. The screen distances vary from less than 10 feet to almost 40 feet, depending on the building construction (see TM 5-853-1). This measure only applies to the low level of protection.

BUILDING ELEMENTS

3-38. Building elements for AT weapons and mortars involve the building's layout. This includes the materials used in the construction.

Layout

3-39. A building's interior layout is only an issue for the low level of protection against a mortar round. The layout issue involves designating sacrificial areas in which unimportant assets are located. The assets to be protected are located in a hardened interior layer. Figure 3-11 includes a plan view (from above). The sacrificial area has to be both around and above the protected area in case a mortar round comes from above. If such a layout is not feasible, other options include going to a higher level of protection and either hardening the entire building or building the facility underground (which are both very expensive).

Walls and Roofs

3-40. Walls and roofs must offer protection against both AT weapons and mortar rounds. The design of walls that protect against AT weapons varies with the level of protection. For the low level of protection where the round is predetonated, the walls can be of conventional construction, varying with the standoff distance from the predetonation screen to the wall. For higher levels of protection, the walls must resist the full effect of the round, requiring the walls to be 24-inch-thick reinforced concrete. Roofs are not an issue in

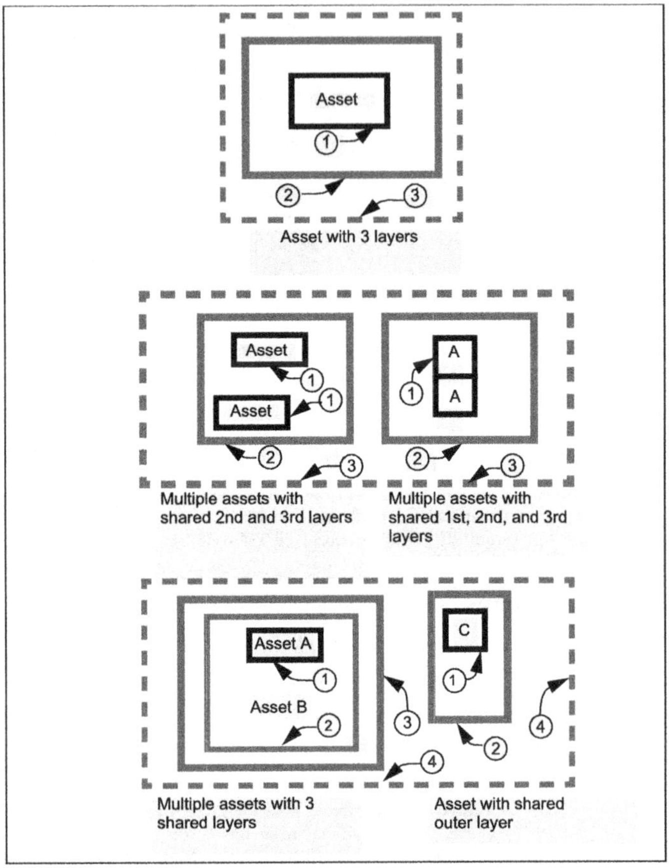

Figure 3-11. Assets Protected by Hardened Interior Layer

protecting against AT weapons because it is difficult to get direct LOSs to roofs. If such LOSs are possible, the roof should be designed like the walls.

3-41. To provide protection against mortar rounds, walls and roofs should be designed to resist the explosive effects in the rounds at the standoff distance that the sacrificial space provides. In the case of sacrificial areas, the walls can be of common construction. The interior protected-area walls are then designed of reinforced concrete or reinforced masonry for the standoff distance those sacrificial walls provide. When the walls must resist the full effect of the rounds (as in the higher level of protection), they are likely to be very thick (up to 30 inches of reinforced concrete for some improvised mortars). Similar considerations should be made for roofs. Roofs are designed to take the direct effects of the round or to take the round at the standoff distance provided by the sacrificial area.

Doors and Windows

3-42. It is impractical to provide doors and windows that are resistant to mortar rounds and AT weapons. Windows should only be used in sacrificial areas where there is a mortar threat. When there is an AT weapon threat, windows can only be used where the round is predetonated. The windows should be narrowed or raised to present a smaller target (see Figures 3-8 and 3-9, page 3-12). Doors should be placed in foyers (see Figure 3-7, page 3-10) for protection against AT rounds and to achieve a low level of protection against mortars. Blast-resistant doors are necessary to achieve a high level of protection against mortar rounds.

BALLISTICS

3-43. In a ballistics tactic, aggressors fire small arms at assets from vantage points outside of the target facility's control. Ballistic attacks cannot be detected reliably before they occur.

GENERAL-DESIGN STRATEGY

3-44. Protective measures to resist these tactics rely on blocking LOSs to protected areas of a facility or by hardening the facility to resist the ballistic effects. This strategy focuses on assets within buildings. Protecting people or property in the open is difficult and can only be addressed through operational measures. Detection measures are not applicable for this tactic.

LEVELS OF PROTECTION

3-45. There are only two levels of protection for this tactic. The low level of protection depends on blocking LOSs to assets. This strategy assumes that the aggressor cannot hit what he cannot see. The risk of an aggressor firing into a building randomly and hitting something is what makes this the low level of protection. The high level of protection involves hardening building components to resist the ballistic effects. These strategies can be thought of as either hardening or hiding.

SITE-WORK ELEMENTS

3-46. Site-work elements are of limited use for the ballistics tactic. When they are applied, they are used to obstruct LOSs from vantage points outside of the site, which is consistent with the low level of protection. The LOSs can be blocked using trees, other buildings, motor pools, or fences.

BUILDING ELEMENTS

3-47. Building elements are the principal means of protecting assets against a ballistics attack. They can be applied to achieve either the low or high level of protection.

Walls and Roofs

3-48. Walls and roofs are inherently opaque, so it is easy to achieve the low level of protection (hiding) with them. Achieving the high level of protection (hardening) for walls and roofs can be done within conventional construction using reinforced concrete, concrete-masonry units (CMUs), or clay brick. The material's required thickness is shown in Table 3-1. The thicknesses of CMUs and clay brick are nominal, meaning they do not represent the actual thickness of the material; they represent the thicknesses at which those materials are commercially available. Steel plates (mild steel and armor steel) and bullet-resistant fiberglass can be used to retrofit existing building components that would not provide the needed bullet resistance.

Table 3-1. Required Thicknesses, in Inches

Ballistics Type	Reinforced Concrete	Grouted CMU*	Clay Brick*	Steel Plate		Bullet-Resistant Fiberglass
				Mild	Armor	
.38 special	2	4	4	1/4	3/16	5/16
9 mm	2 1/2	4	4	5/16	1/4	7/16
7.62 and 5.56 mm	4	6	6	9/16	7/16	1 1/8
7.62-mm AP	6 1/2	8	8	13/16	11/16	N/A
*Nominal thicknesses						

Windows

3-49. Windows can include openings in walls and skylights, although skylights are only an issue where there are LOSs to them. When skylights require protection, treat them like windows. Achieving the low level of protection (hiding) for windows requires making it difficult to see through them, such as installing reflective film on the glass. An aggressor cannot see through the windows during daylight while it is lighter outside than inside, but he may see through them at night when the opposite might be true. Drapes or blinds that can be closed at night address that vulnerability. To achieve the high level of protection requires bullet-resistant window assemblies. These are commercially available for a wide range of ballistics types. They are purchased as manufactured-and-tested assemblies (including glazing and frames, both of which are equally bullet-resistant). The glazing materials and thicknesses and the framing details are proprietary to their manufacturers. The manufacturers make them according to industry test standards to ensure an effective product.

Doors

3-50. Doors without glass easily meet the requirements for the low level of protection. Meeting the high level of protection requires the installation of bullet-resistant door assemblies. Doors can be installed in foyers so that there is no direct LOS into assets within the building (see Figure 3-7, page 3-10).

FORCED ENTRY

3-51. In the forced-entry tactic, an aggressor tries to forcibly gain access to assets. He may use tools or explosives to breach building components or other barriers.

GENERAL-DESIGN STRATEGY

3-52. The general-design strategy for forced entry is to detect the aggressor early in the forced-entry attempt and delay him long enough for a response force to intercept him. The combination of detection and defensive measures must provide sufficient time for a response force to intercept the aggressor before he reaches the asset or before he escapes with it, depending on the protective goals for the asset. The first goal would apply where the asset is likely to be destroyed or where access to it is not acceptable. The second goal would be applied when the idea is to prevent it from being stolen.

LEVELS OF PROTECTION

3-53. Several levels of protection apply to forced entry. These levels vary in terms of system design, delay time, and response-force arrival time.

SITE-WORK ELEMENTS

3-54. Site-work elements do not normally play a major role in protecting against a forced entry. However, the site should be laid out and maintained so that an aggressor does not have a hiding place nearby that will conceal his attempts to break into the building. Another site-work element is the application of perimeter barriers, most commonly fences. Fences are effective at delineating a boundary and at keeping honest people honest, but they are ineffective for preventing a forced entry. The design strategy for forced entry is based on delaying the aggressor, and any serious aggressor could climb a fence in less than 4 seconds or can cut through a fence in less than 10 seconds. Therefore, fences are not used as delay elements, but they are used to establish boundaries and as platforms on which to hang sensors. The final site-work consideration is securing utility-access ports such as manholes. If there are utility tunnels through which aggressors can enter a building, those accesses should be locked using padlocks or locking bolts.

BUILDING ELEMENTS

3-55. Building elements are the principal construction elements of a system for protecting against a forced entry. The building elements are used to provide delay. The process for designing to resist forced entry involves laying out concentric "rings" of delay (called defensive layers). These defensive layers can include the facility's exterior, interior rooms within that layer, and containers within the interior rooms. The individual building components for each of the layers (walls, doors, windows, floors, ceilings, and roofs) provide the delay time (see TM 5-853-1).

DETECTION ELEMENTS

3-56. For a protective system to be effective against a forced entry, the aggressors must be detected at a point of adequate delay. Detection at that point can be achieved by using an IDS. Once a sensor detects an aggressor, the alarm annunciator communicates that event to security personnel, who then dispatch a response force. The alarm can be assessed through a guard response or via CCTV. Chapter 6 and TM 5-853-4 provide detailed discussion of IDSs, CCTV systems, and other elements of ESSs.

COVERT ENTRY AND INSIDER COMPROMISE

3-57. In the covert-entry tactic, an aggressor who is not authorized to be in the facility attempts to enter using false credentials. In the insider-compromise tactic, personnel with legitimate access to a facility try to compromise an asset. The insider may or may not have legitimate access to the asset itself. The purpose of the entry in either case can be to steal or otherwise compromise the asset or to destroy it. In the latter case, the aggressor may bring IEDs or IIDs.

GENERAL-DESIGN STRATEGY

3-58. The general-design strategy for both the insider-compromise and covert-entry tactics is to keep people from entering areas they are not authorized to enter. For covert entry, aggressors are denied access to controlled areas. For insider compromise, aggressors are denied access to assets within controlled areas based on their need to have access to them. The general-design strategy also includes detecting aggressors removing assets from protected areas and detecting aggressors carrying tools, weapons, and explosives into protected areas.

LEVELS OF PROTECTION

3-59. The levels of protection for these tactics address different issues, depending on whether the aggressor's goal is to steal or otherwise compromise an asset or to destroy it. When the goal is to steal or compromise an asset, the levels of protection vary with the number and sophistication of the access controls required to verify personnel access into a controlled area. When the goal is to destroy the assets, the levels of protection vary with the amount of damage the building (and the assets inside) are allowed to sustain and the sophistication of detecting weapons or explosives at entry points.

BUILDING ELEMENTS

3-60. Building elements vary with an aggressor's goal. To protect against theft or compromise of assets, building elements are used to establish and maintain controlled areas into which only authorized personnel can enter. For insider compromise, there may be an additional requirement that access be further limited among personnel otherwise authorized access to the controlled area. That access is based on the need to have access to a specific asset. The result is that the controlled area may be compartmentalized, and each compartmentalized area may have separate access requirements. There are no special construction requirements for these tactics if the goal is theft of compromise. The only requirement is that the building elements of controlled areas should provide enough resistance to require aggressors to force their way through them to gain entry and to provide evidence of the forced entry if it is attempted. Forcing entry would be contrary to the aggressor's assumed goal to be covert. In addition, a common design goal would be to limit the number of entrances into controlled areas because there will need to be access control at each entry.

3-61. To protect against the destruction of assets, building elements are used to shield assets from the effects of explosives going off at access-control points.

The basic approach is to lay out areas at access points in which guards can search for carried-in weapons, explosives, or incendiary devices. The construction of that area is designed to limit damage to the rest of the building if an explosive is detonated in that area. Those levels of damage are similar to those discussed in relation to vehicle bombs. The walls and doors between the access point and the protected area will be hardened, and the walls and doors to the outside will be of lightweight construction so that they may fail and vent the blast pressure away from the building. At the higher level of protection, the access-control area is located in a separate facility and the target building is hardened to resist an explosion in that separate facility.

DETECTION ELEMENTS

3-62. Detection elements for these tactics also vary based on the aggressor's goal. For theft, the detection elements are mainly related to access control. For destruction, the detection elements are used to detect weapons, explosives, or incendiary devices.

3-63. The main detection elements for theft or compromise are access-control devices. These can include procedural systems (such as guards checking ID), mechanical systems (such as keyed or combination locks), or electronic entry-control elements (such as electronic card readers, keypads, and biometric devices). Chapter 6 provides detailed discussion of electronic devices. The sophistication of these elements and the number used varies with the level of protection. For example, achieving the higher levels of protection requires the application of multiple forms of access-control elements such as a card reader and an electronic keypad for electronic-entry control or a badge check and badge exchange for a procedural system.

3-64. When destruction of the assets is the goal, detection is oriented toward detecting weapons, explosives, or incendiary devices. At the lower levels of protection, it is sufficient for guards to search for carried-in items. Achieving higher levels of protection requires the application of such equipment as metal detectors, X-ray machines, and explosive detectors.

SURVEILLANCE AND EAVESDROPPING

3-65. Surveillance and eavesdropping tactics include visual surveillance, acoustic eavesdropping, and electronic-emanations eavesdropping. In these tactics, aggressors remain outside of controlled areas and try to gather information from within those areas. The tools used for these tactics include ocular devices for the visual-surveillance tactic and listening devices and electronic-emanations-eavesdropping equipment for the eavesdropping tactic.

GENERAL-DESIGN STRATEGY

3-66. The general-design strategy for these tactics is to deny aggressors access to information assets. The kind of information (objects, operations, or files; secure conversations; or electronically processed data) and how it can be compromised differs for each tactic as do the specific protective strategies. Therefore, each tactic is addressed separately.

LEVELS OF PROTECTION

3-67. Each of these tactics has only one level of protection. Either one protects or fails to protect against these tactics.

SITE-WORK ELEMENTS

3-68. Site-work elements play a minor role in protecting assets from all surveillance or eavesdropping tactics. The main issue is to eliminate or control vantage points from which aggressors can surveil or eavesdrop on assets or operations. In addition, for the visual-surveillance tactic, a design goal can be to block LOSs from vantage points. Items used to block LOSs include trees, bushes, fences, and other buildings (see Figure 3-12).

BUILDING ELEMENTS

3-69. Building elements are the principal components of the protective strategies for surveillance and eavesdropping tactics. For visual surveillance,

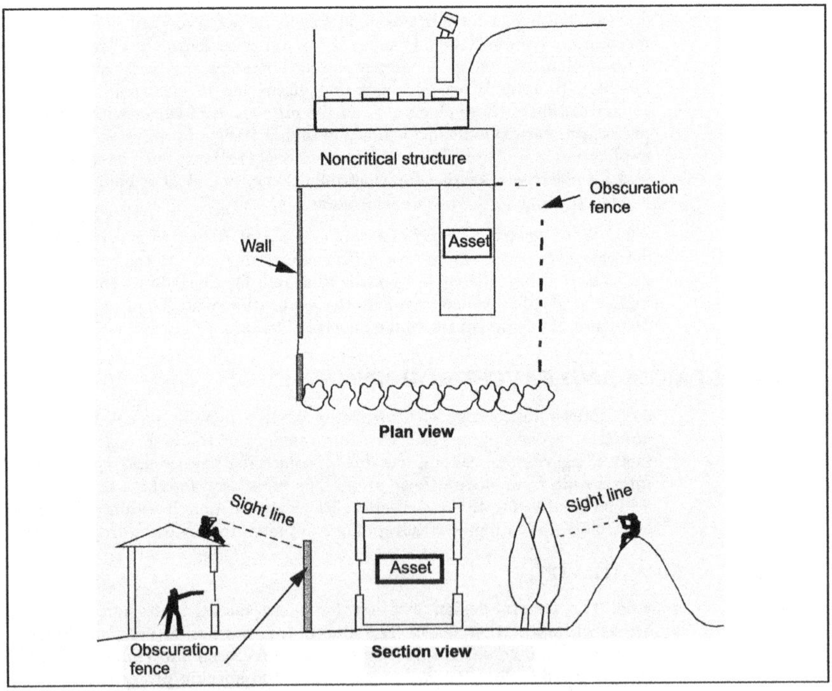

Figure 3-12. LOS Blocked From Potential Vantage Points

the building elements must block LOSs from outside the building. Walls and roofs perform this function effectively. Doors are only a problem when they have windows in them or are made of transparent materials. When this is the case, they can be treated like windows or they can be placed in foyers so that there are no direct LOSs through them. Windows can be treated with reflective film and drapes or blinds as described in the ballistics tactics. When there are LOSs through skylights, they should be treated like windows.

3-70. Building elements for acoustic eavesdropping relate to the construction of areas (preferably separated from the building exterior) that minimize the sound that can be transmitted through them. This requires specialized construction that has a sound-transmission-coefficient (STC) rating. Walls, floors, and ceilings can be constructed to achieve specific STC ratings using conventional construction materials as described in TM 5-853-1. Doors and windows that are STC rated are commonly manufactured and tested as assemblies. This type of design and construction can be expensive.

3-71. Protection against electronic-emanations eavesdropping involves the application of Terminal Electromagnetic-Pulse Emanation Standard (TEMPEST) guidance, most of which is classified. The protection is based on a TEMPEST assessment done for the Army by the US Army Intelligence and Security Command (INSCOM) and on guidance in AR 380-19. The results of a TEMPEST assessment will commonly lead to countermeasures from one or more of the following categories:

- Follow information security policies and procedures recommended during the assessments.
- Provide controlled space both inside and outside the facility.
- Provide TEMPEST-shielded equipment.
- Provide separation between electronic circuits that handle classified information and those that do not. This is commonly called red/black separation.
- Provide TEMPEST-shielded enclosures. This is specialized, metal-shielded construction that is very expensive.

MAIL AND SUPPLY BOMBS

3-72. In mail- and supply-bomb tactics, aggressors place bombs in materials delivered to a facility. Explosives used in supply bombs are significantly larger (briefcase size) than those in mail bombs (pipe bombs or smaller). Mail bombs are usually directed at individuals, while supply bombs may be used to target larger numbers of people. These tactics assume that the facility containing the asset has a mail-handling area or a supplies-handling and -receiving area. These tactics do not apply if mail or supplies are handled and screened in a different facility.

GENERAL-DESIGN STRATEGY

3-73. A bomb exploding within a building has more severe effects than the same size bomb exploding outside of the facility because the blast pressures cannot dissipate inside. Also, there is no standoff distance between the explosive and the facility to mitigate blast effects. The general-design strategy for mail and supply bombs is to detect delivered bombs before they explode and to harden the area where the explosion takes place. This minimizes the damage to the remainder of the facility. Occupants and contents within the mail room or supplies-handling area are likely to be killed or destroyed if an undetected bomb explodes.

LEVELS OF PROTECTION

3-74. The levels of protection for mail and supply bombs are based on the amount of damage allowed to the building and, therefore, the occupants of the building. They also vary based on the sophistication of the detection measures used.

BUILDING ELEMENTS

3-75. The purpose of building elements in relation to these bomb tactics is to shield assets from the effects of explosives going off at supply areas, receiving points, or mail rooms. The basic approach is to lay out either a mail room or a supplies-receiving area in which people can search suspicious packages for explosives or incendiary devices. Constructing this type of area will limit the damage to the rest of the building if an explosive is detonated there. Those levels of damage are similar to those discussed in relation to vehicle bombs.

Mail Rooms

3-76. Mail rooms should be located on the facility's exterior, away from any critical assets. The walls and ceiling between the mail room and the remainder of the building are hardened to keep the blast effects out of the facility. The exterior walls and doors should be of lightweight construction so that they may fail and vent the blast pressure away from the building. There may be an explosives container in the mail room where suspicious packages can be placed. If the package explodes, the container will keep its effects from causing damage or injury. The hardened construction will protect assets outside of the mail room if the explosion occurs outside of the container. Check with EOD personnel to determine the local policy for using explosive containers. At higher levels of protection, the mail room is constructed to completely contain the effects of an explosion either through hardened construction or by using a specialized construction called vented suppressive shielding. Mail rooms should not have windows into protected areas. Doors between the mail room and the rest of the building should be avoided, placed in foyers, or replaced with blast-resistant doors, depending on the desired level of protection.

Supplies-Handling Areas

3-77. Supplies-handling areas should also be on the building's exterior, away from critical areas of the facility. Walls and doors between the handling area and the protected area should be hardened, and the exterior walls and doors should be of lightweight construction so that they may fail and vent the blast pressure away from the building. There should be no windows between the handling area and the protected area. At the higher level of protection, the handling area is located in a separate facility and the target building is hardened to resist an explosion in that separate facility.

DETECTION ELEMENTS

3-78. Detection for these assets varies with the level of protection. At the lower levels of protection, bombs are detected by inspection. As the level of protection goes up, the sophistication of the detection increases. At the higher levels of protection, equipment such as X-ray examining devices, metal detectors, and explosives detectors can be used. Explosive-detection dogs are an alternative to explosive detectors.

CHEMICAL AND BIOLOGICAL CONTAMINATION

3-79. When using chemical- and biological-contamination tactics, aggressors introduce contaminants into the air or water supply to a facility or a group of facilities. Both airborne and waterborne contaminants include chemical, biological, and radiological agents. Aggressors may also forcibly enter a facility to contaminate water or air using the forced-entry tactic.

GENERAL-DESIGN STRATEGY

3-80. Both chemical and biological agents are difficult to detect in water and air supplies. Radiological agents are relatively easy to detect in water, but they are not commonly included in water-quality examinations. It is unlikely that all agents will be detected, so the general-design strategy for these tactics is to filter out suspected airborne contaminants or to shut off suspected waterborne contaminants. Also, because contaminants can easily be entered into the environment from inside a facility, the strategy includes limiting access to the facility (especially mechanical rooms, water intakes, and so forth).

LEVELS OF PROTECTION

3-81. The levels of protection for each of these tactics differ only in the frequency with which some protective measures are exercised. For the low level of protection, they are exercised only in response to a known threat. In the high level of protection, they are exercised continuously.

SITE-WORK ELEMENTS

3-82. Site-work elements are only significant for waterborne contamination. They include protecting water-treatment plants and water-storage structures. This protection may include constructing perimeter barriers (such as chain-link fences) and controlling access to the plant site. These measures are used because most contaminants require quantities on the order of truckloads to contaminate a water supply, so the focus of security is to keep such large vehicles under control. The perimeter barriers do not need to stop the vehicles because the assumption is that the aggressor wants to be covert. An overt act would alert people to avoid the water supply.

BUILDING ELEMENTS

3-83. Building elements for both tactics include controlling access so that aggressors cannot sneak in and plant devices in the building. Protection against airborne contamination at a facility involves making elements of the air-handling system (including air intakes) inaccessible and laying out toxin-free areas for people to be protected. A toxin-free area is an area in which the internal air pressure is higher than the external air pressure. Therefore, if a chemical, biological, or radiological device is set off outside, its contaminant will not be able to penetrate the protected area. Achieving that "net positive pressure" requires a significant air-handling system with air filters to filter contaminants out of the air. It also requires an air-lock entrance into the area so contaminants cannot enter through the door. At the low level of protection, the filters and the air-handling system are only used in response to a credible threat. At the high level of protection, that risk is not acceptable and the filters are run continuously.

3-84. The building-element issues for waterborne contamination are limited to providing protection against forced and covert entries into water-treatment plants and water-storage areas. These methods have been previously described. The only additional issue is the provision for alternative water sources. If it is suspected or detected that the water is contaminated, a backup water source should be in place (such as bottled water). For the high level of protection, bottled water should always be used for drinking.

Chapter 4

Protective Barriers

Protective barriers are used to define the physical limits of an installation, activity, or area. Barriers restrict, channel, or impede access and are fully integrated to form a continuous obstacle around the installation. They are designed to deter the worst-case threat. The barriers should be focused on providing assets with an acceptable level of protection against a threat.

OVERVIEW

4-1. Protective barriers form the perimeter of controlled, limited, and exclusion areas. Utility areas (such as water sources, transformer banks, commercial power and fuel connections, heating and power plants, or air-conditioning units) may require these barriers for safety standards. Protective barriers consist of two major categories—natural and structural.

- Natural protective barriers are mountains and deserts, cliffs and ditches, water obstacles, or other terrain features that are difficult to traverse.
- Structural protective barriers are man-made devices (such as fences, walls, floors, roofs, grills, bars, roadblocks, signs, or other construction) used to restrict, channel, or impede access.

4-2. Barriers offer important benefits to a physical-security posture. They create a psychological deterrent for anyone thinking of unauthorized entry. They may delay or even prevent passage through them. This is especially true of barriers against forced entry and vehicles. Barriers have a direct impact on the number of security posts needed and on the frequency of use for each post.

4-3. Barriers cannot be designed for all situations. Considerations for protective structural barriers include the following:

- Weighing the cost of completely enclosing large tracts of land with significant structural barriers against the threat and the cost of alternate security precautions (such as patrols, MWD teams, ground sensors, electronic surveillance, and airborne sensors).
- Sizing a restricted area based on the degree of compartmentalization required and the area's complexity. As a rule, size should be kept to a minimum consistent with operational efficiency. A restricted area's size may be driven by the likelihood of an aggressor's use of certain tactics. For example, protecting assets from a vehicle bomb often calls for a substantial explosives standoff distance. In these cases, mitigating the vehicle bomb would often be more important than minimizing the restricted area to the extent necessary for operational efficiency. Protective barriers should be established for—
  - Controlling vehicular and pedestrian traffic flow.

- Providing entry-control points where ID can be checked.
- Defining a buffer zone for more highly classified areas.
- Precluding visual compromise by unauthorized individuals.
- Delaying forced entry.
- Protecting individual assets.

4-4. If a secured area requires a limited or exclusion area on a temporary or infrequent basis, it may not be possible to use physical structural barriers. A temporary limited or exclusion area may be established where the lack of proper physical barriers is compensated for by additional security posts, patrols, and other security measures during the period of restriction. Temporary barriers (including temporary fences, coiled concertina wire, and vehicles) may be used. Barriers are not the only restrictive element, and they may not always be necessary. They may not be ideal when working with limited or exclusion areas or when integrated with other controls.

4-5. Because barriers can be compromised through breaching (cutting a hole through a fence) or by nature (berms eroded by the wind and rain), they should be inspected and maintained at least weekly. Guard-force personnel should look for deliberate breaches, holes in and under barriers, sand dunes building up against barriers, and the proper functioning of locks.

FENCING

4-6. Three types of fencing are authorized for use in protecting restricted areas—chain link, barbed wire, and barbed tape or concertina. The type used for construction depends primarily on the threat and the degree of permanence. It may also depend on the availability of materials and the time available for construction. Fencing may be erected for other uses besides impeding personnel access. It can impede observation, can serve as a means to defeat standoff-weapon systems (such as rocket-propelled grenades [RPGs]), and can serve as a barrier to hand-thrown weapons (such as grenades and firebombs).

4-7. Generally, chain-link fencing will be used for protecting permanent limited and exclusion areas. All three types of fencing may be used to augment or increase the security of existing fences that protect restricted areas. Examples would be to create an additional barrier line, to increase existing fence height, or to provide other methods that effectively add to physical security. It is important to recognize that fencing provides very little delay when it comes to motivated aggressors, but it can act as a psychological deterrent.

CHAIN LINK

4-8. Chain-link fence (including gates) must be constructed of 6-foot material, excluding the top guard. Fence heights for conventional arms and ammunition security must be 6 feet for standard chain-link, wire-mesh fencing. Chain-link fences must be constructed with 9-gauge or heavier wire. They must be galvanized with mesh openings not larger than 2 inches per side and have twisted and barbed selvages at the top and the bottom. The wire must be taut and securely fastened to rigid metal or reinforced-concrete posts set in concrete. It must reach within 2 inches of hard ground or pavement. On soft ground, it must reach below the surface deep enough to compensate for shifting soil or sand. Materials and construction must meet with the US Army Corps of Engineers (USACE) guide specifications shown in the USACE Standard (STD) 872-90 series. Weaknesses in the chain-link fence occur as a result of weather (rusting) or failure to keep it fastened to the post that affects the desired tightness. Damage to the fence and fence fabric may be the result of allowing vegetation and trees to grow on or near the fence. The interaction between the fence and the overgrowth often leads to fence damage and reduces the integrity and continuity of the fence as a perimeter boundary and barrier. The perimeter fence is the most obvious protective measure. A well-maintained fence indicates that the asset owner is dedicated to physical security.

BARBED WIRE

4-9. Standard barbed wire is twisted, double-strand, 13.5-gauge wire, with four-point barbs spaced an equal distance apart. Barbed-wire fencing (including gates) intended to prevent human trespassing should not be less than 6 feet high and must be affixed firmly to posts not more than 6 feet apart. The distance between strands should not exceed 6 inches, and at least one wire should be interlaced vertically and midway between posts. The ends must be staggered or fastened together, and the base wire must be picketed to the ground.

BARBED TAPE OR CONCERTINA

4-10. A barbed-taped obstacle (BTO) is fabricated from 0.025-inch stainless steel and is available in 24-, 30-, 40-, and 60-inch-diameter coils. The barbs shall have a minimum length of 1.2 inches, and the barb cluster's width shall be 1.21 inches. A BTO deploys tangle-free for fast installation. It may be recovered and used again. Fifty feet (plus or minus 2 inches) can be covered by 101 coil loops. Handling barbed tape requires the use of heavy barbed-tape gauntlets instead of standard barbed-wire gauntlets.

Barbed-Tape Concertina

4-11. Barbed-tape concertina (standard concertina barbed tape) is a commercially manufactured wire coil of high-strength-steel barbed wire that is clipped together at intervals to form a cylinder. When opened, it is 50 feet long and 3 feet in diameter. When used as the perimeter barrier for a restricted area, the concertina must be laid between poles with one roll on top of another or in a pyramid arrangement (with a minimum of three rolls).

4-12. Reinforced barbed-tape concertina consists of a single strand of spring-steel wire and a single strand of barbed tape. The sections between barbs of the barbed tape are securely clinched around the wire. Each coil is about 37 1/2 inches in diameter and consists of 55 spiral turns connected by steel clips to form a cylindrical diamond pattern when extended to a coil length of 50 feet. One end turn is fitted with four bundling wires for securing the coil when closed and each end turn is fitted with two steel carrying loops. The concertina extends to 50 feet without permanent distortion. When released, it can be retracted into a closed coil.

4-13. When possible, a top guard should be constructed on all perimeter fences and may be added on interior enclosures for additional protection. A top guard is an overhang of barbed wire or tape along the top of a fence, facing outward and upward at about a 45-degree angle. Placing barbed wire or tape above it can further enhance the top guard. Top-guard supporting arms will be permanently affixed to the top of fence posts to increase the overall height of the fence by at least 1 foot. (Due to liability issues in some locations, the top guards will not be allowed to face outward where the fence is adjacent to public areas.) Three strands of barbed wire spaced 6 inches apart must be installed on the supporting arms. The number of strands of wire or tape may be increased when required. The top guard of fencing adjoining gates may range from a vertical height of 18 inches to the normal 45-degree outward protection but only for sufficient distance along the fence to open the gates adequately. Bottom and top tension wires should be used in lieu of fence rails. A concrete sill may be cast at the bottom of the fence to protect against soil erosion. A bottom rail is used on high-security fences to prevent intruders from lifting the fence.

Gates and Entrances

4-14. The number of gates and perimeter entrances must be the minimum required for safe and efficient operation of the facility. Active perimeter entrances must be designed so that the guard force maintains full control. Semiactive entrances, such as infrequently used vehicular gates, must be locked on the inside when not in use. When closed, gates and entrances must provide a barrier structurally comparable to their associated barriers. Care must be afforded against the ability to crawl under gates. Top guards, which may be vertical, are required for all gates.

Triple-Standard Concertina (TSC) Wire

4-15. This type of fence uses three rolls of stacked concertina. One roll will be stacked on top of two rolls that run parallel to each other while resting on the ground, forming a pyramid. In many situations, this fence has been used effectively in place of a chain-link fence. (If perimeter fencing consists of TSC, a top guard is not feasible.)

Tangle-Foot Wire

4-16. Barbed wire or tape may be used in appropriate situations to construct a tangle-foot obstruction either outside a single perimeter fence or in the area between double fences to provide an additional deterrent to intruders. The wire or tape should be supported on short metal or wooden pickets spaced at irregular intervals of 3 to 10 feet and at heights between 6 and 12 inches. The wire or tape should be crisscrossed to provide a more effective obstacle. The space and materials available govern the depth of the field.

AIRCRAFT CABLE

4-17. Although not used very often, aircraft cable can be used as a temporary barrier. Refer to FM 5-34 for information required for determining the barrier's strength. The barrier is created using wire rope. Clips are spaced six times the diameter of the wire rope. Aircraft cable (deployed as described above or attached to a chain-link fence) can also be made to act as a barrier to moving vehicles. To do so, the cable must be anchored into the ground at both ends at about 200-foot intervals (see TM 5-853-1).

UTILITY OPENINGS

4-18. Sewers, air and water intakes and exhausts, and other utility openings of 10 inches or more in diameter that pass through perimeter barriers must have security measures equivalent to that of the perimeter (see TM 5-820-4). Specific requirements of various openings are discussed below:

- Manhole covers 10 inches or more in diameter must be secured to prevent unauthorized opening. They may be secured with locks and hasps, by welding them shut, or by bolting them to their frame. Ensure that hasps, locks, and bolts are made of materials that resist corrosion. Keyed bolts (which make removal by unauthorized personnel more difficult) are also available.
- Drainage ditches, culverts, vents, ducts, and other openings that pass through a perimeter and that have a cross-sectional area greater than 96 square inches and whose smallest dimension is greater than 6 inches will be protected by securely fastened welded bar grilles (refer to TM 5-853-3, Figure 8-1). As an alternative, drainage structures may be constructed of multiple pipes, with each pipe having a diameter of 10 inches or less. Multiple pipes of this diameter may also be placed and secured in the inflow end of a drainage culvert to prevent intrusion into the area. Ensure that any addition of grilles or pipes to culverts or other drainage structures is coordinated with the engineers so that they can compensate for the diminished flow capacity and additional maintenance that will result from the installation.

OTHER PERIMETER BARRIERS

4-19. Buildings less than two stories high that form part of a perimeter must have a top guard along the outside edge to deny access to the roof. When using masonry walls as part of a perimeter barrier, they must be at least 7 feet high and have a barbed-wire top guard. The top guard should be sloped outward at a 45-degree angle and carry at least three strands of barbed wire. This will increase the vertical height of the barrier by at least 1 foot.

4-20. Protect windows, active doors, and other designated openings by securely fastening bars, grilles, or chain-link screens. Fasten window barriers from the inside. If hinged, the hinges and locks must be on the inside. Building elements that provide delay against forced entry have stringent requirements. These elements should be designed according to TM 5-853-1.

SECURITY TOWERS

4-21. It is not acceptable to observe a perimeter from towers only. However, all towers should be located to provide maximum observation and should be constructed for protection from small-arms fire.

4-22. Mobile towers are useful in some temporary situations such as a large, open storage area where receiving and storing activities take place. All facilities using towers must have a support force available for emergencies. Tower personnel should be rotated at frequent intervals.

4-23. The height of a tower increases the range of observation during daylight hours and at night with artificial illumination. However, during inclement weather and during a blackout, towers lose this advantage and must be supplemented by on-ground observation.

4-24. The following considerations should be made when planning for the use of towers:

- Hardening the tower against small-arms effects by using sandbags, salvaged armor, or commercially fabricated bullet-resistant construction. This may require strengthening the tower supports, which should be performed only under the supervision of an engineer. The level of protection required must equate to the threat level identified during the IPB or the military decision-making process (MDMP). The best approach is to design for the worst identified threat rather than to try and modify the tower at a later date on short notice.
- Installing communications and alarm systems, both audible and visual (primary and alternate).
- Using appropriate surveillance, target-acquisition, and night-observation (STANO) equipment with the tower and perimeter barriers being surveilled. Infrared (IR) items may be especially valuable. Considerations for the selection and use of STANO equipment must be made while evaluating the effects of perimeter protective lighting.
- Providing security lighting for route protection to the tower. Security lighting also allows for support of the guard force entering or exiting the perimeter.
- Ensuring that the tower's height is determined according to the area of observation.
- Ensuring that towers have overlapping, mutually supporting fields of observation and fire.
- Providing towers with a backup fortified defensive fighting position, as appropriate.

INSTALLATION ENTRANCES

4-25. The number of installation or activity gates and perimeter entrances in active use should be limited to the minimum number required for safe and efficient operations. When necessary, install vehicle barriers in front of vehicle gates. Security lighting should be considered at entry points (see Chapter 5). Refer to TM 5-853-1 for the application and selection of these barriers.

4-26. Plans to use guards for controlling entry to an installation or activity must be predetermined based on the threat conditions (THREATCON). The construction of the guard post must be included in the security plan.

PERIMETER ENTRANCES

4-27. Active perimeter entrances should be designated so that security forces maintain full control without an unnecessary delay in traffic. This is accomplished by having sufficient entrances to accommodate the peak flow of pedestrian and vehicular traffic and having adequate lighting for rapid and efficient inspection. When gates are not operational during nonduty hours, they should be securely locked, illuminated during hours of darkness, and inspected periodically by a roving patrol. Additionally, warning signs should be used to warn drivers when gates are closed. Doors and windows on buildings that form a part of the perimeter should be locked, lighted, and inspected.

ENTRY-CONTROL STATIONS

4-28. Entry-control stations should be provided at main perimeter entrances where security personnel are present. Considerations for construction and use should be based on the information outlined in USACE STD 872-50-01.

4-29. Entry-control stations should be located as close as practical to the perimeter entrance to permit personnel inside the station to maintain constant surveillance over the entrance and its approaches. Additional considerations at entry-control stations include—

- Establishing a holding area for unauthorized vehicles or those to be inspected further. A turnaround area should be provided to keep from impeding other traffic.
- Establishing control measures such as displaying a decal on the window or having a specially marked vehicle.

4-30. Entry-control stations that are manned 24 hours each day should have interior and exterior lighting, interior heating (where appropriate), and a sufficient glassed area to afford adequate observation for personnel inside. Where appropriate, entry-control stations should be designed for optimum personnel ID and movement control. Each station should also include a telephone, a radio, and badge racks (if required).

4-31. Signs should be erected to assist in controlling authorized entry, to deter unauthorized entry, and to preclude accidental entry. Signs should be plainly displayed and be legible from any approach to the perimeter from a reasonable distance. The size and coloring of a sign, its letters, and the interval of posting must be appropriate to each situation.

4-32. Entry-control stations should be hardened against attacks according to the type of threat. The methods of hardening may include—

- Reinforced concrete or masonry.
- Steel plating.
- Bullet-resistant glass.
- Sandbags, two layers in depth.
- Commercially fabricated, bullet-resistant building components or assemblies.

WARNING SIGNS

4-33. A significant amount of warning signs should be erected to ensure that possible intruders are aware of entry into restricted areas. Warning signs augment control signs. They warn intruders that the area is restricted and that trespassing may result in the use of deadly force.

4-34. Warning signs should be installed along the limited area's physical barriers and at each entry point where they can be seen readily and understood by anyone approaching the perimeter. In areas where English is one of two or more languages commonly spoken, warning signs must contain the local language in addition to English. The wording on the signs will denote warning of a restricted area. The signs should be posted at intervals of no more than 100 feet. They must not be mounted on fences equipped with intrusion-detection equipment. Additionally, the warning signs prescribed in AR 190-13 should be posted at all entrances to limited, controlled, and exclusion areas. See Chapter 7 for more details.

OTHER SIGNS

4-35. Signs setting forth the conditions of entry to an installation or area should be plainly posted at all principal entrances. The signs should be legible under normal conditions at a distance not less than 50 feet from the point of entry. Such signs should inform the entrant of the provisions (search of the person, the vehicle, packages, and so forth) or prohibitions (such as against cameras, matches, and lighters and entry for reasons other than official business) that may be prescribed by the installation commander.

4-36. Signs or notices legibly setting forth the designation of restricted areas and provisions of entry should be plainly posted at all entrances and at other points along the perimeter line as necessary. The wording of these signs or notices is prescribed in AR 190-13.

INSTALLATION PERIMETER ROADS AND CLEAR ZONES

4-37. When the perimeter barrier encloses a large area, an interior all-weather perimeter road should be provided for security-patrol vehicles. Clear zones should be maintained on both sides of the perimeter barrier to provide an unobstructed view of the barrier and the ground adjacent to it. Roads within the clear zone should be as close to the perimeter barrier as possible without interfering with it. The roads should be constructed to allow effective road barriers to deter motor movement of unauthorized personnel during mobilization periods.

4-38. Clear zones should be kept clear of weeds, rubbish, or other material capable of offering concealment or assistance to an intruder attempting to breach the barrier. A clear zone of 20 feet or more should exist between the perimeter barrier and exterior structures, parking areas, and natural or man-made features. When possible, a clear zone of 50 feet or more should exist between the perimeter barrier and structures within the protected area, except when a building's wall constitutes part of the perimeter barrier. Ammunition supply points (ASPs) will have clear zones 12 feet outside of the ASP and 30 feet inside, and the vegetation will not exceed 8 inches (4 inches for high-threat and highly controlled areas). Refer to AR 190-11 and DOD 0-2000.12-H, Appendix EE, for further information.

4-39. When it is impossible to have adequate clear zones because of property lines or natural or man-made features, it may be necessary to increase the height of the perimeter barrier, increase security-patrol coverage, add more security lighting, or install an intrusion-detection device along that portion of the perimeter.

4-40. When considering the construction of a new site or perimeter, ensure that the plans include a fence located well inside the property line, thus permitting control of enough space outside the fence to maintain at least a minimal clear zone. The following considerations apply:

- On a large installation (such as a proving ground), it is unreasonable to construct an expensive perimeter fence and keep it under constant observation. Such an installation is usually established in a sparsely inhabited area. Its comparative isolation and the depth of the installation give reasonable perimeter protection. Under these circumstances, it is usually sufficient to post warning signs or notices, reduce access roads to a minimum, and periodically patrol the area between the outer perimeter and the conventionally protected vital area of the installation.
- An alternative to erecting new or replacing old chain-link fence involving an entire installation perimeter is to relocate or isolate the sensitive area or item by—
  - Relocating the item within a safe perimeter.
  - Consolidating the item with other items.
  - Erecting a chain-link fence (regulations permitting) around individual assets rather than the installation's perimeter.

ARMS-FACILITY STRUCTURAL STANDARDS

4-41. It is next to impossible to build a protective barrier that cannot be penetrated by a human or heavy armor. Therefore, as opposed to protecting a facility using only one barrier, enhance security by using a combination of barriers to increase delay. Multiple barriers also cause aggressors to expend more energy trying to breach all of the barriers. They also provide the appearance of additional security and may further deter some aggressors.

4-42. The interest of security must be kept in mind when constructing walls, ceilings, floors, and roofs. Facilities that house arms and ammunition are constructed as security barriers in the interest of deterring and delaying penetration. Construction guidelines for arms facilities are outlined in AR 190-11. AR 190-11 requires coordination with the engineer office, the safety office, the provost marshal office (PMO), or the security-force office when definitive drawings and specifications for new construction or upgrades or modifications of AA&E storage structures are proposed. This coordinated effort ensures that safety and physical-security requirements are met. AR 190-11 also addresses waivers and exceptions for AA&E storage structures, as well as the requirements for a tactical (training or operational) or shipboard environment. Waivers and exceptions are not discussed in this manual. The following guidelines are provided for securing AA&E in tactical and shipboard environments:

- The criteria and standards for protecting AA&E will be developed by the major Army command (MACOM) according to AR 190-11.
- The deploying commander will establish and enforce procedures for securing deployed AA&E based on the assessment of the threat, the objectives, the location, and the duration of the deployment.
- The AA&E in the tactical environment will be secured at all times.
- The AA&E will be under continuous positive control.
- Persons charged with the custody of AA&E will have the capability to sound the alarm if a forceful theft is attempted.
- A response force will be available to protect the AA&E.
- A system of supervisory checks will be established to ensure that all personnel comply with security measures. Supervisory checks of the AA&E holding area will be made to ensure that the AA&E being guarded have not been tampered with.
- All officers, noncommissioned officers (NCOs), or civilian equivalents will closely monitor the control of ammunition and explosives during field training or range firing.
- Selection of personnel to perform guard duties at AA&E holding areas will be closely monitored by commanders to ensure that only responsible individuals are assigned duties.

Chapter 5

Physical-Security Lighting

Security lighting allows security personnel to maintain visual-assessment capability during darkness. When security-lighting provisions are impractical, additional security posts, patrols, MWD patrols, NVDs, or other security means are necessary.

OVERVIEW

5-1. Security lighting should not be used as a psychological deterrent only. It should also be used along perimeter fences when the situation dictates that the fence be under continuous or periodic observation.

5-2. Lighting is relatively inexpensive to maintain and, when properly used, may reduce the need for security forces. It may also enhance personal protection for forces by reducing the advantages of concealment and surprise for a determined intruder.

5-3. Security lighting is desirable for those sensitive areas or structures within the perimeter that are under observation. Such areas or structures include pier and dock areas, vital buildings, storage areas, motor pools, and vulnerable control points in communication and power- and water-distribution systems. In interior areas where night operations are conducted, adequate lighting facilitates the detection of unauthorized persons approaching or attempting malicious acts within the area. Security lighting has considerable value as a deterrent to thieves and vandals and may make the job of the saboteur more difficult. It is an essential element of an integrated physical-security program.

5-4. A secure auxiliary power source and power-distribution system for the facility should be installed to provide redundancy to critical security lighting and other security equipment. During deployed operations, primary power may not exist or may be subject to constraints or interruptions due to poor infrastructure or hostile activity. Auxiliary power sources must be available for critical electrical loads and must be secured against direct and indirect fires as well as sabotage. If automatic-transfer switches are not installed, security procedures must designate the responsibility for the manual start of the source.

COMMANDER'S RESPONSIBILITY

5-5. Commanders determine perimeter lighting needs based on the threat, site conditions along the perimeter, surveillance capabilities, and available guard forces. Commanders ensure that security lighting is designed and used to discourage unauthorized entry and to facilitate the detection of intruders approaching or attempting to gain entry into protected areas.

PLANNING CONSIDERATIONS

5-6. Security lighting usually requires less intensity than working lights, except for ID and inspection at entry-control points. Each area of a facility presents its own unique set of considerations based on physical layout, terrain, atmospheric and climatic conditions, and security requirements. Information is available from the manufacturers of lighting equipment and from the installation's director of public works, who will assist in designing a lighting system. This information includes—

- Descriptions, characteristics, and specifications of various lighting fixtures, arc, and gaseous-discharge lamps.
- Lighting patterns of various fixtures.
- Typical layouts showing the most efficient height and spacing of equipment.
- Minimum levels of illumination and lighting uniformity required for various applications.

5-7. In planning a security-lighting system, the physical-security manager considers the—

- Cost of replacing lamps and cleaning fixtures, as well as the cost of providing the required equipment (such as ladders and mechanical buckets) to perform this maintenance.
- Provision of manual-override capability during a blackout, including photoelectric controls. These controls may be desirable in a peacetime situation but undesirable when a blackout is a possibility.
- Effects of local weather conditions on lighting systems.
- Fluctuating or erratic voltages in the primary power source.
- Grounding requirements.
- Provisions for rapid lamp replacement.
- Use of lighting to support a CCTV system.
- Limited and exclusion areas. Specific lighting requirements are referenced in AR 190-59 and TM 5-853-2. TM 5-853-4 provides guidance for facility applications that include CCTV cameras.
  - Lighting in these areas must be under the control of the guard force.
  - For critical areas (such as weapons storage areas), instantaneous lighting with a backup source is required. Any period without lighting in a critical area is unacceptable. Therefore, these areas generally have a requirement for backup power (such as diesel-engine generators, uninterrupted power supplies, and batteries) in case of power loss.
  - Security-lighting systems are operated continuously during hours of darkness.
  - Protective lights should be used so that the failure of one or more lights will not affect the operation of the remaining lights.
- Lighting requirements for adjoining properties and activities.
- Restrike time (the time required before the light will function properly after a brief power interruption).

- Color accuracy.
- Other facilities requiring lighting, such as parking areas.

PRINCIPLES OF SECURITY LIGHTING

5-8. Security lighting enables guard-force personnel to observe activities around or inside an installation while minimizing their presence. An adequate level of illumination for all approaches to an installation will not discourage unauthorized entry; however, adequate lighting improves the ability of security personnel to assess visually and intervene on attempts at unauthorized entry. Lighting is used with other security measures (such as fixed security posts or patrols, fences, and ESSs) and should never be used alone. Other principles of security lighting include the following:

- Optimum security lighting is achieved by adequate, even light on bordering areas; glaring lights in the eyes of an intruder; and little light on security-patrol routes. In addition to seeing long distances, security forces must be able to see low contrasts (such as indistinct outlines of silhouettes) and must be able to detect an intruder who may be exposed to view for only a few seconds. Higher levels of illumination improve these abilities.
- High brightness contrast between an intruder and the background should be the first consideration when planning for security lighting. With predominantly dark, dirty surfaces or camouflage-type painted surfaces, more light is needed to produce the same brightness around installations and buildings than when clean concrete, light brick, and grass predominate. When the same amount of light falls on an object and its background, the observer must depend on contrasts in the amount of light reflected. His ability to distinguish poor contrasts is significantly improved by increasing the illumination level.
- The observer primarily sees an outline or a silhouette when the intruder is darker than his background. Using light finishes on the lower parts of buildings and structures may expose an intruder who depends on dark clothing and darkened face and hands. Stripes on walls have also been used effectively, as they provide recognizable breaks in outlines or silhouettes. Providing broad-lighted areas around and within the installation against which intruders can be seen can also create good observation conditions.

5-9. To be effective, two basic systems or a combination of both may be used to provide practical and effective security lighting. The first method is to light the boundaries and approaches; the second is to light the area and structures within the property's general boundaries. Protective lighting should—

- Discourage or deter attempts at entry by intruders. Proper illumination may lead a potential intruder to believe detection is inevitable.
- Make detection likely if entry is attempted.
- Prevent glare that may temporarily blind the guards.

TYPES OF LIGHTING

5-10. The type of lighting system used depends on the installation's overall security requirements. Four types of lighting units are used for security-lighting systems—continuous, standby, movable (portable), and emergency.

5-11. Continuous lighting is the most common security-lighting system. It consists of a series of fixed lights arranged to flood a given area continuously during darkness with overlapping cones of light. Two primary methods of using continuous lighting are glare projection and controlled lighting.

- The glare security-lighting method is used when the glare of lights directed across the surrounding territory will not be annoying nor interfere with adjacent operations. It is a strong deterrent to a potential intruder because it makes it difficult to see inside of the area. Guards are protected by being kept in comparative darkness and being able to observe intruders at a considerable distance beyond the perimeter.
- Controlled lighting is best when it limits the width of the lighted strip outside the perimeter, such as along highways. In controlled lighting, the width of the lighted strip is controlled and adjusted to fit the particular need. This method of lighting may illuminate or silhouette security personnel.

5-12. Standby lighting has a layout similar to continuous lighting. However, the luminaries are not continuously lit but are either automatically or manually turned on when suspicious activity is detected or suspected by the security force or alarm systems.

5-13. Movable lighting consists of manually operated, movable searchlights that may be lit during hours of darkness or only as needed. The system normally is used to supplement continuous or standby lighting.

5-14. Emergency lighting is a system of lighting that may duplicate any or all of the above systems. Its use is limited to times of power failure or other emergencies that render the normal system inoperative. It depends on an alternative power source such as installed or portable generators or batteries.

FENCED PERIMETERS

5-15. Fenced perimeters require the lighting specifications indicated in TM 5-853-2. Specific lighting requirements are based on whether the perimeter is isolated, semi-isolated, or nonisolated.

- Isolated fenced perimeters are fence lines around areas where the fence is 100 feet or more from buildings or operating areas. The approach area is clear of obstruction for 100 or more feet outside of the fence. Other personnel do not use the area. Use glare projection for these perimeters and keep patrol routes unlit.
- Semi-isolated fenced perimeters are fence lines where approach areas are clear of obstruction for 60 to 100 feet outside of the fence. The general public or installation personnel seldom have reason to be in the area. Use controlled lighting for these perimeters and keep patrol routes in relative darkness.

- Nonisolated fenced perimeters are fence lines immediately adjacent to operating areas. These areas may be within an installation or public thoroughfares. Outsiders or installation personnel may move about freely in this approach area. The width of the lighted strip depends on the clear zones inside and outside the fence. Use controlled lighting for these perimeters. It may not be practical to keep the patrol area dark.

ENTRANCES

5-16. Entrances for pedestrians will have two or more lighting units providing adequate illumination for recognition of persons and examination of credentials. Vehicle entrances will have two lighting units located to facilitate the complete inspection of passenger cars, trucks, and freight cars as well as their contents and passengers. Semiactive and inactive entrances will have the same degree of continuous lighting as the remainder of the perimeter, with standby lighting to be used when the entrance becomes active. Gatehouses at entrances should have a low level of interior illumination, enabling guards to see approaching pedestrians and vehicles.

OTHER

5-17. Areas and structures within the installation's property line consist of yards; storage spaces; large, open working areas; piers; docks; and other sensitive areas and structures.

- Open yards (unoccupied land only) and outdoor storage spaces (material storage areas, railroad sidings, motor pools, and parking areas) should be illuminated. An open yard adjacent to a perimeter (between guards and fences) will be illuminated according to the perimeter's illumination requirements. Where lighting is necessary in other open yards, illumination will not be less than 0.2 foot-candle at any point.
- Lighting units are placed in outdoor storage spaces to provide an adequate distribution of light in aisles, passageways, and recesses to eliminate shadowed areas where unauthorized persons may hide.
- Illuminating both water approaches and the pier area safeguards piers and docks located on an installation. Decks on open piers will be illuminated to at least 1 foot-candle and the water approaches (extending to a distance of 100 feet from the pier) to at least 0.5 foot-candle. The area beneath the pier floor will be lit with small wattage floodlights arranged on the piling. Movable lighting is recommended as a part of the protective lighting system for piers and docks. The lighting must not in any way violate marine rules and regulations (it must not be glaring to pilots). Consult the US Coast Guard (USCG) for approval of protective lighting adjacent to navigable waters.

WIRING SYSTEMS

5-18. The wiring circuit should be arranged so that failure of any one lamp will not leave a large portion of the perimeter line or a major segment of a critical or vulnerable position in darkness. Feeder lines will be placed underground (or sufficiently inside the perimeter in the case of overhead wiring) to minimize the possibility of sabotage or vandalism from outside the perimeter. Another advantage to underground wiring is reduced effects from adverse weather conditions.

MAINTENANCE

5-19. Periodic inspections will be made of all electrical circuits to replace or repair worn parts, tighten connections, and check insulation. Keep fixtures clean and properly aimed.

POWER SOURCES

5-20. Primary and alternate power sources must be identified. The following is a partial list of considerations:

- The primary source is usually a local public utility.
- An alternate source (standby batteries or diesel-fuel-driven generators may be used) is provided where required and should—
  - Start automatically upon failure of primary power.
  - Be adequate to power the entire lighting system.
  - Be equipped with adequate fuel storage and supply.
  - Be tested under load to ensure efficiency and effectiveness.
  - Be located within a controlled area for additional security.

CCTV-CAMERA LIGHTING REQUIREMENTS

5-21. TM 5-853-4 provides a detailed discussion of CCTV-camera lighting requirements and guidelines for minimum lighting levels and lighting uniformity. The following considerations apply when lighting systems are intended to support CCTV assessment or surveillance:

- The camera's field of view.
- Lighting intensity levels.
- Maximum light-to-dark ratio.
- Scene reflectance.
- Daylight-to-darkness transitions.
- Camera mounting systems relative to lighting.
- The camera's spectral response.
- The cold-start time.
- The restrike time.

Chapter 6

Electronic Security Systems

An overall site-security system is comprised of three major subelements—detection, delay, and response. The detection subelement includes intrusion detection, assessment, and entry control. An ESS is an integrated system that encompasses interior and exterior sensors; CCTV systems for assessing alarm conditions; electronic entry-control systems (EECSs); data-transmission media (DTM); and alarm reporting systems for monitoring, controlling, and displaying various alarm and system information. Interior and exterior sensors and their associated communication and display subsystems are collectively called IDSs.

OVERVIEW

6-1. Many Army and DOD regulations specify protective measures, policies, and operations related to security. Although the regulations specify minimum requirements, it is possible that more stringent requirements will be necessary at specific sites. A designer will use a previously performed site survey to determine which regulations apply and to determine whether circumstances require more stringent measures. Refer to TM 5-853-4 for additional detailed information.

6-2. AR 190-13 requires the use of a standardized ESS, if practical and available. The receiving element must determine whether a standardized system can meet the requirements and whether it is available. After coordinating with the product manager for physical-security equipment to verify that a standardized system is available, the associated MACOM can issue approval to procure a commercial system in lieu of a standardized system.

USE OF ESS

6-3. An ESS is used to provide early warning of an intruder. This system consists of hardware and software elements operated by trained security personnel.

6-4. A system is configured to provide one or more layers of detection around an asset. Each layer is made up of a series of contiguous detection zones designed to isolate the asset and to control the entry and exit of authorized personnel and materials.

GENERAL ESS DESCRIPTION

6-5. An ESS consists of sensors interfaced with electronic entry-control devices, CCTV, alarm reporting displays (both visual and audible), and security lighting. The situation is assessed by sending guards to the alarm point or by using CCTV. Alarm reporting devices and video monitors are located in the security center. The asset's importance will determine whether multiple or redundant security centers are required and, ultimately, the required sophistication of all elements in the ESS. Digital and analog data are transmitted from local (field) interior and exterior locations to the security center for processing. Reliability and accuracy are important functional requirements of the data-transmission system.

ESS IMPLEMENTATION PROCESS

6-6. The ESS implementation process is shown in Figure 6-1. Implementing an ESS is based on general requirements tailored to a site-specific mission and physical profile. The process begins with a site survey that includes a top-down view of basic needs and classic configurations that are tailored to such site-specific characteristics as terrain, site geography, climatic conditions, the type of asset, and priorities. This data is used to determine the hardware and software requirements, taking into account the additional capacity that should be factored into the design system for future expansion. Once the requirements for an ESS have been identified, the user must determine whether an existing standardized system is suitable for the application. (AR 190-13 outlines the process for gaining approval to use nonstandard equipment.) The user must also secure funding for the equipment (refer to Appendix J). Depending on the current funding regulations, operation-and-maintenance, procurement, or other funds may be required. For example, operations and procurement, Army (OPA) funds may be required for IDS devices; and operations and maintenance, Army (OMA) funds may be required for installation items. A contract is normally awarded to procure and install the equipment. The procurement or installation must be overseen. This may be accomplished by reviewing submittals, inspecting the contractor's work, or responding to the contractor's requests for information. Once the equipment is installed, the acceptance-testing activities must be witnessed and verified. Site conditions during acceptance testing affect the demonstrated detection capability of an exterior IDS. As feasible, acceptance testing should be designed to determine a sensor system's probability of detection (PD) under a range of conditions. For some types of sensor systems, this may be as straightforward as conducting both daytime and nighttime trials to experience differences in temperature and solar heating. After the ESS has been accepted, it must be operated and maintained throughout the remainder of its life cycle. Planning for manpower to operate the system and forecasting the funding and personnel to properly maintain the system is critical for success.

ESS DESIGN CONSIDERATIONS

6-7. A facility may require interior and exterior ESS elements, depending on the level of protection required. The applicable regulations, threat, and design criteria will define the ESS's general requirements. For an existing ESS, hardware and software may need to be supplemented, upgraded, or completely replaced. A site layout (in which all assets are identified and located) is required. It is a useful design tool for such tasks as configuring the DTM.

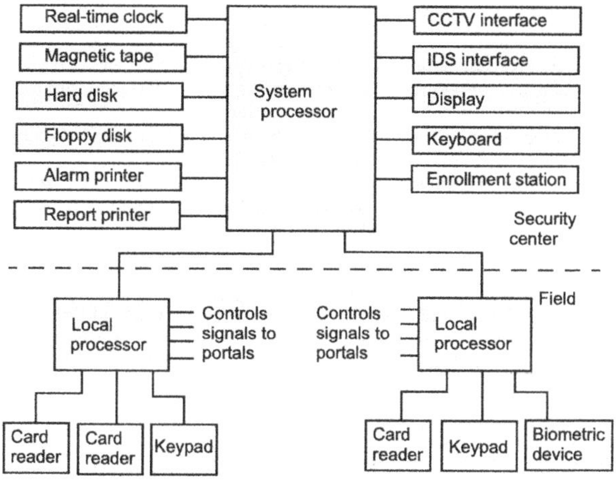

Figure 6-1. Entry-Control System Configured With Distributed Control

6-8. The exterior and interior IDSs should be configured as layers of unbroken rings concentrically surrounding the asset. These rings should correspond to defensive layers that constitute the delay system. The first detection layer is located at the outermost defensive layer necessary to provide the required delay. Detection layers can be on a defensive layer, in the area between two defensive layers, or on the asset itself, depending on the delay required. For example, if a wall of an interior room provides sufficient delay for effective response to aggression, detection layers could be between the facility exterior and interior-room wall or on the interior-room wall. These would detect the intruder before penetration of the interior wall is possible.

RESPONSE AND DELAY

6-9. When dealing with an ESS, the response time is defined as the time it takes the security force to arrive at the scene after an initial alarm is received at the security center. The total delay time is defined as the sum of all of the barriers' delay times, the time required to cross the areas between barriers after an intrusion alarm has been reported, and the time required to accomplish the mission and leave the protected area.

6-10. An ESS's basic function is to notify security personnel that an intruder is attempting to penetrate, or has penetrated, a protected area in sufficient time to allow the response force to intercept and apprehend him. To accomplish this, there must be sufficient physical delay between the point where the intruder is first detected and his objective. This provides delay time equal to or greater than the response time (refer to TM 5-853-1).

6-11. When dealing with interior sensors, boundary sensors that detect penetration (such as structural-vibration sensors or passive ultrasonic sensors) provide the earliest warning of an attempted intrusion. This alarm is usually generated before the barrier is penetrated. This gives the security force advance notification of an attempted penetration, thus allowing the barrier's delay time to be counted as part of the total delay time. Door-position sensors and glass-breakage sensors do not generate an alarm until the barrier has been breached; therefore, the delay time provided by the barrier cannot be counted as part of the total delay time.

6-12. Volumetric motion sensors do not generate an alarm until the intruder is already inside the area covered by the sensors. Therefore, if these sensors are to be used to provide additional response time, additional barriers must be placed between the volumetric motion sensors and the protected asset. Point sensors, such as capacitance sensors and pressure mats, provide warning of attempted penetration only if they detect the intruder before access is gained to the protected area.

BASIC GUIDANCE

6-13. An IDS is deployed in and around barriers (as detailed in TM 5-853-1). Voice communication links (radio, intercom, and telephone) with the response force are located in the security center. Security personnel will man the center and will alert and dispatch response forces in case of an alarm.

6-14. The barrier should always be deployed behind the IDS to ensure that integrity is maintained against intruders. An intruder will then activate the alarm sensor before penetrating or bypassing the barriers, thus providing delay for alarm assessment and response. The delay time is the determining factor in whether an assessment is conducted by dispatching a guard or by observing the CCTV. Normally, an intruder can climb a fence before a guard can be dispatched; therefore, a CCTV is usually required with an exterior IDS. Barriers can be located ahead of an alarm sensor as a boundary demarcation and can serve to keep people and animals from causing nuisance alarms by inadvertently straying into a controlled area. These barriers provide no additional response time because the barrier could be breached before the IDS sensors could be activated.

6-15. Data for monitoring and controlling an ESS are gathered and processed in the security center where the operator interacts with information from the ESS components located at remote facilities. The ESS's alarm-annunciation computer and its DTM line-termination equipment should be located in a controlled area and provided with tamper protection. Supervisory personnel should permit changes to software only, and these changes should be documented. If redundant DTM links connect the central computer to the local processor, diverse paths should be used to route these links.

6-16. The preferred medium for transmitting data in an ESS is a dedicated fiber-optics system. It provides for communications not susceptible to voltage transients, lightning, electromagnetic interference, and noise. Additionally, the fiber optics will provide a measure of communication-line security and wide bandwidth for video signals and increased data-transmission rates.

ESS EFFECTIVENESS

6-17. An ESS has a degree-of-protection effectiveness that is based on its probability of detecting intruders attempting to go over, under, around, or through the physical-security system. The intruder may use forced-entry, covert-entry, or insider-compromise tactics. A well-designed system will minimize the possibility of a successful penetration through covert entry or insider compromise. Interior and exterior alarm sensors have a PD based on the capability to detect an intruder passing through a sensing field. An intruder disturbs the steady-state quiescent condition of a sensor for a finite period. Sensors are designed to detect a person of minimum stature moving within a specific range of speeds and distances from the sensor, and any target outside of those parameters will probably not be detected. The PD for a specific sensor is usually specified at 0.9 or greater, but the designer must be aware that the PD is based on certain constraints and environmental conditions.

6-18. Manufacturer specifications usually do not discuss environmental or nuisance alarms that can be caused by climatic conditions (such as wind or rain) or by the intrusion of animals (including birds). The alarm annunciation is valid because the sensor's thresholds have been exceeded; however, the alarm does not represent a valid penetration attempt. If the assessment system is slow, the operator may not be able to determine the cause of the alarm and must, therefore, treat an environmental or nuisance alarm as real.

6-19. Another type of false alarm is caused by electronic-circuit tolerances being exceeded, resulting in the sensor's actuation. False alarms may also result from improper installation of the sensor or from effects of other equipment in the immediate area.

6-20. After an alarm is sensed and information is displayed in the security center, the console operator must determine the cause of the alarm (intrusion, nuisance, environmental, or false). Timely assessment is required when determining its cause. For example, if an intruder scales a fence in 10 seconds and runs 20 feet per second, the intruder will have overcome the barrier and be 2,200 feet from the point of penetration in 2 minutes. To conduct an accurate assessment of the alarm after 2 minutes, guards will have to search an area of about 200 acres. A fixed-television camera properly located and integrated with the alarm processor can assess the situation while the intruder is still in the controlled area.

6-21. For a CCTV camera to be effective, the area it views must be adequately lighted. To correlate the alarms and cameras in a large system (more than 10 cameras) in a timely manner, a computer-based processing system must be used to select and display alarms and camera scenes for the operator. A complex ESS has the following basic components:

- Intrusion-detection sensors.
- Electronic entry-control devices.
- CCTV.

- Alarm-annunciation system.
- DTM.

6-22. The intrusion-detection sensors are normally deployed in a series of concentric layers. The overall PD improves with each added layer of sensors. The layers (interior and exterior) should be functionally uniform; however, their overall effectiveness and cost are different. The exterior zones significantly differ from the interior zones due to the following considerations:

- The consistency of the PD.
- The PD.
- The cost per detection zone.
- The number of zones.
- The overall sensor coverage.

6-23. Exterior IDSs usually have PDs equal to those of interior IDSs. However, exterior sensors are more likely to experience weather-related situations that cause the system's PD to vary. Sensor phenomenology (passive infrared [PIR], microwave radar, and so forth) determines which environmental factors may alter the system's PD. The frequency of occurrence, severity, and duration of a weather event jointly determine whether it represents security vulnerability with the IDS in use. Typically, sophisticated intruders will attempt their penetration and challenge an ESS under conditions most favorable to themselves. Inclement weather (fog, snow, and rain) affects the usefulness of CCTVs and security lighting such that the capability for remote assessment of alarm events may be lost. Exterior IDSs are not necessarily less likely to detect a penetration attempt during fog, rain and snow; the effect of such site conditions on the IDS depends on sensor phenomenology. For example, fence motion caused by rain impact may drive the response of a fence-mounted sensor closer to satisfying the system's alarm criteria, with the result that the margin of disturbance available to the intruder is less. Also, certain buried sensors are more likely to detect an intruder when the ground is wet because of rain or melting snow. Since interior sensor systems are less influenced by environmental conditions, their PD is typically more consistent than that of some types of exterior sensor systems. Other considerations in comparing an interior and exterior ESS are the cost, the number and size of detection zones required, and the detection height.

- Because of environmental conditions, the exterior electronics must be designed and packaged for extremes of temperature, moisture, and wind. The result is that exterior electronic packages are more costly than equivalent packages for interior applications.
- State-of-the-art exterior sensors do not detect penetration attempts above the height of a fence (typically 8 feet). Fence-mounted sensors are usually limited to this height because the fence fabric or poles are used to support the sensor. For aboveground sensors in the controlled area between the fences, the sensor's mounting brackets and posts limit the detection height. In some applications of field sensors (especially buried sensors), the detection height is no more than 3 feet. For a facility, interior sensors can be deployed on walls, floors, or ceilings, thus permitting complete protection of the asset.

6-24. An interior ESS may be far less costly than that of a comparable exterior ESS. This comparison indicates to the designer the value of selecting and deploying a well-planned, well-designed, layered system. The basic rule in overall design of an ESS is to design from the inside out; that is, layered from the asset to the site boundary.

INTERIOR ESS CONSIDERATIONS

6-25. An interior ESS is typically deployed within a boundary in the immediate vicinity of the asset being protected. If the interior ESS operates in a controlled environment, its PD will be independent of any weather-induced variation in exterior conditions. Also, the physical-security system's effectiveness is enhanced by the interior barriers (walls, ceiling, and floor) that inherently impose a longer delay than exterior barriers (fences and gates).

6-26. Functionally, an interior asset should be viewed as being contained within a cube with sensors protecting all six faces. Interior sensors can be deployed at the cube's perimeter, in its interior space, or in the space immediately outside of the cube.

6-27. If an increased level of protection is dictated by the threat, and if the building is large enough, multiple layers of interior sensors may be deployed for a given asset. A multilayered interior IDS will improve the overall PD. Tamper protection and access-/secure-mode capabilities must be considered when planning and laying out interior sensors.

TAMPER PROTECTION

6-28. To minimize the possibility of someone tampering with circuitry and associated wiring, all sensor-related enclosures must be equipped with tamper switches. These switches must be positioned so that an alarm is generated before the cover has been moved enough to permit access to the circuitry of adjustment controls. In addition, several types of sensors should be equipped with tamper switches to protect against being repositioned or removed. Security screens containing grid-wire sensors and vibration sensors that can be easily removed from a wall are examples of sensors that require tamper switches.

ACCESS/SECURE MODE

6-29. During regular working hours, many of the interior sensors must be deactivated by placing the area in the access mode. For example, door-position sensors and volumetric sensors in occupied areas must be deactivated to prevent multiple nuisance alarms caused by the normal movement of people. This can be done locally or remotely. With local control, a switch is used to bypass or shunt alarm contacts when the sensor is placed in the access mode. When done remotely, the security-center operator usually enters a command that causes the processor software to ignore incoming alarms from those sensors placed in the access mode. However, when a sensor is placed in the access mode, its tamper-protection circuitry must remain in the activated or secure mode. During nonworking hours when the facility is unoccupied, all sensors must be placed in the secure mode. Certain devices (such as duress-alarm switches, tamper switches, grid-wire sensors covering vent openings, and glass-breakage sensors) should never be placed in the access mode. The designer must ensure that selected sensors can be placed in an access mode (if required) and that certain types of sensors (such as duress and tamper switches) are configured so that they cannot be put in the access mode under any condition.

EXTERIOR ESS CONSIDERATIONS

6-30. An exterior ESS is typically deployed at a site's boundary or some other significant boundary such as the demarcation fence for a group of bunkers. An exterior ESS has the advantage that it remains in the secure mode at all times.

6-31. The ideal configuration for an exterior ESS is a rectangle or a polygon, with all sides being straight. The ESS is located in and around barriers that typically include a dual fence. The outside fence is used for demarcation, and the interior fence is used to aid in detection and provide some delay. If dual fences are not used, the sensors should be deployed on the fence or inside it.

DESIGN GUIDELINES

6-32. The general-design criteria of a perimeter IDS involves primarily the selection and layout of exterior sensors that are compatible with the physical and operational characteristics of a specific site. Important factors to consider during the selection process include physical and environmental conditions at the site, the sensor's performance, and the overall cost of the system. Refer to TMs 5-853-1 and 5-853-2 for additional guidance on the requirements for and placement of exterior sensor systems. Since exterior barriers provide very little delay, exterior sensor systems generally do not provide a significant increase in the available response time.

Physical and Environmental Considerations

6-33. Physical and environmental considerations are often the determining factors for selecting exterior sensors. The site's characteristics can significantly affect a sensor's operational performance, both in terms of PD and the susceptibility to nuisance alarms. Exterior sensor systems should be selected on the basis of the frequency and duration of weather-related periods of poor detection capability. An exterior IDS may have an unacceptably low PD during a particular weather event or site condition, yet otherwise be superior to other IDSs in terms of good detection capability and a low nuisance-alarm rate. It may be appropriate to select that IDS in spite of its known vulnerability, precisely because the circumstances of its vulnerability are known and precautionary measures can be taken at those times. The overall performance of that IDS, together with its cost, may justify its selection.

6-34. Weather and climatic conditions at a specific site can significantly influence sensor selection. For example, IR detectors are not very effective in heavy rain, fog, dust, or snow. Deep snow can affect detection patterns and performance of both IR and microwave sensors. High winds can cause numerous false alarms in fence-mounted sensors. Electrical storms can cause alarms in many types of sensors and may also damage the equipment.

6-35. Vegetation can be a significant cause of nuisance alarms. Tall grass or weeds can disturb the energy pattern of microwave and both thermal IR and near-IR beam-break sensors. Vegetation growing near electric-field sensors and capacitance sensors can cause nuisance alarms. Large weeds or bushes rubbing against a fence can produce nuisance alarms from fence-mounted sensors. Large trees and bushes moving within the field of view of video motion sensors can cause nuisance or environmental alarms. A clear area must be established for exterior sensors. This area must be void of vegetation or contain vegetation of carefully controlled growth.

6-36. Topographic features are extremely important. Ideally, perimeter terrain should be flat, although gently sloping terrain is acceptable. Irregular terrain with steep slopes may preclude the use of LOS sensors and make CCTV assessment difficult. Gullies and ditches crossing the perimeter represent a vulnerability to LOS sensors and may be a source of false alarms (from flowing water) for buried line sensors. Large culverts can provide an intruder with an entry or exit route across the perimeter without causing an alarm. Likewise, overhead power and communication lines may permit an intruder to bridge the perimeter without causing an alarm.

6-37. Large animals (such as cows, horses, and deer) can cause nuisance alarms in both aboveground and buried sensors. Sensors sensitive enough to detect a crawling or rolling intruder are susceptible to nuisance alarms from small animals such as rabbits, squirrels, cats, and dogs. To minimize the interference from animals, a dual chain-link-fence configuration may be established around the site perimeter with the sensors installed between the fences.

Sensor Performance

6-38. Exterior sensors must have a high PD for all types of intrusion and have a low unwanted-alarm rate for all expected environmental and site conditions. Unfortunately, no single exterior sensor that is presently available meets both these criteria. All are limited in their detection capability, and all have high susceptibility to nuisance and environmental conditions. Table 6-1, page 6-10, provides estimates of PDs for various types of intrusions. Table 6-2, page 6-10, lists the relative susceptibility of various types of sensors to nuisance and environmental alarms.

Economic Considerations

6-39. Exterior sensor costs are usually given in cost per linear foot per detection zone (typically 300 feet). These costs include both equipment and installation. Fence-mounted sensors (such as strain-sensitive cable, electromechanical, and mechanical) are generally less costly than stand-alone and buried line sensors. Installation costs can vary significantly, depending on the type of sensor. Table 6-3, page 6-11, provides a comparison of relative costs for procuring and installing various types of exterior sensor systems. It should be remembered that the sensor system's cost is only a portion of the total cost for employing a perimeter IDS. Additional costs include fencing, site preparation, CCTV assessment, and perimeter lighting.

Table 6-1. Estimate of PD by Exterior Sensors

Type of Sensor	Intruder Technique: Slow walk	Walking	Running	Crawling	Rolling	Jumping	Tunnelling	Trenching	Bridging	Cutting	Climbing	Lifting
Fence mounted	N/A	N/A	N/A	N/A	N/A	VH	VL	L	VL	M/H	H	M/H
Taut wire	N/A	N/A	N/A	N/A	N/A	VH	VL	VL	VL	H	H	H
Electric field	VH	VH	VH	H	VH	VH	VL	L	L	N/A	N/A	N/A
Capacitance	VH	VH	VH	H	H	VH	VL	L	L	N/A	N/A	N/A
Ported cable	H	VH	VH	VH	VH	H	M	VH	L	N/A	N/A	N/A
Seismic	H	VH	H	M	M	M	L	M	L	N/A	N/A	N/A
Seismic/magnetic	H	VH	H	M	M	M	L	M	L	N/A	N/A	N/A
Microwave	H	VH	H	M/H	M/H	M/H	VL	L/M	L	N/A	N/A	N/A
IR	VH	VH	VH	M/H	M/H	H	VL	L	VL	N/A	N/A	N/A
Video motion	H	VH	VH	H	H	H	VL	L/M	M	N/A	N/A	N/A
VL = very low, L = low, M = medium, H = high, VH = very high, N/A = not applicable												

Table 6-2. Relative Susceptibility of Exterior Sensors to False Alarms

Type of Sensor	Intruder Technique: Wind	Rain	Standing water/runoff	Snow	Fog	Small animals	Large animals	Small birds	Large birds	Lightning	Overhead power lines	Buried power lines
Fence mounted	H	M	L	L	VL	L	M	L	L	L	VL	VL
Taut wire	VL	VL	VL	VL	VL	VL	L	VL	VL	VL	VL	VL
Electric field	M	L/H	VL	M	VL	M	VH	L	M	M	L	VL
Capacitance	M	M	VL	M	VL	M	VH	L	M	M	L	VL
Ported cable	VL	M	H	L	VL	VL	M	VL	VL	M	VL	L
Seismic	M	L	L	L	VL	L	VH	VL	VL	L	L	M
Seismic/magnetic	M	L	L	L	VL	L	VH	VL	VL	H	M	H
Microwave	L	L	M/H	L/M	L	M/H	VH	VL	M	L/M	L	VL
IR	L	L	L	M	M	M	VH	L	M	L	VL	VL
Video motion	M	L	L	L	M/H	L	VH	VL	M	L	L	VL
VL = very low, L = low, M = medium, H = high, VH = very high												

Table 6-3. Exterior IDS Sensor Cost Comparison

Type of Sensor	Equipment	Installation	Maintenance
Fence mounted	L	L	L
Taut wire	H	H	M
Electric field	H	M	M
Capacitance	M	L	M
Ported cable	H	M	M
Seismic	M	M	L
Seismic/magnetic	H	M	L
Microwave	M	M	L
IR	M	L	M
Video motion	M	L	M
L = low, M = medium, H = high			

PERIMETER LAYOUT AND ZONING

6-40. A protected area's perimeter is usually defined by an enclosing wall or fence or a natural barrier such as water. For exterior sensors to be effective, the perimeter around which they are to be deployed must be precisely defined. In most applications, a dual chain-link-fence configuration will be established around the perimeter. Typically, fences should be between 30 and 50 feet apart; as the distance increases, it is harder for an intruder to bridge the fences. If fence separation is less than 30 feet, some microwave and ported-coax sensors cannot be used. The area between fences (called the controlled area or isolation zone) may need to be cleared of vegetation and graded, depending on the type of sensor used. Proper drainage is required to preclude standing water and to prevent the formation of gullies caused by running water after a heavy rain or melting snow. Cleared areas are required inside and outside of the controlled area. These areas enhance routine observation, as well as sensor-alarm assessment, and minimize the protective cover available to a would-be intruder.

6-41. After the perimeter has been defined, the next step is to divide it into specific detection zones. The length of each detection zone is determined by evaluating the contour, the existing terrain, and the operational activities along the perimeter. Detection zones should be long and straight to minimize the number of sensors or cameras necessary and to aid guard assessment if cameras are not used. It may be more economical to straighten an existing fence line than to create numerous detection zones in accommodating a crooked fence line. If the perimeter is hilly and LOS sensors or CCTV assessment are used, the length of individual detection zones will be commensurate with sensor limitations. Entry points for personnel and vehicles must be configured as independent zones. This enables deactivation of the sensors in these zones; that is, placing them in the access mode during customary working hours (assuming the entry points are manned) without having to deactivate adjacent areas.

6-42. The specific length of individual zones can vary around the perimeter. Although specific manufacturers may advertise maximum zone lengths exceeding 1,000 feet, it is not practical to exceed a zone length of 300 feet. If the zone is longer, it will be difficult for an operator using CCTV assessment or for the response force to identify the location of an intrusion or the cause of a false alarm.

6-43. When establishing zones using multiple sensors, the designer should establish coincident zones where the length and location of each individual sensor will be identical for all sensors within a given zone. If an alarm occurs in a specific zone, the operator can readily determine its approximate location by referring to a map of the perimeter. This also minimizes the number of CCTV cameras required for assessment and simplifies the interface between the alarm-annunciation system and the CCTV switching system.

ESS ALARM-ANNUNCIATION SYSTEM

6-44. Status information from the various intrusion-detection sensors and entry-control terminal devices must be collected from the field and transmitted to the alarm-annunciation system in the security center, where it is processed, annunciated, and acted on by security personnel. The alarm-annunciation system may also interface with a CCTV system. There are typically two types of alarm-annunciation configurations available. The simplest configuration, which is suitable for small installations, is the point-to-point configuration. With this configuration, a separate transmission line is routed from the protected area to the security center (see Figure 6-2). The Joint-Service Interior Intrusion-Detection System (J-SIIDS) is typical of this type of configuration but will not be further discussed in this manual. The second, and more popular type, is a digital multiplexed configuration that allows multiple protected areas to communicate with the security center over a common data line. A block diagram of a typical multiplexed alarm-annunciation system is shown in Figure 6-3.

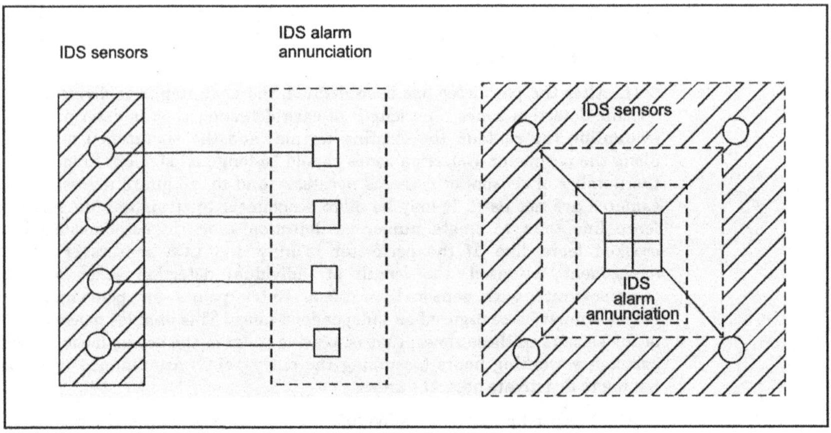

Figure 6-2. Typical Point-to-Point IDS

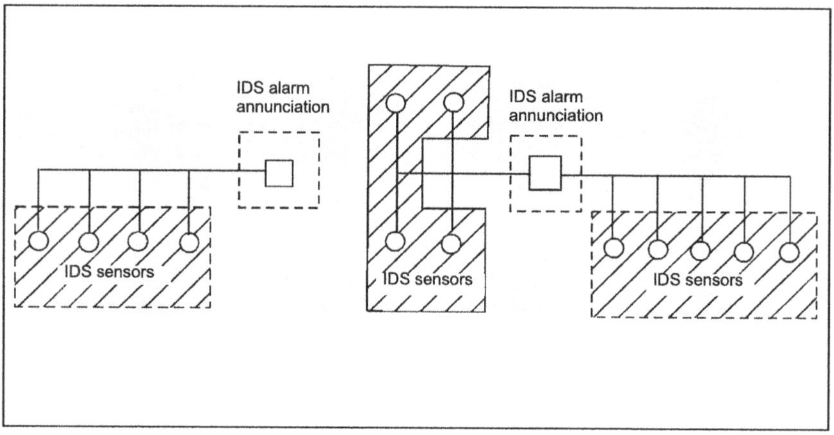

Figure 6-3. Typical Multiplexed IDS

ALARM-ANNUNCIATION CONFIGURATION

6-45. A block diagram of a typical alarm-annunciation system is shown in Figure 6-4, page 6-14. As shown in the figure, the central computer is the hub of the information flow. The central computer receives and displays alarm and device status information and sends operator-control commands to the ESS's local processors. It also interfaces with the CCTV system. For larger facilities, the management of the DTM communications tasks may be delegated to a separate communication processor so that the central computer can turn its full attention to interpreting the incoming information and updating the control and display devices located at the security console (display, logging, control, and storage devices).

6-46. The central computer may consist of one or more digital computers. The real-time clock is usually integral to the central computer and provides a time stamp for alarms and other events. It allows for time synchronization with the CCTV and other systems, if included. The console operator must be able to set the clock, which should include a battery backup. All system events must be properly time-correlated. For example, there will be an exact time correlation for an ESS alarm event reported on the alarm printer and the corresponding video scene recorded by the CCTV's video processor.

DATA STORAGE

6-47. Computer-based systems are required to store large amounts of information such as system software, application programs, data structures, and system events (alarm transactions and status changes). Therefore, a large amount of nonvolatile memory is required. The semiconductor memory provided with a central computer is designed for rapid storage and retrieval

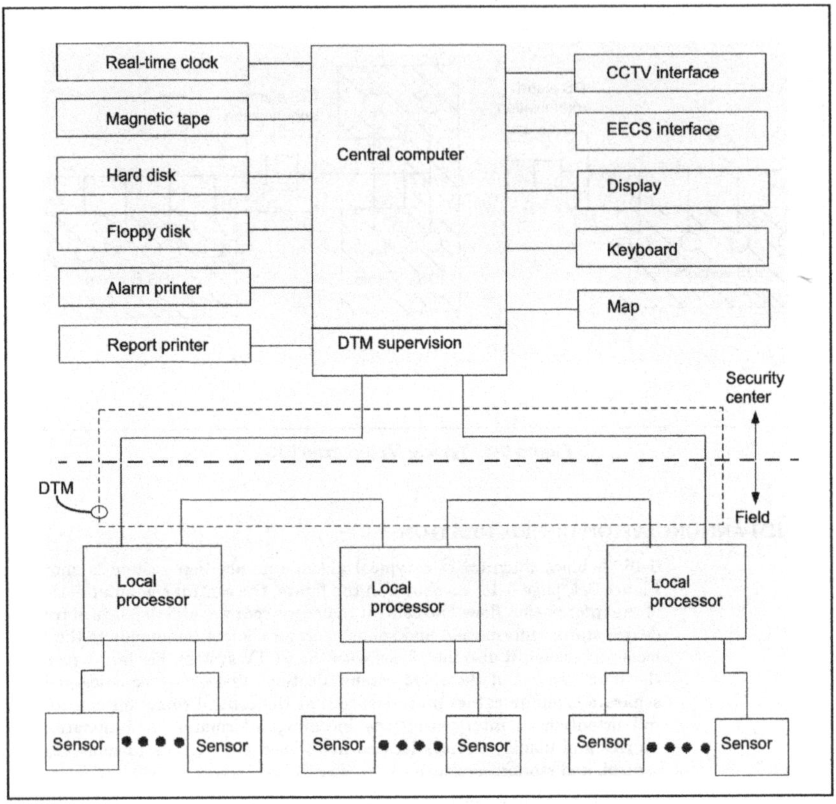

Figure 6-4. Typical IDS Alarm-Annunciation System

and possesses extremely fast access times. The most commonly used media for archival storage are magnetic tape; compact-disk, read-only memory (CD-ROM); and magnetic disk. These media are capable of economically storing large amounts of data.

OPERATOR INTERFACE

6-48. The operator interacts with the alarm-annunciation system through devices that can be seen, heard, or touched and manipulated. Visual displays and printers can be used to inform the operator of an alarm or the equipment's status. Audible devices are used to alert an operator to an alarm or the equipment's failure. Devices such as push buttons and keyboards permit an operator to acknowledge and reset alarms, as well as change operational parameters.

- **Visual displays.** The type of display used to inform the operator visually of the ESS's status is determined primarily by the system's complexity. Status information is usually displayed on monitors. Alphanumeric displays and map displays are seldom used. Monitors provide great flexibility in the type and format of alarm information that may be displayed. Both text and graphic information can be displayed in a variety of colors. Multiple alarms may also be displayed. If alarms are prioritized, higher-priority alarms may be highlighted by blinking, by using bold print or reverse video, or by changing colors. To assist the operator in determining the correct response, alarm-specific instructions may be displayed adjacent to the alarm information.
- **Audible alarm devices.** In conjunction with the visual display of an alarm, the alarm-annunciation system must also generate an audible alarm. The audible alarm may be produced by the ringing of a bell or by the generation of a steady or pulsating tone from an electronic device. In any case, the audible alarm serves to attract the operator's attention to the visual-alarm display. A silence switch is usually provided to allow the operator to silence the bell or tone before actually resetting the alarm.
- **Logging devices.** All alarm-system activity (such as a change of access/secure status, an alarm event, an entry-control transaction, or a trouble event) should be logged and recorded. Logged information is important not only for security personnel investigating an event, but also for maintenance personnel checking equipment performance for such causes as false and nuisance alarms. Most alarm-annunciation systems are equipped with logging and alarm printers.
- **Alarm printers.** Alarm printers are typically of the high-speed, continuous-feed variety. The printer provides a hard-copy record of all alarm events and system activity, as well as limited backup in case the visual display fails.
- **Report printers.** Most ESSs include a separate printer (report printer) for generating reports using information stored by the central computer. This printer will usually be typical of those found in modern office environments.
- **Operator control.** A means is required to transmit information from the operator to the system. The type of controls provided usually depends on the type of display provided. The following are consistent with the controls:
  - Keypads consist of a numeric display system that will generally be provided with a 12-digit keypad and several function keys such as access, secure, acknowledge, and reset. The keypad enables an operator to key in numeric requests for the status of specific zones.
  - Monitor-based systems are usually provided with a typewriter-type keyboard that enables an operator to enter more information using a combination of alphanumeric characters and function keys.
  - An ESS may be equipped with enhancement hardware/devices to help the operator enter information or execute commands quickly. A mouse or a trackball are typical examples.

FIELD-DATA COLLECTION

6-49. Sensor and terminal device data must be transmitted to the central alarm monitor located in the security center using a selected DTM. The following are DTM methods that may be used:

Local Processors

6-50. Multiplexing techniques can be used to minimize the number of data links needed to communicate field-device status to the security center. This is done through devices called local processors. The following is descriptive of a local processor's capabilities:

- A local processor may have very few device inputs, or it may have many (depending on the manufacturer). Rather than having a fixed number of inputs, many local processors are expandable. For example, a basic local processor may be provided with eight device inputs with additional blocks of eight inputs available by using plug-in modules.
- The local processor must provide line supervision for all communication links to sensors, terminal devices, and so forth. Usually, direct-current (DC) line supervision is supplied as the standard with more secure techniques available as options. The data communication links between the local processor and the central alarm monitor must also be supervised.
- Local processors can also provide output signals that can be used for such functions as activating sensor remote test features, light control, or portal control or activating a deterrent (such as a loud horn).
- The local processor contains a microprocessor, solid-state memory, and appropriate software. It has the capability to perform a number of functions locally (such as access-/secure-mode selection, alarm reset, card or keypad electronic-entry control, portal control, and device testing). If the communication link to the security center is temporarily lost, local processors can continue to operate in a stand-alone mode, storing data for transmission after the link is restored.
- The number of local processors required for a specific site depends on the number of protected areas and their proximity to each other and the number of sensors within a protected area. For example, a small building may require one local processor, whereas a large building may require one or more for each floor. An exterior IDS perimeter with two or three different sensors may require one local processor for every two perimeter zones. All local processors may be linked to the central computer using one common DTM link, or the DTM may consist of several links. The designer should note that the temporary loss of a DTM link would render all local processors on that link inactive for the duration of the loss.

Central Computer and Local-Processor Data Exchange

6-51. When the ESS is powered up or reset at the security center, the central computer will download all necessary operational information over the DTM to all local processors. After the download is complete, the central computer will automatically begin polling the local processors for ESS device status. In addition to alarm status, tamper indications, and local-processor status, the DTM may be required to convey security-center console-operator commands to field devices. Examples include security-area access-/secure-mode changes and initiation of the intrusion-sensor self test.

CCTV Interface

6-52. If a CCTV assessment system is deployed with the ESS, an interface between the two is required. This interface allows CCTV system alarms (such as loss of video) to be displayed by the ESS's alarm-annunciation system. The interface also provides IDS alarm signals to the CCTV's video switcher so that the correct CCTV camera will be displayed on the CCTV monitors to allow real-time alarm assessment and video recording as required.

ESS SOFTWARE

6-53. The software provided with computer-based ESS alarm-annunciation systems consists of three types—a standard operating system (such as the Microsoft®-disk operating system [MS-DOS]); vendor-developed application programs; and user-filled, site-specific data structures.

- **System software.** The designer will ensure that system software provided by the vendor conforms to accepted industry standards so that standard, follow-on maintenance and service contracts can be negotiated to maintain the central computer system.
- **Application software.** The vendor-developed application programs are typically proprietary and include ESS monitoring, display, and entry-control capabilities.
- **User-filled data structures.** These data structures are used to populate the site-specific database. Specific electronic address information, personnel access schedules, and normal duty hours are typically included in the site-specific database. The information may include preferred route descriptions for the response force, the phone number of the person responsible for the alarmed area, and any hazardous material that may be located in the alarmed area.

6-54. ESS software functions typically include the following:

- **Alarm monitoring and logging.** The software should provide for monitoring all sensors, local processors, and data communication links and notifying the operator of an alarm condition. All alarm messages should be printed on the alarm printer, archived, and displayed at the console. As a minimum, printed alarm data should include the date and time (to the nearest second) of the alarm and the location and type of alarm.
- **Alarm display.** The software should be structured to permit several alarms to be annunciated simultaneously. A buffer or alarm queue should be available to store additional alarms until they are annunciated and, subsequently, acted upon and reset by the console operator.
- **Alarm priority.** A minimum of five alarm-priority levels should be available. Higher-priority alarms should always be displayed before

lower-priority alarms. This feature permits an operator to respond quickly to the more important alarms before those of lesser importance. For example, the priority of alarm devices may be as follows:

- Duress.
- Intrusion detection.
- Electronic-entry control.
- Tamper.
- CCTV alarms and equipment-malfunction alarms.

- **Reports.** The application software should provide for generating, displaying, printing, and storing reports.

PASSWORDS

6-55. Software security will be provided by limiting access to personnel with authorized passwords assigned by a system manager. A minimum of three password levels shall be provided. Additional security can be provided by programmed restrictions that limit the keyboard actions of logged-in passwords to the user ranks of system managers, supervisors, and console operators, as appropriate.

OPERATOR INTERFACE

6-56. The software should enable an operator with the proper password to enter commands and to obtain displays of system information. As a minimum, an operator should be able to perform the following functions through the keyboard or the keypad:

- Log on by password to activate the keyboard.
- Log off to deactivate the keyboard.
- Request display of all keyboard commands that are authorized for the logged-in password.
- Request display of detailed instructions for any authorized keyboard command.
- Acknowledge and clear alarm messages.
- Display the current status of any device in the system.
- Command a status change for any controlled device in the system.
- Command a mode change for any access/secure device in the system.
- Command printouts of alarm summaries, status summaries, or system activity on a designated printer.
- Add or delete ESS devices or modify parameters associated with a device.

INTERIOR INTRUSION-DETECTION SENSORS

6-57. Interior intrusion-detection sensors are devices used to detect unauthorized entry into specific areas or volumetric spaces within a building. These sensors are usually not designed to be weatherproof or rugged enough to survive an outdoor environment. Therefore, this type of sensor should not be used outdoors unless described by the manufacturer as suitable for outdoor use.

6-58. Interior intrusion-detection sensors generally perform one of three detection functions—detection of an intruder penetrating the boundary of a protected area, detection of intruder motion within a protected area, and detection of an intruder touching or lifting an asset within a protected area. Therefore, interior sensors are commonly classified as boundary-penetration sensors, volumetric motion sensors, and point sensors. Although duress switches are not intrusion-detection sensors, they are included in this discussion because they are usually wired to the same equipment that monitors the interior intrusion-detection sensors.

BOUNDARY-PENETRATION SENSORS

6-59. Boundary-penetration sensors are designed to detect penetration or attempted penetration through perimeter barriers. These barriers include walls, ceilings, duct openings, doors, and windows.

Structural-Vibration Sensors

6-60. Structural-vibration sensors detect low-frequency energy generated in an attempted penetration of a physical barrier (such as a wall or a ceiling) by hammering, drilling, cutting, detonating explosives, or employing other forcible methods of entry. A piezoelectric transducer senses mechanical energy and converts it into electrical signals proportional in magnitude to the vibrations. To reduce false alarms from single accidental impacts on the barrier, most vibration sensors use a signal processor that has an adjustable pulse-counting accumulator in conjunction with a manual sensitivity adjustment. The count circuit can be set to count a specific number of pulses of specific magnitude within a predefined time interval before an alarm is generated. However, the circuitry is usually designed to respond immediately to large pulses, such as those caused by an explosion. The sensitivity adjustment is used to compensate for the type of barrier and the distance between transducers. Typically, several transducers can be connected together and monitored by one signal processor. Figure 6-5, page 6-20, shows an example of wall-mounted, structural-vibration sensors.

Glass-Breakage Sensors

6-61. Glass-breakage sensors detect the breaking of glass. The noise from breaking glass consists of frequencies in both the audible and ultrasonic range. Glass-breakage sensors use microphone transducers to detect the glass breakage. The sensors are designed to respond to specific frequencies only, thus minimizing such false alarms as may be caused by banging on the glass.

Passive Ultrasonic Sensors

6-62. Passive ultrasonic sensors detect acoustical energy in the ultrasonic frequency range, typically between 20 and 30 kilohertz (kHz). They are used to detect an attempted penetration through rigid barriers (such as metal or masonry walls, ceilings, and floors). They also detect penetration through windows and vents covered by metal grilles, shutters, or bars if these openings are properly sealed against outside sounds.

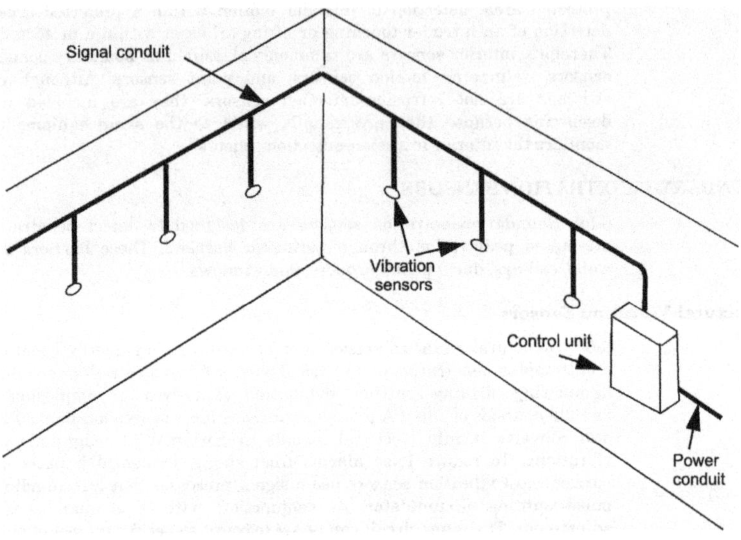

Figure 6-5. Wall-Mounted, Structural-Vibration Sensors

6-63. **Detection Transducer.** The detection transducer is a piezoelectric crystal that produces electrical signals proportional to the magnitude of the vibrations. A single transducer provides coverage of an area about 15 by 20 feet in a room with an 8- to 12-foot ceiling. A typical detection pattern is shown in Figure 6-6. Ten or more transducers can be connected to a signal processor. As with vibration sensors, the signal processor for a passive ultrasonic sensor has manual sensitivity adjustment and an adjustable pulse-counting accumulator.

6-64. **Sensors.** Passive ultrasonic sensors detect ultrasonic energy that results from the breaking of glass, the snipping of bolt cutters on metal barriers, the hissing of an acetylene torch, and the shattering of brittle materials (such as concrete or cinderblock). However, the sensors will not reliably detect drilling through most material nor attacks against soft material such as wallboard. Their effective detection range depends largely on the barrier material, the method of attempted penetration, and the sensitivity adjustment of the sensor. Examples of maximum detection distances for a typical sensor for different types of attempted penetration are shown in Table 6-4.

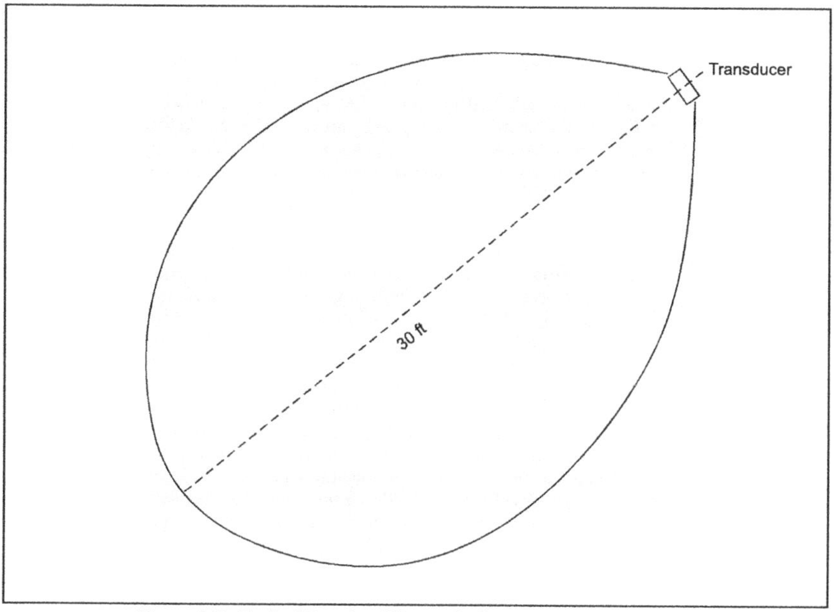

Figure 6-6. Typical Passive-Ultrasonic-Sensor Detection Pattern

Table 6-4. Detection Range for Passive Ultrasonic Sensors

Penetration	Distance (in Feet)
Cut 1/4-inch-thick expanded metal with bolt cutters	55
Cut 5/8-inch reinforcing bar with bolt cutters	45
Use acetylene cutting torch	39
Cut wood with circular saw	30
Cut 5/8-inch reinforcing bar with hacksaw	19
Drill through brick	15
Drill through 1/8-inch steel plate	6
Cut 1/8-inch steel plate with hacksaw	4
Drill through cinderblock	3

6-65. **Balanced Magnetic Switches.** Balanced magnetic switches (BMSs) are typically used to detect the opening of a door. These sensors can also be used on windows, hatches, gates, or other structural devices that can be opened to gain entry. When using a BMS, mount the switch mechanism on the door frame and the actuating magnet on the door. Typically, the BMS has a three-position reed switch and an additional magnet (called the bias magnet) located adjacent to the switch. When the door is closed, the reed switch is held in the balanced or center position by interacting magnetic fields. If the door is opened or an external magnet is brought near the sensor in an attempt to defeat it, the switch becomes unbalanced and generates an alarm. A BMS must be mounted so that the magnet receives maximum movement when the door or window is opened. Figure 6-7 shows several configurations for mounting BMSs.

Grid-Wire Sensors

6-66. The grid-wire sensor consists of a continuous electrical wire arranged in a grid pattern. The wire maintains an electrical current. An alarm is generated when the wire is broken. The sensor detects forced entry through walls, floors, ceilings, doors, windows, and other barriers. An enamel-coated number 24 or 26 American wire gauge (AWG) solid-copper wire typically forms the grid. The grid's maximum size is determined by the spacing between the wires, the wire's resistance, and the electrical characteristics of the source providing the current. The grid wire can be installed directly on the barrier, in a grille or screen that is mounted on the barrier, or over an opening that requires protection. The wire can be stapled directly to barriers made of wood or wallboard. Wood panels should be installed over the grid to protect it from day-to-day abuse and to conceal it. When used on cinder, concrete, and

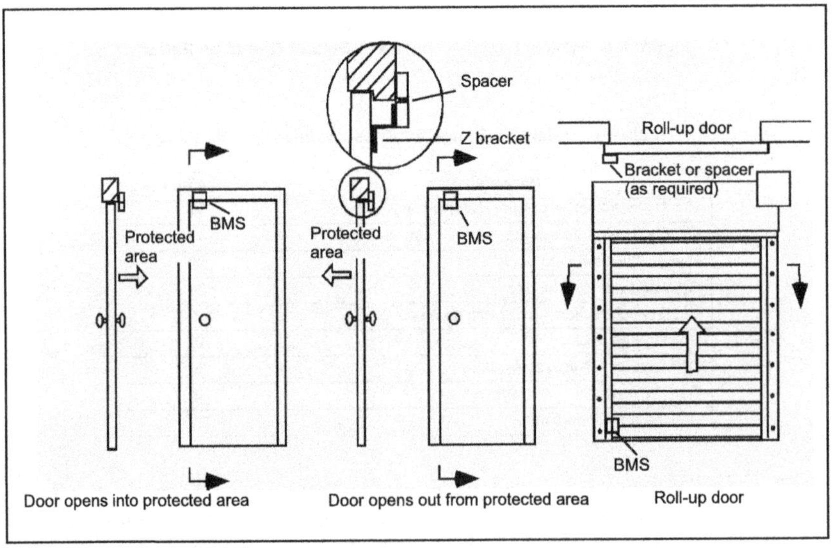

Figure 6-7. BMS Mounting Configurations

masonry surfaces, these surfaces must first be covered with plywood or other material to which the wire can be stapled. An alternative method is to staple the wire grid to the back side of a panel and install the panel over the surface.

VOLUMETRIC MOTION SENSORS

6-67. Volumetric motion sensors are designed to detect intruder motion within the interior of a protected volume. Volumetric sensors may be active or passive. Active sensors (such as microwave) fill the volume to be protected with an energy pattern and recognize a disturbance in the pattern when anything moves within the detection zone. Whereas active sensors generate their own energy pattern to detect an intruder, passive sensors (such as IR) detect energy generated by an intruder. Some sensors, known as dual-technology sensors, use a combination of two different technologies, usually one active and one passive, within the same unit. If CCTV assessment or surveillance cameras are installed, video motion sensors can be used to detect intruder movement within the area. Since ultrasonic motion sensors are seldom used, they will not be discussed here.

Microwave Motion Sensors

6-68. With microwave motion sensors, high-frequency electromagnetic energy is used to detect an intruder's motion within the protected area. Interior or sophisticated microwave motion sensors are normally used.

6-69. **Interior Microwave Motion Sensors.** Interior microwave motion sensors are typically monostatic; the transmitter and the receiver are housed in the same enclosure (transceiver). They may each be provided with a separate antenna or they may share a common antenna. The high-frequency signals produced by the transmitter are usually generated by a solid-state device, such as a gallium arsenide field-effect transistor. The power generated is usually less than 10 milliwatts, but it is sufficient to transmit the signal for distances up to about 100 feet. The shape of the transmitted beam is a function of the antenna configuration. The range of the transmitted beam can be controlled with a range adjustment. A variety of detection patterns can be generated (see Figure 6-8, page 6-24). The frequency of the transmitted signal is compared with the frequency of the signal reflected back from objects in the protected area. If there is no movement within the area, the transmitted and received frequencies will be equal and no alarm will be generated. Movement in the area will generate a Doppler frequency shift in the reflected signal and will produce an alarm if the signal satisfies the sensor's alarm criteria. The Doppler shift for a human intruder is typically between 20 and 120 hertz (Hz). Microwave energy can pass through glass doors and windows as well as lightweight walls or partitions constructed of plywood, plastic, or fiberboard. As a result, false alarms are possible because of the reflection of the microwave signals from the movement of people or vehicles outside of the protected area. The designer can sometimes take advantage of this when the protected area is large and contains a number of partitions, but this is not normally done.

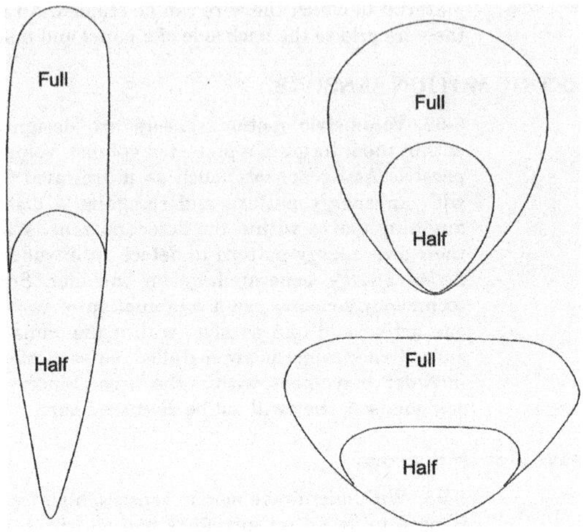

Figure 6-8. Typical Detection Patterns for Microwave Motion Sensors

6-70. **Sophisticated Microwave Motion Sensors.** Sophisticated microwave motion sensors may be equipped with electronic range gating. This feature allows the sensor to ignore the signals reflected beyond the settable detection range. Range gating may be used to effectively minimize unwanted alarms from activity outside the protected area.

PIR Motion Sensors

6-71. PIR motion sensors detect a change in the thermal energy pattern caused by a moving intruder and initiate an alarm when the change in energy satisfies the detector's alarm criteria. These sensors are passive devices because they do not transmit energy; they monitor the energy radiated by the surrounding environment.

6-72. All objects with temperatures above absolute zero radiate thermal energy. The wavelengths of the IR energy spectrum lie between 1 and 1,000 microns. Because the human body radiates thermal energy of between 7 and 14 microns, PIR motion sensors are typically designed to operate in the far IR wavelength range of 4 to 20 microns.

6-73. The IR energy must be focused onto a sensing element, somewhat as a camera lens focuses light onto a film. Two techniques are commonly used. One technique uses reflective focusing; parabolic mirrors focus the energy. The other uses an optical lens. Of the various types of optical lenses, Fresnel lenses are preferred because they can achieve short focal lengths with minimal thickness. Because IR energy is severely attenuated by glass, lenses are usually made of plastic.

6-74. The sensor's detection pattern is determined by the arrangement of lenses or reflectors. The pattern is not continuous but consists of a number of rays or fingers, one for each mirror or lens segment. Numerous detection patterns are available, several of which are shown in Figure 6-9. The PIR is not provided with a range adjustment, but the range can be adjusted

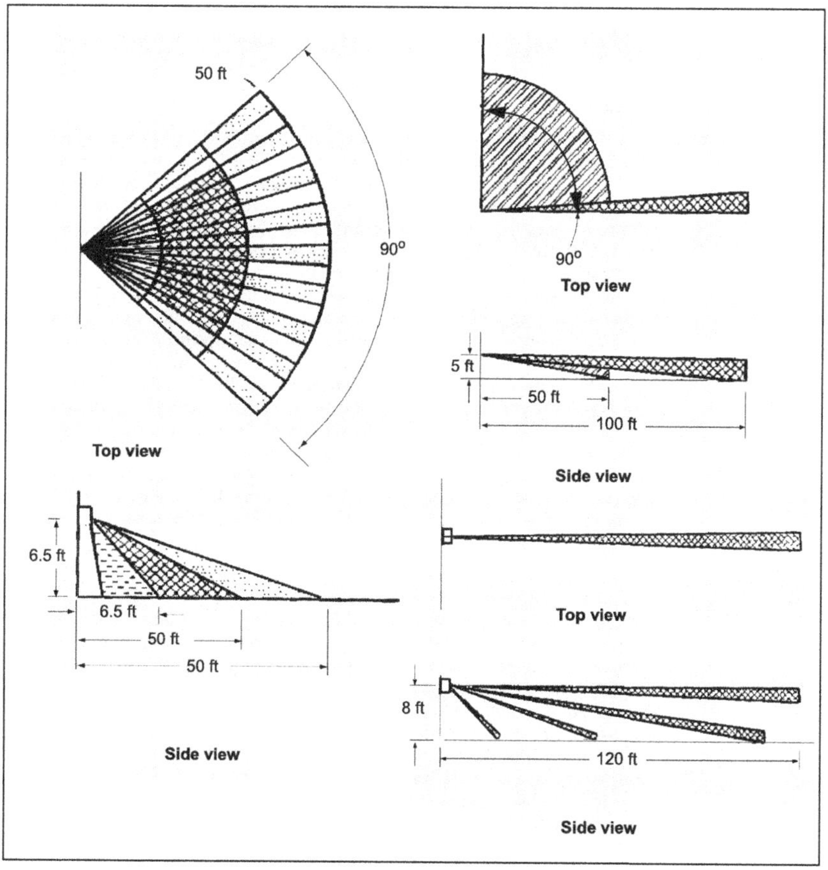

Figure 6-9. Typical Detection Patterns for a PIR Motion Detector

somewhat by manipulating the sensor's position; therefore, careful selection of the appropriate detection pattern is critical to proper sensor performance.

6-75. Most manufacturers use a pyroelectric material as the thermal sensing element. This material produces a change in electric charge when exposed to changes in temperature. To minimize false alarms caused by changes in ambient temperature, most manufacturers use a dual-element sensor. The sensing element is split into halves, one that produces a positive voltage pulse and the other a negative pulse when a change in temperature changes. An intruder entering one of the detection fingers produces an imbalance between the two halves, resulting in an alarm condition. Quadelement sensors that combine and compare two dual-element sensors are also in use. Pulse-count activation, a technique in which a predefined number of pulses within a specific interval of time must be produced before an alarm is generated, is also used.

Dual-Technology Sensors

6-76. To minimize the generation of alarms caused by sources other than intruders, dual-technology sensors combine two different technologies in one unit. Ideally, this is achieved by combining two sensors that individually have a high PD and do not respond to common sources of false alarms. Available dual-technology sensors combine an active ultrasonic or microwave sensor with a PIR sensor. The alarms from each sensor are logically combined in an "and" configuration; that is, nearly simultaneous alarms from both active and passive sensors are needed to produce a valid alarm. Although combined technology sensors have a lower false-alarm rate than individual sensors, the PD is also reduced. For example, if each individual sensor has a PD of 0.95, the PD of the combined sensors is the product of individual probabilities (0.9). Also, ultrasonic and microwave motion sensors have the highest probability of detecting movement directly toward or away from the sensor, whereas PIR motion sensors have the highest probability of detecting movement across the detection pattern. Therefore, the PD of sensors combined in a single unit is less than that obtainable if the individual sensors are mounted perpendicular to each other with overlapping detection patterns. Because of the lower false-alarm rate, the reduced PD can be somewhat compensated for by increasing the sensitivity or detection criteria of each individual sensor.

Video Motion Sensors

6-77. A video motion sensor generates an alarm when an intruder enters a selected portion of a CCTV camera's field of view. The sensor processes and compares successive images between the images against predefined alarm criteria. There are two categories of video motion detectors—analog and digital. Analog detectors generate an alarm in response to changes in a picture's contrast. Digital devices convert selected portions of the analog video signal into digital data that are compared with data converted previously; if differences exceed preset limits, an alarm is generated. The signal processor usually provides an adjustable window that can be positioned anywhere on the video image. Available adjustments permit changing horizontal and vertical window size, window position, and window sensitivity. More sophisticated units provide several adjustable windows that can be individually sized and positioned. Multiple windows permit concentrating on several specific areas of an image while ignoring others. For example, in a scene containing six doorways leading into a long hallway, the sensor can be set to monitor only two critical doorways.

POINT SENSORS

6-78. Point sensors are used to protect specific objects within a facility. These sensors (sometimes referred to as proximity sensors) detect an intruder coming in close proximity to, touching, or lifting an object. Several different types are available, including capacitance sensors, pressure mats, and pressure switches. Other types of sensors can also be used for object protection.

Capacitance Sensors

6-79. Capacitance sensors detect an intruder approaching or touching a metal object by sensing a change in capacitance between the object and the ground. A capacitor consists of two metallic plates separated by a dielectric medium. A change in the dielectric medium or electrical charge results in a change in capacitance. In practice, the metal object to be protected forms one plate of the capacitor and the ground plane surrounding the object forms the second plate. The sensor processor measures the capacitance between the metal object and the ground plane. An approaching intruder alters the dielectric value, thus changing the capacitance. If the net capacitance change satisfies the alarm criteria, an alarm is generated.

6-80. The maximum capacitance that can be monitored by this type of sensor is usually between 10,000 and 50,000 picofarads. The minimum detectable change in capacitance can be as low as 20 picofarads. The signal processor usually has a sensitivity adjustment that can be set to detect an approaching intruder several feet away or to require that the intruder touch the object before an alarm is generated.

6-81. Because air forms most of the dielectric of the capacitor, changes in relative humidity will affect the sensor's sensitivity. An increase in humidity causes the conductivity of the air to increase, lowering the capacitance. Moving a metal object (such as a file cabinet) closer to or away from the protected object can also affect the sensitivity of a capacitance sensor. Figure 6-10, page 6-28, illustrates a typical application using a capacitance sensor.

Pressure Mats

6-82. Pressure mats generate an alarm when pressure is applied to any part of the mat's surface, as when someone steps on the mat. One type of construction uses two layers of copper screening separated by soft-sponge rubber insulation with large holes in it. Another type uses parallel strips of ribbon switches made from two strips of metal separated by an insulating material and spaced several inches apart. When enough pressure is applied to the mat, either the screening or the metal strips make contact, generating an alarm. Pressure mats can be used to detect an intruder approaching a protected object, or they can be placed by doors or windows to detect entry. Because pressure mats are easy to bridge, they should be well concealed, such as placing them under a carpet.

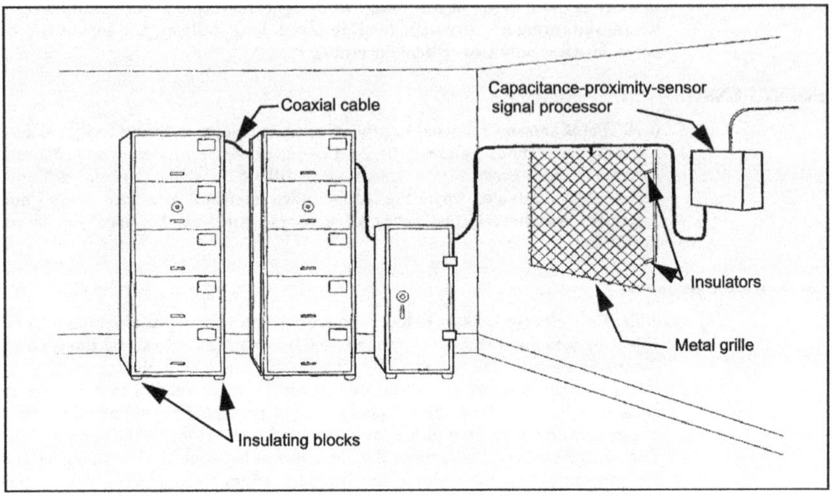

Figure 6-10. Capacitance-Proximity-Sensor Application

Pressure Switches

6-83. Mechanically activated contact switches or single ribbon switches can be used as pressure switches. Objects that require protection can be placed on top of the switch. When the object is moved, the switch actuates and generates an alarm. In this usage, the switch must be well concealed. The interface between the switch and the protected object should be designed so that an adversary cannot slide a thin piece of material under the object to override the switch while the object is removed.

DURESS-ALARM DEVICES

6-84. Duress-alarm devices may be fixed or portable. Operations and security personnel use them to signal a life-threatening emergency. Activation of a duress device will generate an alarm at the alarm-monitoring station. Because of the nature of the alarm, duress devices should never annunciate at the point of threat. These devices are customarily manually operated.

6-85. Fixed duress devices are mechanical switches permanently mounted in an inconspicuous location, such as under a counter or desk. They can be simple push-button switches activated by the touch of a finger or hand or foot-operated switches attached to the floor.

6-86. Portable duress devices are wireless units consisting of a transmitter and a receiver. The transmitter is portable and small enough to be conveniently carried by a person. The receiver is mounted in a fixed location within the facility. Either ultrasonic or RF energy can be used as the communication medium. When activated, the transmitter generates an alarm that is detected (within range) by the receiver. The receiver then activates a relay that is hardwired to the alarm-monitoring system.

EXTERIOR INTRUSION-DETECTION SENSORS

6-87. Exterior intrusion-detection sensors are customarily used to detect an intruder crossing the boundary of a protected area. They can also be used in clear zones between fences or around buildings, for protecting materials and equipment stored outdoors within a protected boundary, or in estimating the PD for buildings and other facilities.

6-88. Exterior sensors are designed to operate in outdoor environmental conditions. The detection function must be performed with a minimum of unwanted alarms such as those caused by wind, rain, ice, standing water, blowing debris, animals, and other sources. Important criteria for selecting an exterior sensor are the PD, the sensor's susceptibility to unwanted alarms, and the sensor's vulnerability to defeat.

6-89. The PD of an exterior sensor is much more vulnerable to the physical and environmental conditions of a site than that of an interior sensor. Many uncontrollable forces (such as wind, rain, ice, frozen soil, standing or running water, falling and accumulated snow, and blowing dust and debris) may affect an exterior sensor's performance. Although attention generally is directed to circumstances that cause a dramatic drop in the PD, environmental factors can also cause short-term increases in the PD. If controlled intrusions (intrusions by security personnel to verify the current detection capability of an IDS) are done while an IDS temporarily has a higher than usual PD as the result of current site conditions, the results may give a false indication of the general effectiveness of that IDS.

6-90. Because of the nature of the outdoor environment, exterior sensors are also more susceptible to nuisance and environmental alarms than interior sensors. Inclement weather conditions (heavy rain, hail, and high wind), vegetation, the natural variation of the temperature of objects in the detection zone, blowing debris, and animals are major sources of unwanted alarms.

6-91. As with interior sensors, tamper protection, signal-line supervision, self-test capability, and proper installation make exterior sensors less vulnerable to defeat. Because signal-processing circuitry for exterior sensors is generally more vulnerable to tampering and defeat than that for interior sensors, it is extremely important that enclosures are located and installed properly and that adequate physical protection is provided. Several different types of exterior intrusion-detection sensors are available. They can be categorized as—

- Fence sensors.
- Buried line sensors.
- LOS sensors.
- Video motion sensors.

FENCE SENSORS

6-92. Fence sensors detect attempts to penetrate a fence around a protected area. Penetration attempts (such as climbing, cutting, or lifting) generate mechanical vibrations and stresses in fence fabric and posts that are usually different than those caused by natural phenomena like wind and rain. The basic types of sensors used to detect these vibrations and stresses are strain-sensitive cable, taut wire, and fiber optics. Other types of fence sensors detect penetration attempts by sensing changes in an electric field or in capacitance. Mechanical and electromechanical fence sensors are seldom used and will not be discussed here.

Strain-Sensitive Cable

6-93. Strain-sensitive cables are transducers that are uniformly sensitive along their entire length. They generate an analog voltage when subject to mechanical distortions or stress resulting from fence motion. Strain-sensitive cables are sensitive to both low and high frequencies. The signal processor usually has a band-pass filter that passes only those signals characteristic of fence-penetration actions. An alarm is initiated when the signal's frequency, amplitude, and duration characteristics satisfy the processor's criteria. Because the cable acts like a microphone, some manufacturers offer an option that allows the operator to listen to fence noises causing the alarm. Operators can then determine whether the noises are naturally occurring sounds from wind or rain or are from an actual intrusion attempt. This feature is relatively costly to implement because it requires additional cable from each signal processor to the security center and, if CCTV is being used, it may be of limited benefit. Strain-sensitive cable is attached to a chain-link fence about halfway between the bottom and top of the fence fabric with plastic ties. One end of the cable is terminated at the signal processor and the other end with a resistive load. The DC through the cable provides line supervision against cutting or electrically shorting the cable or disconnecting it from the processor. A typical installation is shown in Figure 6-11.

Taut-Wire Sensor

6-94. A taut-wire sensor combines a physically taut-wire barrier with an intrusion-detection sensor network. The taut-wire sensor consists of a column of uniformly spaced horizontal wires up to several hundred feet in length and securely anchored at each end. Typically, the wires are spaced 4 to 8 inches apart. Each is individually tensioned and attached to a detector located in a sensor post. Two types of detectors are commonly used—mechanical switches and strain gauges.

- The mechanical switch consists of a specially designed switch mechanism that is normally open. The tensioned wires are mechanically attached to the switch, and movement of the wire beyond a preset limit causes the switch to close. To counteract small gradual movements of a wire (such as that caused by settling of the fence or by freezing or thawing of soil) switches are usually supported in their housing by a soft plastic material. This material allows the switch to self-adjust when acted upon by gradual external forces and wire effects

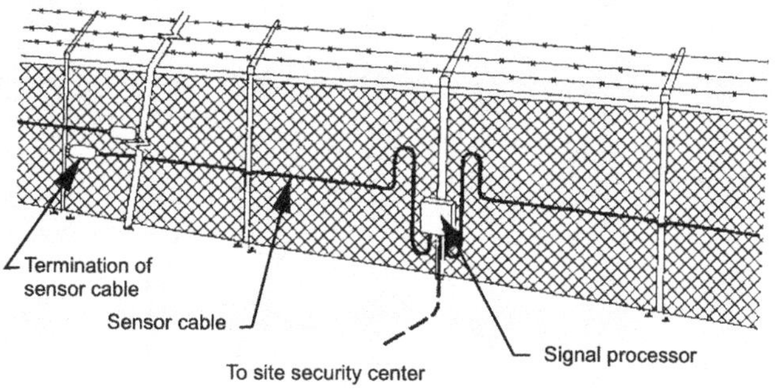

Figure 6-11. Typical Strain-Sensor Cable Installation

such as the relaxation of the wire with time and its thermal expansion or contraction.

- Strain-gauge detectors are attached to the taut wire with a nut on a threaded stud. When a force is applied to the taut wire, the resulting deflection is converted by the strain gauge into a change in electrical output that is monitored by a signal processor.

6-95. With sensors that use mechanical switches as detectors, the switches in a single sensor-post assembly are wired in parallel and are connected directly to the alarm-annunciation system. Pulse-count circuitry is not used because a single switch closure, such as that caused by an intruder moving or cutting one wire, is indicative of an intrusion attempt. Strain-gauge detectors in a sensor post are monitored by a signal processor. When the signal from one or more strain gauges satisfies the processor's criteria, an alarm is initiated.

6-96. The taut-wire sensor can be installed as a freestanding fence or can be mounted on an existing fence or wall. Figure 6-12, page 6-32, shows a freestanding configuration.

Fiber-Optic Cable Sensors

6-97. Fiber-optic cable sensors are functionally equivalent to the strain-sensitive cable sensors previously discussed. However, rather than electrical signals, modulated light is transmitted down the cable and the resulting received signals are processed to determine whether an alarm should be initiated. Since the cable contains no metal and no electrical signal is present, fiber-optic sensors are generally less susceptible to electrical interference from lightning or other sources.

Electric-Field Sensors

6-98. Electric-field sensors consist of an alternating-current (AC) field generator, one or more field wires, one or more sense wires, and a signal

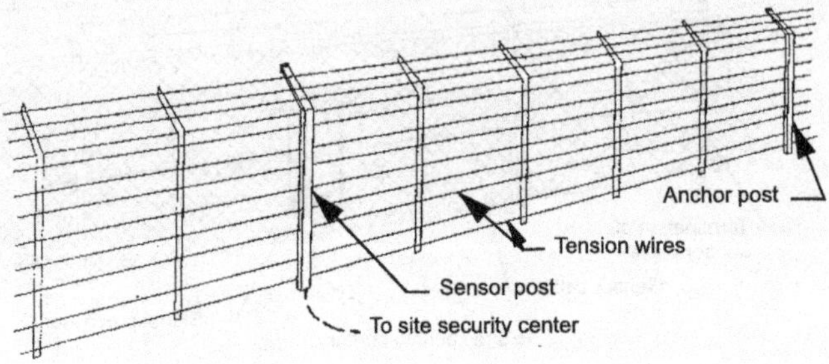

Figure 6-12. Typical Taut-Wire Installation

processor. The generator excites the field wires around which an electrostatic-field pattern is created. The electrostatic field induces electrical signals in the sense wires, which are monitored by the signal processor. Under normal operating conditions, the induced signals are constant. However, when an intruder approaches the sensor, the induced electrical signals are altered, causing the signal processor to generate an alarm.

6-99. Several different field- and sense-wire configurations are available. They range from one field wire and one sense wire to as many as four field wires and one sense wire or four field wires and four sense wires. Figure 6-13 shows the detection pattern produced by vertical three-wire (one field and two sense wires) configurations. The three-wire system has a wider detection envelope and is less costly (one less field wire and associated hardware). However, because of the tighter coupling between wires, the four-wire system is less susceptible to nuisance alarms caused by extraneous noise along the length of the zone.

6-100. A signal processor monitors the signals produced by the sense wires. The processor usually contains a band-pass filter that rejects high-frequency signals such as those caused by objects striking the wires. Additional criteria that must be satisfied before the processor initiates an alarm include signal amplitude and signal duration. By requiring the signal to be present for a preset amount of time, false alarms (such as those caused by birds flying through the detection pattern) can be minimized.

6-101. As with taut-wire sensors, electric-field sensors can be freestanding (mounted on their own posts) or attached by standoffs to an existing fence. They can also be configured to follow contours of the ground. The area under the sensor must be clear of vegetation, since vegetation near or touching sense wires can cause false alarms. These sensors can also be installed on the walls and roof of a building.

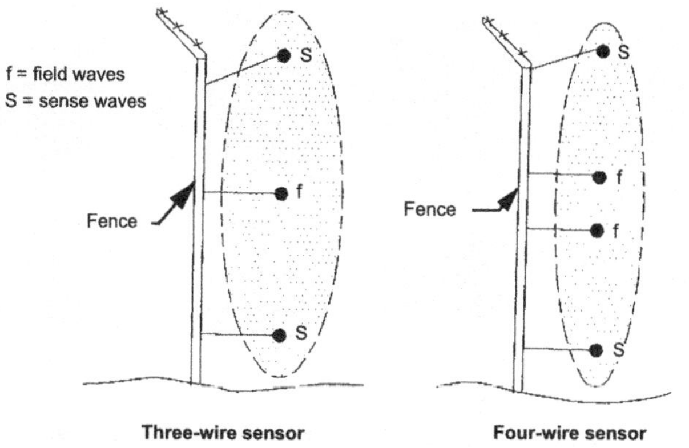

Figure 6-13. Typical Electric-Field-Sensor Detection Patterns

Capacitance Proximity Sensors

6-102. Capacitance proximity sensors measure the electrical capacitance between the ground and an array of sense wires. Any variations in capacitance, such as that caused by an intruder approaching or touching one of the sense wires, initiates an alarm. These sensors usually consist of two or three wires attached to outriggers along the top of an existing fence, wall, or roof edge. Figure 6-14, page 6-34, shows a typical capacitance sensor consisting of three sensor wires attached to the outrigger of a fence. To minimize environmental alarms, the capacitance sensor is divided into two arrays of equal length. The signal processor monitors the capacitance of each array. Changes in capacitance common to both arrays (such as produced by wind, rain, ice, fog, and lightning) are canceled within the processor. However, when changes occur in one array and not the other because of an intruder, the processor initiates an alarm.

BURIED-LINE SENSORS

6-103. A buried-line sensor system consists of detection probes or cable buried in the ground, typically between two fences that form an isolation zone. These devices are wired to an electronic processing unit. The processing unit generates an alarm if an intruder passes through the detection field. Buried-line sensors have several significant features:

- They are hidden, making them difficult to detect and circumvent.
- They follow the terrain's natural contour.
- They do not physically interfere with human activity, such as grass mowing or snow removal.

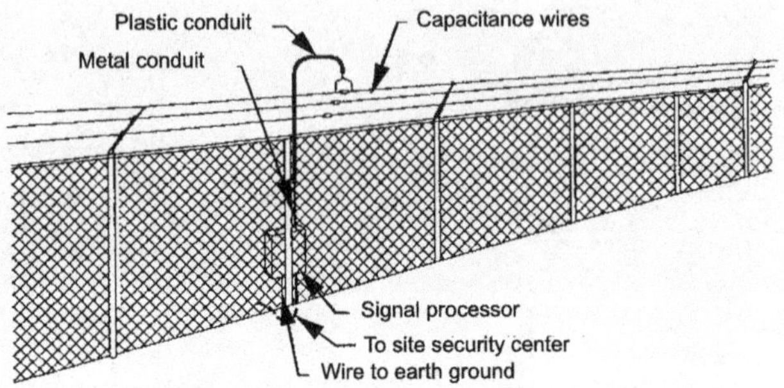

Figure 6-14. Typical Capacitance-Sensor Configuration

- They are affected by certain environmental conditions, such as running water and ground freeze/thaw cycles. (Seismic, seismic/magnetic, magnetic, and balanced pressure sensors are seldom used and will not be discussed here.)

6-104. The ported-coax cable sensor consists of two coax cables buried in the ground parallel to each other. An RF transmitter is connected to one cable and a receiver to the other. The outer conductor of each cable is ported (fabricated with small holes or gaps in the shield). The transmitter cable radiates RF energy into the medium surrounding the cables. A portion of this energy is coupled into the receiver cable through its ported shield. (Because of the ported shields, these cables are frequently referred to as leaky cables.) When an intruder enters the RF field, the coupling is disturbed, resulting in a change of signal monitored by the receiver, which then generates an alarm. Two basic types of ported-coax sensors are available—pulse and continuous wave.

- Pulse-type sensors transmit a pulse of RF energy down one cable and monitors the received signal on the other. The cables can be up to 10,000 feet long. The signal processor initiates an alarm when the electromagnetic field created by the pulse is disturbed and identifies the disturbance's approximate location.
- Continuous-wave sensors apply continuous RF energy to one cable. The signal received on the other cable is monitored for electromagnetic-field disturbances that indicate an intruder's presence. Cable lengths are limited to 300 to 500 feet. Additionally, the sensor is available in a single-cable configuration as well as two separate cables. The pattern typically extends 2 to 4 feet above the ground and can be 5 to 13 feet wide, depending on cable spacing and soil composition. Figure 6-15 represents a typical cross-section of a detection pattern created by a ported-cable sensor.

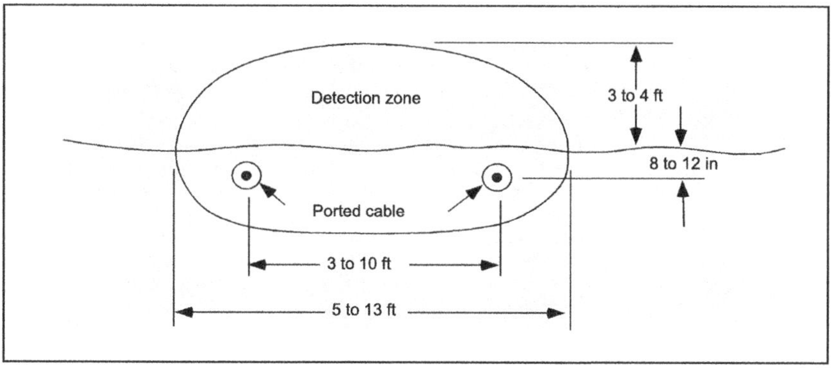

Figure 6-15. Typical Ported-Cable Detection Pattern

6-105. Sensor performance depends on properties of the medium surrounding the cables. Velocity and attenuation of the RF wave that propagates along the cables and the coupling between the cables are functions of the dielectric constant of the soil and its conductivity which, in turn, depends on its moisture content. For example, the velocity is greater and the attenuation is less for cables buried in dry, low-loss soil than in wet, conductive soil. Freeze/thaw cycles in the soil also affect the sensor's performance. When wet soil freezes, the wave velocity and the cable coupling increase and the attenuation decreases, resulting in greater detection sensitivity. Seasonal sensitivity adjustments may be necessary to compensate for changing ground conditions.

6-106. Although usually buried in soil, ported cables can also be used with asphalt and concrete. If the asphalt or concrete pavement area is relatively small and only a few inches thick (such as a pedestrian pavement crossing the perimeter), the ported cables can be routed under the pavement. However, for the large and deep pavements, slots must be cut into the asphalt or concrete to accept the cable.

6-107. A portable ported-coax sensor is available that can be rapidly deployed and removed. The cables are placed on the surface of the ground rather than buried. This sensor is useful for temporary perimeter detection coverage for small areas or objects (such as vehicles or aircraft).

LOS SENSORS

6-108. The LOS sensors, which are mounted above ground, can be either active or passive. Active sensors generate a beam of energy and detect changes in the received energy that an intruder causes by penetrating the beam. Each sensor consists of a transmitter and a receiver and can be in a monostatic or bistatic configuration. Passive sensors generate no beam of energy; they simply look for changes in the thermal characteristics of their field of view. For effective detection, the terrain within the detection zone must be flat and free of obstacles and vegetation.

Microwave Sensors

6-109. Microwave intrusion-detection sensors are categorized as bistatic or monostatic. Bistatic sensors use transmitting and receiving antennas located at opposite ends of the microwave link, whereas monostatic sensors use the same antenna.

- A bistatic system uses a transmitter and a receiver that are typically separated by 100 to 1,200 feet and that are within direct LOS with each other. The signal picked up by the receiver is the vector sum of the directly transmitted signal and signals that are reflected from the ground and nearby structures. Detection occurs when an object (intruder) moving within the beam pattern causes a change in net-vector summation of the received signals, resulting in variations of signal strength.
  - The same frequency bands allocated by the Federal Communications Commission (FCC) for interior microwave sensors are also used for exterior sensors. Because high-frequency microwave beams are more directive than low-frequency beams and the beam pattern is less affected by blowing grass in the area between the transmitter and the receiver, most exterior sensors operate at the next to highest allowable frequency, 10.525 gigahertz (GHz).
  - The shape of the microwave beam and the maximum separation between the transmitter and the receiver are functions of antenna size and configuration. Various antenna configurations are available, including parabolic-dish arrays, strip-line arrays, and slotted arrays. The parabolic antenna uses a microwave-feed assembly located at the focal point of a metallic parabolic reflector. A conical beam pattern is produced (see Figure 6-16). A strip-line antenna configuration produces a nonsymmetrical beam that is higher than its height. Larger antenna configurations generally produce narrower beam patterns.

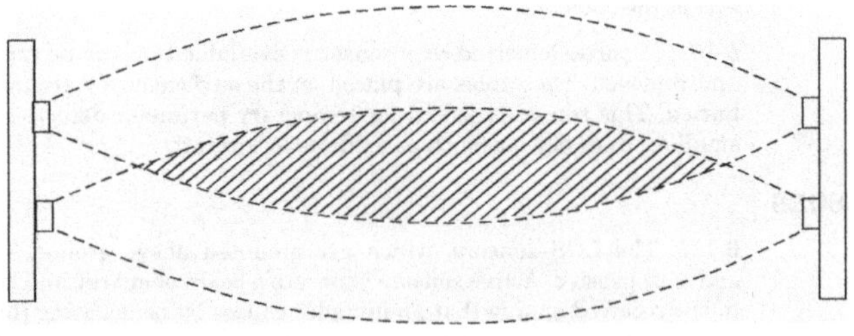

Figure 6-16. Stacked Microwave Configuration

- Monostatic microwave sensors use the same antenna or virtually coincident antenna arrays for the transmitter and receiver, which are usually combined into a single package. Two types of monostatic sensors are available. Amplitude-modulated (AM) sensors detect changes in the net-vector summation of reflected signals similar to bistatic sensors. Frequency-modulated (FM) sensors operate on the Doppler principle similar to interior microwave sensors. The detection pattern is typically shaped like a teardrop (see Figure 6-17). Monostatic sensors can provide volumetric coverage of localized areas, such as in corners or around the base of critical equipment.

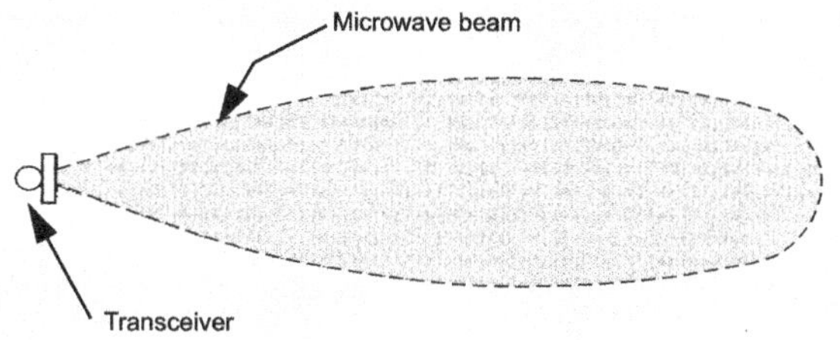

Figure 6-17. Typical Monostatic-Microwave-Sensor Detection Pattern

IR Sensors

6-110. The IR sensors are available in both active and passive models. An active sensor generates one or more near-IR beams that generate an alarm when interrupted. A passive sensor detects changes in thermal IR radiation from objects located within its field of view.

6-111. Active sensors consist of transmitter/receiver pairs. The transmitter contains an IR light source (such as a gallium arsenide light-emitting diode [LED]) that generates an IR beam. The light source is usually modulated to reduce the sensor's susceptibility to unwanted alarms resulting from sunlight or other IR light sources. The receiver detects changes in the signal power of the received beam. To minimize nuisance alarms from birds or blowing debris, the alarm criteria usually require that a high percentage of the beam be blocked for a specific interval of time.

6-112. Active sensors can be single- or multiple-beam systems. Because single-beam sensors can be easily bypassed, multiple-beam systems are generally used in perimeter applications. There are two basic types of multiple-beam configurations—one type uses all transmitters on one post and all receivers on the other post; the second type uses one transmitter and several receivers on each post. Both types are illustrated in Figure 6-18, page 6-38.

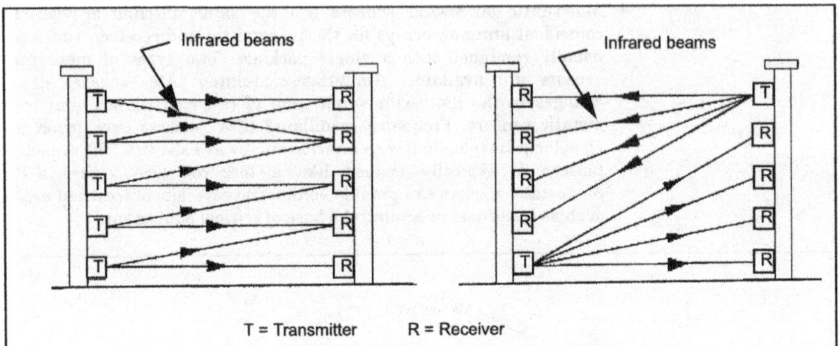

Figure 6-18. Typical IR-Sensor Beam Patterns

6-113. The spacing between transmitters and receivers can be as great as 1,000 feet when operation is under good weather conditions. However, conditions such as heavy rain, fog, snow, or blowing dust particles attenuate the IR energy, reducing its effective range to 100 to 200 feet or less.

VIDEO MOTION SENSORS

6-114. A video motion sensor generates an alarm whenever an intruder enters a selected portion of a CCTV camera's field of view. The sensor processes and compares successive images from the camera and generates an alarm if differences between the images satisfy predefined criteria. Digital devices convert selected portions of the analog video signal into digital data that are compared with data converted previously; if differences exceed preset limits, an alarm is generated.

6-115. The signal processor usually provides an adjustable window that can be positioned anywhere on the video image. Available adjustments permit changing the window's horizontal and vertical sizes, its position, and its sensitivity. More sophisticated units provide several adjustable windows that can be individually sized and positioned. Multiple windows permit concentrating on several specific areas of an image while ignoring others. For example, in a scene that contains several critical assets and multiple sources of nuisance alarms (such as large bushes or trees), the sensor can be adjusted to monitor only the assets and ignore the areas that contain the nuisance-alarm sources.

6-116. The use of video motion-detection systems for exterior applications has been limited, primarily because of difficulties with uncontrolled exterior environments. Lighting variations caused by cloud movement and shadows of slow-moving objects, birds and animals moving within the camera's field of view, camera motion and moving vegetation during windy conditions, and severe weather conditions have traditionally caused a multitude of unwanted alarms in this type of system. Systems using more advanced signal-processing algorithms have improved motion-detection capability and nuisance-alarm rejection; however, they are still subject to high unwanted-alarm rates under certain conditions and should be used with due caution and extreme care.

ELECTRONIC ENTRY CONTROL

6-117. The function of an entry-control system is to ensure that only authorized personnel are permitted into or out of a controlled area. Entry can be controlled by locked fence gates, locked doors to a building or rooms within a building, or specially designed portals.

6-118. These means of entry control can be applied manually by guards or automatically by using entry-control devices. In a manual system, guards verify that a person is authorized to enter an area, usually by comparing the photograph and personal characteristics of the individual requesting entry. In an automated system, the entry-control device verifies that a person is authorized to enter or exit. The automated system usually interfaces with locking mechanisms on doors or gates that open momentarily to permit passage. Mechanical hardware (such as locking mechanisms, electric door strikes, and specially designed portal hardware) and equipment used to detect contraband material (such as metal detectors, X-ray baggage-search systems, explosives detectors, and special nuclear-material monitors) are described in other documentation. Refer to TM 5-853-1 for additional information on determining entry-control requirements and integrating manual electronic-entry control into a cohesive system.

6-119. All entry-control systems control passage by using one or more of three basic techniques—something a person knows, something a person has, or something a person is or does. Automated entry-control devices based on these techniques are grouped into three categories—coded, credential, and biometric devices.

CODED DEVICES

6-120. Coded devices operate on the principle that a person has been issued a code to enter into an entry-control device. This code will match the code stored in the device and permit entry. Depending on the application, a single code can be used by all persons authorized to enter the controlled area or each authorized person can be assigned a unique code. Group codes are useful when the group is small and controls are primarily for keeping out the general public. Individual codes are usually required for control of entry to more critical areas. Coded devices verify the entered code's authenticity, and any person entering a correct code is authorized to enter the controlled area. Electronically coded devices include electronic and computer-controlled keypads.

Electronic Keypad Devices

6-121. The common telephone keypad (12 keys) is an example of an electronic keypad. This type of keypad consists of simple push-button switches that, when depressed, are decoded by digital logic circuits. When the correct sequence of buttons is pushed, an electric signal unlocks the door for a few seconds.

Computer-Controlled Keypad Devices

6-122. These devices are similar to electronic keypad devices, except they are equipped with a microprocessor in the keypad or in a separate enclosure at a different location. The microprocessor monitors the sequence in which the keys are depressed and may provide additional functions such as personal ID and digit scrambling. When the correct code is entered and all conditions are satisfied, an electric signal unlocks the door.

CREDENTIAL DEVICES

6-123. A credential device identifies a person having legitimate authority to enter a controlled area. A coded credential (plastic card or key) contains a prerecorded, machine-readable code. An electric signal unlocks the door if the prerecorded code matches the code stored in the system when the card is read. Like coded devices, credential devices only authenticate the credential; it assumes a user with an acceptable credential is authorized to enter. Various technologies are used to store the code upon or within a card. Hollerith, optically coded, magnetic-spot, capacitance, and electric-circuit cards are seldom used and will not be discussed here. The most commonly used types of cards are described as follows:

Magnetic-Stripe Card

6-124. A strip of magnetic material located along one edge of the card is encoded with data (sometimes encrypted). The data is read by moving the card past a magnetic read head.

Wiegand-Effect Card

6-125. The Wiegand-effect card contains a series of small-diameter, parallel wires about one-half inch long, embedded in the bottom half of the card. The wires are manufactured from ferromagnetic materials that produce a sharp change in magnetic flux when exposed to a slowly changing magnetic field. This type of card is impervious to accidental erasure. The card reader contains a small read head and a tiny magnet to supply the applied magnetic field. It usually does not require external power.

Proximity Card

6-126. A proximity card is not physically inserted into a reader; the coded pattern on the card is sensed when it is brought within several inches of the reader. Several techniques are used to code cards. One technique uses a number of electrically tuned circuits embedded in the card. Data are encoded by varying resonant frequencies of the tuned circuits. The reader contains a transmitter that continually sweeps through a specified range of frequencies and a receiver that senses the pattern of resonant frequencies contained in the card. Another technique uses an integrated circuit embedded in the card to generate a code that can be magnetically or electrostatically coupled to the reader. The power required to activate embedded circuitry can be provided by a small battery embedded in the card or by magnetically coupling power from the reader.

Laser Card

6-127. The optical memory card, commonly called the laser card, uses the same technology developed for recording video and audio disks for entertainment purposes. Data is recorded on the card by burning a microscopic hole (using a laser) in a thin film covering the card. Data is read by using a laser to sense the hole locations. The typical laser card can hold several megabytes of user data.

Smart Card

6-128. A smart card is embedded with a microprocessor, memory, communication circuitry, and a battery. The card contains edge contacts that enable a reader to communicate with the microprocessor. Entry-control information and other data may be stored in the microprocessor's memory.

Bar Code

6-129. A bar code consists of black bars printed on white paper or tape that can be easily read with an optical scanner. This type of coding is not widely used for entry-control applications because it can be easily duplicated. It is possible to conceal the code by applying an opaque mask over it. In this approach, an IR scanner is used to interpret the printed code. For low-level security areas, the use of bar codes can provide a cost-effective solution for entry control. Coded strips and opaque masks can be attached to existing ID badges, alleviating the need for complete badge replacement.

BIOMETRIC DEVICES

6-130. The third basic technique used to control entry is based on the measurement of one or more physical or personal characteristics of an individual. Because most entry-control devices based on this technique rely on measurements of biological characteristics, they have become commonly known as biometric devices. Characteristics such as fingerprints, hand geometry, voiceprints, handwriting, and retinal blood-vessel patterns have been used for controlling entry. Typically, in enrolling individuals, several reference measurements are made of the selected characteristic and then stored in the device's memory or on a card. From then on, when that person attempts entry, a scan of the characteristic is compared with the reference data template. If a match is found, entry is granted. Rather then verifying an artifact, such as a code or a credential, biometric devices verify a person's physical characteristic, thus providing a form of identity verification. Because of this, biometric devices are sometimes referred to as personnel identity-verification devices. The most common biometric devices are discussed below.

Fingerprints

6-131. Fingerprint-verification devices use one of two approaches. One is pattern recognition of the whorls, loops, and tilts of the referenced fingerprint, which is stored in a digitized representation of the image and compared with the fingerprint of the prospective entrant. The second approach is minutiae comparison, which means that the endings and branching points of ridges and valleys of the referenced fingerprint are compared with the fingerprint of the prospective entrant.

Hand Geometry

6-132. Several devices are available that use hand geometry for personnel verification. These devices use a variety of physical measurements of the hand, such as finger length, finger curvature, hand width, webbing between fingers, and light transmissivity through the skin to verify identity. Both two- and three-dimensional units are available.

Retinal Patterns

6-133. This type of technique is based on the premise that the pattern of blood vessels on the human eye's retina is unique to an individual. While the eye is focused on a visual target, a low-intensity IR light beam scans a circular area of the retina. The amount of light reflected from the eye is recorded as the beam progresses around the circular path. Reflected light is modulated by the difference in reflectivity between blood-vessel pattern and adjacent tissue. This information is processed and converted to a digital template that is stored as the eye's signature. Users are allowed to wear contact lenses; however, glasses should be removed.

Device Combinations

6-134. Frequently, an automated entry-control system uses combinations of the three types of entry-control devices. Combining two different devices can significantly enhance the system's security level. In some cases, combining devices results in reduced verification times.

APPLICATION GUIDELINES

6-135. The primary function of an automated entry-control system is to permit authorized personnel to enter or exit a controlled area. Features available to the designer are described below.

- **Enrollment.** All entry-control systems must provide a means of entering, updating, and deleting information about authorized individuals into the system's database files. This is usually accomplished with a dedicated enrollment station for enrolling and disenrolling purposes that is directly connected to the central-processing unit. When credential devices are used, all authorized users must be provided with an appropriate credential. A means should also be provided to disenroll a person quickly without having to retrieve a credential. When using biometric devices, additional enrollment equipment will be required.
- **Entry-control techniques.** Some entry-control functions require additional hardware, while others are accomplished with software. Those features accomplished with software require that the appropriate database be available for every portal affected by them. Typically, these techniques include—
  - Area zones.
  - Time zones.
  - Team zones.
  - Anti-pass back.

  - Antitailgate.
  - Guard tour.
  - Elevator control.
- **Alarms.** Several types of alarms can be used with an entry-control system. These alarms must annunciate audibly and visually in the security center.
- **Entry denial.** Most entry-control devices are configured to permit the user three entry attempts. If more than three unsuccessful entry attempts are made within a specified period, the device generates an alarm. An alarm is also generated if an invalid credential is used or attempted entries are detected that violate specified area, time, or team zoning requirements.
- **Communication failure.** This alarm is generated when a loss of communication between the central processor and the local equipment is detected.
- **Portal open.** If a portal door remains open longer than a predefined time, an alarm is generated.
- **Duress.** This alarm is generated when a special duress code is entered at a keypad.
- **Guard overdue.** This duress alarm is generated when a security guard is determined to be overdue at a checkpoint during a predefined guard tour.
- **Software tamper.** This type of alarm is generated when unauthorized persons are detected attempting to invoke certain system commands or modify database files.

PERFORMANCE CRITERIA

6-136. The overall performance of an entry-control system can be evaluated by examining the verification error rate and the throughput rate. An entry-control system can produce two types of errors—denial of admission of a person who should be admitted or admission of a person who should not be admitted. These are commonly referred to as false-reject errors (type I errors) and false-accept errors (type II errors). Although a false-reject error does not constitute a breach of security, it does create an operational problem that must be handled by an alternative method. False-accept errors constitute a breach of security. Ideally, both false-reject and false-accept error rates should be zero; in practice, however, they are not. In fact, they tend to act in opposition to each other. When the system is adjusted to minimize the false-accept error rate, the false-reject error rate usually increases. Verification error rates are typically measured in percent (number of errors/number of attempts x 100 percent). These error rates are typically very low for coded and credential devices, but many become significant if biometric devices are used.

6-137. The throughput rate is the number of persons that can pass through an entry point in a given unit of time and is usually expressed in persons per minute. It is the time required to approach the entry-control device and for the device to verify information (verification time) and the time required passing through the entry point. Typically, an individual can approach the device and pass through in 3 to 5 seconds. Verification time depends on the type of device and may vary from 3 to 15 seconds. Table 6-5 provides a list of typical verification times for different types of entry-control devices.

Table 6-5. Typical Verification Times of Entry-Control Devices

Device	Verification Time
Keypad	3 seconds
Card reader	3 seconds
Keypad/card reader	6 seconds
Biometric/keypad	6 to 15 seconds
Biometric/card reader	6 to 15 seconds
Biometric	2 minutes

DATA TRANSMISSION

6-138. A critical element in an integrated ESS is the DTM that transmits information from sensors, entry-control devices, and video components to display and assessment equipment. A DTM link is a path for transmitting data between two or more components (such as a sensor and alarm reporting system, a card reader and controller, a CCTV camera and monitor, or a transmitter and receiver). The DTM links connect remote ESS components to the security center. An effective DTM link ensures rapid and reliable transmission media, transmission technique, associated transmission hardware, and degree of security to be provided for the communication system.

6-139. A number of different media are used in transmitting data between elements of an IDS, an EECS, and a CCTV system. These include wire lines, coaxial cable, fiber-optic cable, and RF transmission.

- **Wire line.** Wire lines are twisted pairs that consist of two insulated conductors twisted together to minimize interference by unwanted signals.
- **Coaxial cable.** Coaxial cable consists of a center conductor surrounded by a shield. The center conductor is separated from the shield by a dielectric. The shield protects against electromagnetic interference.
- **Fiber optics.** Fiber optics uses the wide bandwidth properties of light traveling through transparent fibers. Fiber optics is a reliable communication medium best suited for point-to-point, high-speed data transmission. Fiber optics is immune to RF electromagnetic interference and does not produce electromagnetic radiation emission. The preferred DTM for an ESS is fiber-optic cables unless there are justifiable economic or technical reasons for using other types of media.
- **RF transmission.** Modulated RF can be used as a DTM with the installation of radio receivers and transmitters. An RF transmission system does not require a direct physical link between the points of communication, and it is useful for communicating over barriers such as bodies of water and heavily forested terrain. A disadvantage is that the signal power received depends on many factors (including transmission power, antenna pattern, path length, physical

obstructions, and climatic conditions). Also, RF transmission is susceptible to jamming and an adversary with an appropriately tuned receiver has access to it. The use of RF will be coordinated with the communications officer to avoid interference with other existing or planned facility RF systems.

6-140. There are two basic types of communication links—point-to-point and multiplex lines. A point-to-point link is characterized by a separate path for each pair of components. This approach is cost effective for several component pairs or when a number of scattered remote areas communicate with a single central location. The multiplex link, commonly referred to as a multidrop or multipoint link, is a path shared by a number of components. Depending on the number and location of components, this type of configuration can reduce the amount of cabling required. However, the cost reduction from reduced cabling is somewhat offset by costs of equipment required to multiplex and demultiplex data.

6-141. Data links used to communicate the status of ESS devices or other sensitive information to the security center must be protected from possible compromise. Attempts to defeat the security system may range from simple efforts to cut or short the transmission line to more sophisticated undertakings, such as tapping and substituting bogus signals. Data links can be made more secure by physical protection, tamper protection, line supervision, and encryption.

CCTV FOR ALARM ASSESSMENT AND SURVEILLANCE

6-142. A properly integrated CCTV assessment system provides a rapid and cost-effective method for determining the cause of intrusion alarms. For surveillance, a properly designed CCTV system provides a cost-effective supplement to guard patrols. For large facilities, the cost of a CCTV system is more easily justified. It is important to recognize that CCTV alarm-assessment systems and CCTV surveillance systems perform separate and distinct functions. The alarm-assessment system is designed to respond rapidly, automatically, and predictably to the receipt of ESS alarms at the security center. The surveillance system is designed to be used at the discretion of and under the control of the security center's console operator. When the primary function of the CCTV system is to provide real-time alarm assessment, the design should incorporate a video-processing system that can communicate with the alarm-processing system.

6-143. A candidate site for a CCTV assessment system will typically have the following characteristics:

- Assets requiring ESS protection.
- A need for real-time alarm assessment.
- Protected assets spaced some distance apart.

6-144. Figure 6-19, page 6-46, shows a typical CCTV system configuration. A typical site will locate CCTV cameras—

- Outdoors, along site-perimeter isolation zones.
- Outdoors, at controlled access points (sally ports).

- Outdoors, within the protected area, and at viewing approaches to selected assets.
- Indoors, at selected assets within the protected area.

6-145. The security console is centrally located in the security center. The CCTV monitors and the ancillary video equipment will be located at this console, as will the ESS alarm-processing and -annunciation equipment.

CCTV CAMERA COMPONENTS

6-146. An optical-lens system that captures and focuses reflected light from the scene being viewed onto an image target is common to all CCTV cameras. The image target converts reflected light energy into electrical impulses in a two-dimensional array of height and width. An electronic scanning system (reading these impulses in a predetermined order) creates a time-sensitive voltage signal that is a replica of optical information captured by the lens and focused on the target. This voltage signal is then transmitted to a location where it is viewed and possibly recorded. For components and technical information regarding CCTV cameras, see the appropriate TMs.

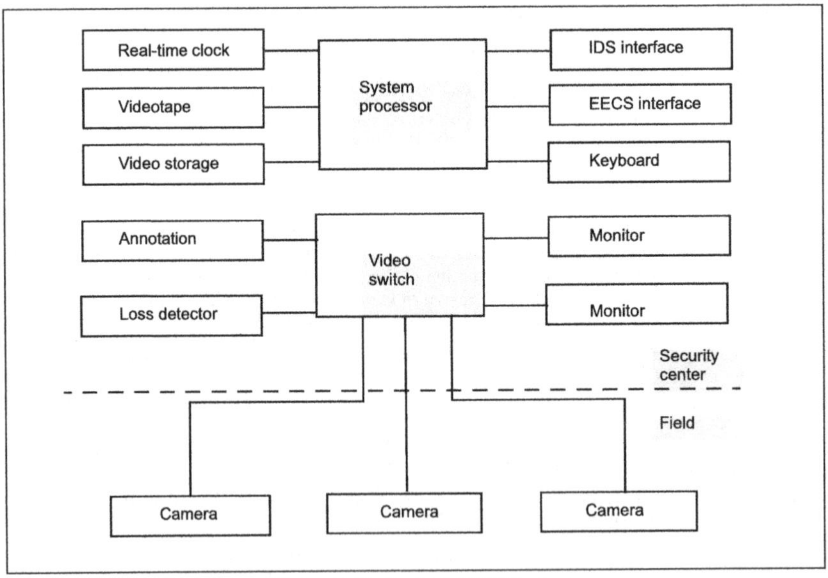

Figure 6-19. Typical CCTV System

VIDEO SIGNAL AND CONTROL LINKS

6-147. A CCTV transmission system is needed to convey video signals from various facility cameras to the security center and to carry commands from the security center to the cameras. Information may be sent via metallic cable, RF, or optical transmission.

Metallic Cable

6-148. Metallic video cables are electrical conductors manufactured specifically for the transmission of frequencies associated with video components. Coaxial cable is a primary example of this type of transmission media. Devices such as video-equalization amplifiers, ground loop correctors, and video-distribution amplifiers may be required.

RF Transmission

6-149. For a system that has widely separated nodes, RF transmission may be a good alternative to metallic cable and associated amplifiers. The information can be transmitted over a microwave link. A microwave link can be used for distances of about 50 miles, as long as the receiver and the transmitter are in the LOS.

Fiber-Optic Cable

6-150. In fiber-optic cable systems, electrical video signals are converted to optical light signals that are transmitted down the optical fiber. The signal is received and reconverted into electrical energy. An optic driver and a receiver are required per fiber. The fiber-optic transmission method provides a low-loss, high-resolution transmission system with usable length three to ten times that of traditional metallic in cable systems. Fiber-optic cable is the transmission media favored by DA.

CCTV-SYSTEM SYNCHRONIZATION

6-151. Timing signals are processed within the image-scan section of the CCTV camera. These signals may be generated internally from a crystal clock, derived from the camera's AC power source, or supplied by an external signal source. The camera should be capable of automatic switchover to its internal clock in case of external signal loss. When CCTV cameras are supplied by a common external (master) signal source or are all powered from the same AC power source, all cameras scan in synchronism. In this case, a console CCTV monitor will display a smooth transition when switched from one video source to another. Without this feature, the monitor display breaks up or rolls when switched between video sources. The rolling occurs for as long as it takes the monitor to synchronize its scan with that of the new video source, typically one second. The resynchronization delay will be experienced by all system components that receive video information, including recorders. To avoid this delay, the designer must specify that all cameras are powered from the AC power phase or must specify master synchronization for the design.

VIDEO PROCESSING AND DISPLAY COMPONENTS

6-152. As shown in Figure 6-19, page 6-46, CCTV camera signals propagate through the video transmission system and through coverage at the security center. In very simple configurations with only a few cameras and monitors, a hardwired connection between each camera and console monitor is adequate. As the number of cameras increases, the need to manage and add supplemental information to camera signals also increases. Psychological testing has demonstrated that the efficiency of console-operator assessment improves as the number of console monitors is reduced, with the optimum number being four to six monitors. Effectiveness is also enhanced by the use of alarm-correlated video. Major components of the video-processor system are the video switcher, the video-loss detector, the alarm-processor communication path, the master video-sync generator, video recorders, and monitors.

- **Video switchers.** Video switchers are required when the number of cameras exceeds the number of console monitors or when a monitor must be capable of selecting video from one of many sources. Video switchers are capable of presenting any of multiple video images to various monitors, recorders, and so forth.
- **Video-loss detector.** Video-loss detectors sense the continued integrity of incoming camera signals.
- **ESS interface and communication path.** There must be a means of rapid communication between the ESS alarm-annunciation and video-processor systems. The alarm processor must send commands that cause the video switcher to select the camera appropriate for the sensor reporting an alarm. The video-processor system must report system tampering or failures (such as loss of video) to the alarm processor. The path should also pass date-and-time synchronizing information between processors so that recorded video scenes and printed alarm logs are properly correlated.
- **Master video-sync generation and distribution.** Master video sync includes a crystal-controlled timing generator, distribution amplifiers, and a transmission link to each camera.
- **Video recorders.** Video recorders provide the means to record alarm-event scenes in real time for later analysis. A recorder typically receives its input through dedicated video-switcher outputs. To support recorder playback, the recorder output is connected to a dedicated switcher input and must be compatible with the switcher-signal format. In addition, the recorder receives start commands from the switcher, and compatibility must exist at this interface. Videocassette recorders should be used when alarm events are to be recorded for later playback and analysis. The cassettes can record in time lapse for up to 240 hours (depending on the user-selected speed) and will change to real-time recording on command. The cassettes can be erased and reused or archived if required.
- **Monitors.** Monitors are required to display the individual scenes transmitted from the cameras or from the video switcher. In alarm-assessment applications, the monitors are driven by dedicated outputs of the video switcher and the monitors display video sources selected by the switcher. For security-console operations, the 9-inch monitor is the

smallest screen that should be used for operator recognition of small objects in a camera's field of view. Two 9-inch monitors can be housed side by side in a standard 19-inch console. If the monitors are to be mounted in freestanding racks behind the security console, larger units will be used.

6-153. Video-processor equipment will be specified to append the following alphanumeric information so that it appears on both monitors and recordings. The equipment must allow the operator to program the annotated information and dictate its position on the screen. This information includes—

- Time and date information.
- Video-source or alarm-zone identification.
- Programmable titles.

CCTV APPLICATION GUIDELINES

6-154. Site-specific factors must be taken into consideration in selecting components that comprise a particular CCTV system. The first is the system's size in terms of the number of cameras fielded, which is the minimum number needed to view all ESS sensor-detection fields and surveillance cameras. Another factor is that some CCTV cameras may require artificial light sources. Finally, there are CCTV-system performance criteria and physical, environmental, and economic considerations. Each is discussed in detail in TM 5-853-4.

Scene Resolution

6-155. The level to which video details can be determined in a CCTV scene is referred to as resolving ability or resolution. It is generally accepted that for assessment purposes, three resolution requirements can be defined. In order of increasing resolution requirements, they are detection, recognition, and identification.

- Detection is the ability to detect the presence of an object in a CCTV scene.
- Recognition is the ability to determine the type of object in a CCTV scene (animal, blowing debris, or crawling human).
- Identification is the ability to determine object details (a particular person, a large rabbit, a small deer, or tumbleweed).

6-156. A CCTV assessment system should provide sufficient resolution to recognize human presence and to detect small animals or blowing debris. Given an alarmed intrusion sensor, it is crucial that the console operator be able to determine if the sensor detected an intruder or if it is simply responding to a nuisance condition. (Refer to TM 5-853-4 for detailed design applications.)

Illumination Levels

6-157. For interior applications where the same camera type is used in several different areas and the scene illumination in each area is constant, specify the manually adjustable iris. This allows a manual iris adjustment appropriate for each particular area's illumination level at the time of installation. If the camera must operate in an area subject to a wide dynamic range of illumination levels (such as would be found outdoors), specify the automatically adjusted iris feature.

Cost Considerations

6-158. The cost of a CCTV system is usually quoted as cost-per-assessment zone. When estimating the total system cost, video-processor equipment costs and the video-transmission system's costs must be included. Other potentially significant costs are outdoor lighting system upgrades and the site preparation required to support the CCTV cameras. The CCTV systems are expensive compared to other electronic security subsystems and should be specified with discretion.

DESIGN GUIDELINES

6-159. The design and application of CCTV systems are quite complex and should be left to professionals who are abreast of the current state-of-the-art systems. Some of the general design guidelines include the following:

- **System familiarity.** Before designing an effective CCTV assessment system, the designer must be familiar with the ESS's sensor placement and the detection field's shape.
- **CCTV camera placement and lighting.** The placement of exterior cameras requires more attention than that of interior cameras because of weather and illumination extremes. The field-of-view alignment, illumination range, and balanced lighting are major design factors. Exterior CCTV design considerations include environmental housings, camera mounting heights, system types, and so forth. Indoor design considerations include the mounting location and tamper detection. The layout for indoor alarm-assessment cameras is subject to three constraints—
  - The camera's location should enclose the complete sensor detection field in the camera's field of view.
  - Lighting that is adequate to support alarm assessment will be provided.
  - Protection from tampering and inadvertent damage by collision during normal area operations will be provided.

Chapter 7

Access Control

Perimeter barriers, intrusion-detection devices, and protective lighting provide physical-security safeguards; however, they alone are not enough. An access-control system must be established and maintained to preclude unauthorized entry. Effective access-control procedures prevent the introduction of harmful devices, materiel, and components. They minimize the misappropriation, pilferage, or compromise of materiel or recorded information by controlling packages, materiel, and property movement. Access-control rosters, personal recognition, ID cards, badge-exchange procedures, and personnel escorts all contribute to an effective access-control system.

DESIGNATED RESTRICTED AREAS

7-1. The installation commander is responsible for designating and establishing restricted areas. A restricted area is any area that is subject to special restrictions or controls for security reasons. This does not include areas over which aircraft flight is restricted. Restricted areas may be established for the following:

- The enforcement of security measures and the exclusion of unauthorized personnel.
- Intensified controls in areas requiring special protection.
- The protection of classified information or critical equipment or materials.

DEGREE OF SECURITY

7-2. The degree of security and control required depends on the nature, sensitivity, or importance of the security interest. Restricted areas are classified as controlled, limited, or exclusion areas.

- A controlled area is that portion of a restricted area usually near or surrounding a limited or exclusion area. Entry to the controlled area is restricted to personnel with a need for access. Movement of authorized personnel within this area is not necessarily controlled since mere entry to the area does not provide access to the security interest. The controlled area is provided for administrative control, for safety, or as a buffer zone for in-depth security for the limited or exclusion area. The commander establishes the control of movement.
- A limited area is a restricted area within close proximity of a security interest. Uncontrolled movement may permit access to the item. Escorts and other internal restrictions may prevent access within limited areas.

- An exclusion area is a restricted area containing a security interest. Uncontrolled movement permits direct access to the item.

7-3. The security protection afforded by a restricted area pertains particularly to subversive-activity control; that is, protection against espionage, sabotage, or any such action adversely affecting national defense. Within this context, the designation "restricted area" is not applicable to an area solely for protection against common pilferage or misappropriation of property or material that is not classified or not essential to national defense. For example, an area devoted to the storage or use of classified documents, equipment, or materials should be designated as a restricted area to safeguard against espionage. An installation communications center should also be so designated to safeguard against sabotage. On the other hand, a cashier's cage or an ordinary mechanic's tool room should not be so designated, although the commander may impose controls to access. This may be a simple matter of posting an "off limits to unauthorized personnel" sign. The PM or the physical-security manager acts as an advisor to the commander. In his recommendations, he must consider evaluating the purpose of designating a restricted area and coordinating with the intelligence officer and the staff judge advocate (SJA).

7-4. A restricted area must be designated in writing by the commander and must be posted with warning signs according to AR 190-13. In areas where English is one of two or more languages commonly spoken, warning signs will be posted in English and in the local language (see Figure 7-1).

7-5. An installation may have varying degrees of security. It may be designated in its entirety as a restricted area, with no further restrictions; or it may be subdivided into controlled, limited, or exclusion areas with restrictions of movement and specific clear zones. Figure 7-2 depicts a simplified restricted area and the degrees of security.

CONSIDERATIONS

7-6. There are other important considerations concerning restricted areas and their lines of division. These considerations include the following:

- A survey and analysis of the installation, its missions, and its security interests. This can determine immediate and anticipated needs that require protection. Anticipated needs are determined from plans for the future.
- The size and nature of the security interest being protected. Safes may provide adequate protection for classified documents and small items; however, large items may have to be placed within guarded enclosures.
- Some security interests are more sensitive to compromise than others. Brief observation or a simple act by an untrained person may constitute a compromise in some cases. In others, detailed study and planned action by an expert may be required.
- All security interests should be evaluated according to their importance. This may be indicated by a security classification such as confidential, secret, or top secret.

RESTRICTED AREA

THIS (INSTALLATION OR ACTIVITY) HAS BEEN DECLARED A RESTRICTED AREA BY AUTHORITY OF (TITLE, COMMANDING GENERAL OR COMMANDING OFFICER) IN ACCORDANCE WITH THE PROVISIONS OF THE DIRECTIVE ISSUED BY THE SECRETARY OF DEFENSE ON 20 AUGUST 1954, PURSUANT TO THE PROVISIONS OF SECTION 21, INTERNAL SECURITY ACT OF 1950. UNATHORIZED ENTRY IS PROHIBITED.

ALL PERSONS AND VEHICLES ENTERING HEREIN ARE LIABLE TO SEARCH. PHOTOGRAPHING OR MAKING NOTES, DRAWINGS, MAPS, OR GRAPHIC REPRESENTATIVES OF THIS AREA OR ITS ACTIVITIES ARE PROHIBITED UNLESS SPECIFICALLY AUTHORIZED BY THE COMMANDER. ANY SUCH MATERIAL FOUND IN THE POSSESSION OF UNAUTHORIZED PERSONS WILL BE CONFISCATED.

Figure 7-1. Sample Restricted-Area Warning

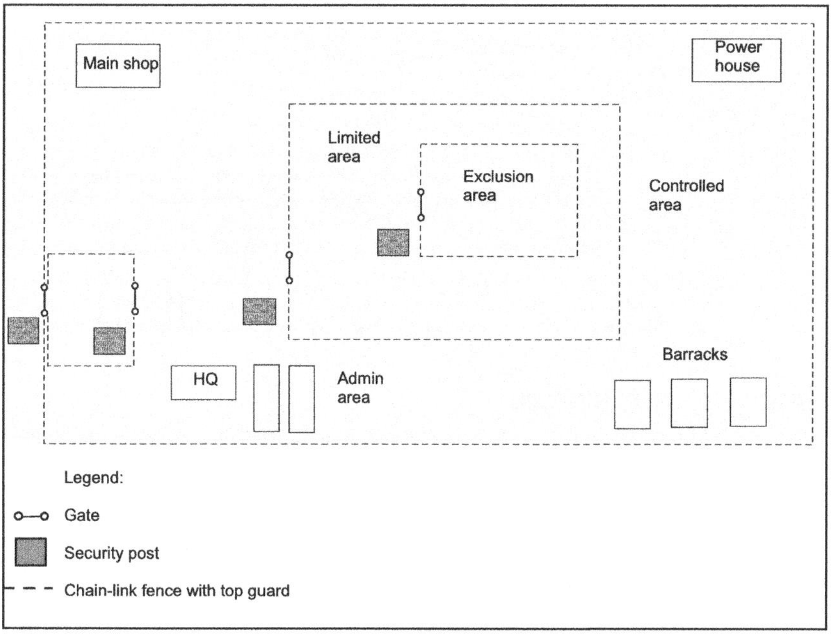

Figure 7-2. Schematic Diagram of a Simplified Restricted Area and the Degrees of Security

- Parking areas for privately owned vehicles (POVs) are established outside of restricted areas. Vehicle entrances must be kept at a minimum for safe and efficient control.
- Physical protective measures (such as fences, gates, and window bars) must be installed.

EMPLOYEE SCREENING

7-7. Screening job applicants to eliminate potential acts of espionage and sabotage and other security risks is important in peacetime and is critical during a national emergency. Personnel screenings must be incorporated into standard personnel policies.

7-8. An applicant should be required to complete a personnel security questionnaire, which is then screened for completeness and used to eliminate undesirable applicants. A careful investigation should be conducted to ensure that the applicant's character, associations, and suitability for employment are satisfactory. The following sources may be helpful in securing employment investigative data:

- State and local police (including national and local police in overseas areas).
- Former employers.
- Public records.
- Credit agencies.
- Schools (all levels).
- References. (These references should include those names not furnished by the applicant. These are known as throw offs, and they are obtained during interviews of references furnished by applicants.)
- Others as appropriate. (These may include the FBI, the US Army Criminal Records Repository, and the Defense Investigative Agency).

7-9. Medical screening considerations should be made (based on an applicant's position [such as a guard]) to evaluate physical and mental stamina. Once an applicant has been identified for employment, he is placed on an access-control roster.

IDENTIFICATION SYSTEM

7-10. An ID system is established at each installation or facility to provide a method of identifying personnel. The system provides for personal recognition and the use of security ID cards or badges to aid in the control and movement of personnel activities.

7-11. Standard ID cards are generally acceptable for access into areas that are unrestricted and have no security interest. Personnel requiring access to restricted areas should be issued a security ID card or badge as prescribed in AR 600-8-14. The card's/badge's design must be simple and provide for adequate control of personnel.

7-12. A security ID card/badge system must be established for restricted areas with 30 or more employees per shift. Commanders may (at their discretion) authorize a card/badge system in restricted areas for less than 30 people.

ID METHODS

7-13. Four of the most commonly used access-control ID methods are the personal-recognition system, the single-card or -badge system, the card- or badge-exchange system, and the multiple-card or -badge system.

Personal-Recognition System

7-14. The personal-recognition system is the simplest of all systems. A member of the security force providing access control visually checks the person requesting entry. Entry is granted based on—

- The individual being recognized.
- The need to enter has been established.
- The person is on an access-control roster.

Single-Card or -Badge System

7-15. This system reflects permission to enter specific areas by the badge depicting specific letters, numbers, or particular colors. This system lends to comparatively loose control and is not recommended for high-security areas. Permission to enter specific areas does not always go with the need to know. Because the ID cards/badges frequently remain in the bearer's possession while off duty, it affords the opportunity for alteration or duplication.

Card- or Badge-Exchange System

7-16. In this system, two cards/badges contain identical photographs. Each card/badge has a different background color, or one card/badge has an overprint. One card/badge is presented at the entrance to a specific area and exchanged for the second card/badge, which is worn or carried while in that area. Individual possession of the second card/badge occurs only while the bearer is in the area for which it was issued. When leaving the area, the second card/badge is returned and maintained in the security area. This method provides a greater degree of security and decreases the possibility of forgery, alteration, or duplication of the card/badge. The levels of protection described in TM 5-853-1 require multiple access-control elements as the levels of protection increase. In the case of the badge exchange, this system counts as two access-control elements.

Multiple-Card or -Badge System

7-17. This system provides the greatest degree of security. Instead of having specific markings on the cards/badges denoting permission to enter various restricted areas, the multiple card/badge system makes an exchange at the entrance to each security area. The card/badge information is identical and allows for comparisons. Exchange cards/badges are maintained at each area only for individuals who have access to the specific area.

MECHANIZED/AUTOMATED SYSTEMS

7-18. An alternative to using guards or military police (MP) to visually check cards/badges and access rosters is to use building card-access systems or biometric-access readers. These systems can control the flow of personnel entering and exiting a complex. Included in these systems are—

- Coded devices such as mechanical or electronic keypads or combination locks.
- Credential devices such as magnetic-strip or proximity card readers.
- Biometric devices such as fingerprint readers or retina scanners.

7-19. Access-control and ID systems base their judgment factor on a remote capability through a routine discriminating device for positive ID. These systems do not require guards at entry points; they identify an individual in the following manner:

- The system receives physical ID data from an individual.
- The data is encoded and compared to stored information.
- The system determines whether access is authorized.
- The information is translated into readable results.

7-20. Specialized mechanical systems are ideal for highly sensitive situations because they use a controlled process in a controlled environment to establish the required database and accuracy. One innovative technique applied to ID and admittance procedures involves dimension comparisons. The dimension of a person's full hand is compared to previously stored data to determine entry authorization. Other specialized machine readers can scan a single fingerprint or an eye retina and provide positive ID of anyone attempting entry.

7-21. An all-inclusive automated ID and access-control system reinforces the security in-depth ring through its easy and rapid change capability. The computer is able to do this through its memory. Changes can be made quickly by the system's administrator.

7-22. The commercial security market has a wide range of mechanized and automated hardware and software systems. Automated equipment is chosen only after considering the security needs and the environment in which it operates. These considerations include whether the equipment is outdoors or indoors, the temperature range, and weather conditions. Assessment of security needs and the use of planning, programming, and budgeting procedures greatly assist a security manager in improving the security posture.

CARD/BADGE SPECIFICATIONS

7-23. Security cards/badges should be designed and constructed to meet the requirements of AR 600-8-14. Upon issuing a card/badge, security personnel must explain to the bearer the wear required and the authorizations allowed with the card/badge. This includes—

- Designation of the areas where an ID card/badge is required.

- A description of the type of card/badge in use and the authorizations and limitations placed on the bearer.
- The required presentation of the card/badge when entering or leaving each area during all hours of the day.
- Details of when, where, and how the card/badge should be worn, displayed, or carried.
- Procedures to follow in case of loss or damage of the card.
- The disposition of the card/badge upon termination of employment, investigations, or personnel actions.
- Prerequisites for reissuing the card/badge.

VISITOR IDENTIFICATION AND CONTROL

7-24. Procedures must be implemented to properly identify and control personnel. This includes visitors presenting their cards/badges to guards at entrances of restricted areas. Visitors are required to stay with their assigned escort. Guards must ensure that visitors stay in areas relating to their visit; an uncontrolled visitor, although conspicuously identified, could acquire information for which he is not authorized. Foreign-national visitors should be escorted at all times.

7-25. Approval for visitors should be obtained at least 24 hours in advance (if possible). Where appropriate, the installation should prepare an agenda for the visitor and designate an escort officer. Measures must be in place to recover visitor cards/badges on the visit's expiration or when they are no longer required.

7-26. Physical-security precautions against pilferage, espionage, and sabotage require the screening, ID, and control of visitors. Further information about visiting requirements and procedures are found in ARs 12-15 and 381-20. Visitors are generally classed in the following categories:

- Persons with whom every installation or facility has business (such as suppliers, customers, insurance inspectors, and government inspectors).
- Individuals or groups who desire to visit an installation or facility for personal or educational reasons. Such visits may be desired by educational, technical, or scientific organizations.
- Individuals or groups specifically sponsored by the government (such as foreign nationals visiting under technical cooperation programs and similar visits by US nationals). Requests for visits by foreign nationals must be processed according to AR 380-10.
- Guided tours to selected portions of the installation in the interest of public relations.

7-27. The ID and control mechanisms for visitors must be in place. They may include the following:

- Methods of establishing the authority for admitting visitors and any limitations relative to access.

- Positive ID of visitors by personal recognition, visitor permit, or other identifying credentials. Contact the employer, supervisor, or officer in charge to validate the visit.
- The use of visitor registration forms. These forms provide a record of the visitor and the time, location, and duration of his visit.
- The use of visitor ID cards/badges. The cards/badges bear serial numbers, the area or areas to which access is authorized, the bearer's name, and escort requirements.

7-28. Individual groups entering a restricted area must meet specific prerequisites before being granted access. The following guidance is for group access into a restricted area:

Visitors

7-29. Before allowing visitors into a restricted area, contact the person or activity being visited. After verifying the visitor's identity, issue a badge, complete the registration forms, and assign an escort (if required). Visitors may include public-utility and commercial-service representatives.

Very Important Persons

7-30. The procedures for admitting very important persons (VIPs) and foreign nationals into restricted areas are contained in AR 12-15. Special considerations and coordination with the protocol office are necessary. A 24-hour advance notice is desirable for these requests, along with an agenda for the visit and the designation of an escort, if appropriate.

Civilians Working on Jobs Under Government Contract

7-31. To allow these personnel to conduct business in restricted areas, the security manager must coordinate with the procurement office. The security manager must also identify movement-control procedures for these employees.

Cleaning Teams

7-32. Supervisors using cleaning teams must seek technical advice from the physical-security office on internal controls for each specific building. This may include providing escorts.

DOD Employees in Work Areas After Normal Operating Hours

7-33. Supervisors establish internal controls based on coordination with the security manager. They also notify security personnel of the workers' presence, type, and duration of work.

ENFORCEMENT MEASURES

7-34. The most vulnerable link in any ID system is its enforcement. Security forces must be proactive in performing their duties. A routine performance of duty will adversely effect even the most elaborate system. Positive enforcement measures must be prescribed to enhance security. Some of these measures may include—

- Designating alert and tactful security personnel at entry control points.
- Ensuring that personnel possess quick perception and good judgment.
- Requiring entry-control personnel to conduct frequent irregular checks of their assigned areas.
- Formalizing standard procedures for conducting guard mounts and posting and relieving security personnel. These measures will prevent posting of unqualified personnel and a routine performance of duty.
- Prescribing a uniform method of handling or wearing security ID cards/badges. If carried on the person, the card must be removed from the wallet (or other holder) and handed to security personnel. When worn, the badge will be worn in a conspicuous position to expedite inspection and recognition from a distance.
- Designing entry and exit control points of restricted areas to force personnel to pass in a single file in front of security personnel. In some instances, the use of turnstiles may be advisable to assist in maintaining positive control.
- Providing lighting at control points. The lighting must illuminate the area to enable security personnel to compare the bearer with the ID card/badge.
- Enforcing access-control measures by educating security forces and employees. Enforcement of access-control systems rests primarily with the security forces; however, it is essential that they have the full cooperation of the employees. Employees must be instructed to consider each unidentified or improperly identified individual as a trespasser. In restricted areas where access is limited to a particular zone, employees must report unauthorized individuals to the security force.
- Positioning ID card/badge racks or containers at entry control points so that they are accessible only to guard-force personnel.
- Appointing a responsible custodian to accomplish control procedures of cards/badges according to AR 600-8-14. The custodian is responsible for the issue, turn in, recovery, and renewal of security ID cards/badges.

7-35. The degree of compromise tolerable in the ID system is in direct proportion to the degree of security required. The following control procedures are recommended for preserving the integrity of a card/badge system:

- Maintenance of an accurate written record or log listing (by serial number) all cards and badges and showing those on hand, to whom they are issued, and their disposition (lost, mutilated, or destroyed).
- Authentication of records and logs by the custodian.
- A periodic inventory of records by a commissioned officer.
- The prompt invalidation of lost cards/badges.
- The conspicuous posting at security control points of current lists of lost or invalidated cards/badges.
- The establishment of controls within restricted areas to enable security personnel to determine the number of persons within the area.
- The establishment of the two-person rule (when required).

- The establishment of procedures to control the movement of visitors. A visitor-control record will be maintained and located at entry control points.

SIGN/COUNTERSIGN AND CODE WORD

7-36. This method of verifying identity is primarily used in a tactical environment. According to the local SOP, the sign/countersign or code-word procedures must be changed immediately if compromised.

DURESS CODE

7-37. The duress code is a simple word or phrase used during normal conversation to alert other security personnel that an authorized person is under duress. A duress code requires planning and rehearsal to ensure an appropriate response. This code is changed frequently to minimize compromise.

ACCESS-CONTROL ROSTERS

7-38. Admission of personnel to a restricted area is granted to those identified and listed on an access-control roster. Pen-and-ink changes may be made to the roster. Changes are published in the same manner as the original roster.

7-39. Rosters are maintained at access control points. They are kept current, verified, and accounted for by an individual designated by the commander. Commanders or their designated representatives authenticate the rosters. Admission of persons other than those on the rosters is subject to specific approval by the security manager. These personnel may require an escort according to the local SOP.

METHODS OF CONTROL

7-40. There are a number of methods available to assist in the movement and control of personnel in limited, controlled, and restricted areas. The following paragraphs discuss the use of escorts and the two-person rule:

ESCORTS

7-41. Escorts are chosen because of their ability to accomplish tasks effectively and properly. They possess knowledge of the area being visited. Escorts may be guard-force personnel, but they are normally personnel from the area being visited. Local regulations and SOPs determine if a visitor requires an escort while in the restricted area. Personnel on the access list may be admitted to restricted areas without an escort.

TWO-PERSON RULE

7-42. The two-person rule is designed to prohibit access to sensitive areas or equipment by a lone individual. Two authorized persons are considered present when they are in a physical position from which they can positively detect incorrect or unauthorized procedures with respect to the task or operation being performed. The team is familiar with applicable safety and security requirements, and they are present during any operation that affords access to sensitive areas or equipment that requires the two-person rule. When application of the two-person rule is required, it is enforced constantly by the personnel who constitute the team.

7-43. The two-person rule is applied in many other aspects of physical-security operations, such as the following:

- When uncontrolled access to vital machinery, equipment, or materiel might provide opportunity for intentional or unintentional damage that could affect the installation's mission or operation.
- When uncontrolled access to funds could provide opportunity for diversion by falsification of accounts.
- When uncontrolled delivery or receipt for materials could provide opportunity for pilferage through "short" deliveries and false receipts.
- When access to an arms or ammunition storage room could provide an opportunity for theft. Keys should be issued so that at least two people must be present to unlock the locks required under the provisions of AR 190-11.

7-44. The two-person rule is limited to the creativity of the PM and the physical-security manager. They should explore every aspect of physical-security operations in which the two-person rule would provide additional security and assurance and include all appropriate recommendations and provisions of the physical-security plan. An electronic-entry control system may be used to enforce the two-person rule. The system can be programmed to deny access until two authorized people have successfully entered codes or swiped cards.

SECURITY CONTROLS OF PACKAGES, PERSONAL PROPERTY, AND VEHICLES

7-45. A good package-control system helps prevent or minimize pilferage, sabotage, and espionage. The local SOP may allow the entry of packages with proper authorization into restricted areas without inspection. A package-checking system is used at the entrance gate. When practical, inspect all outgoing packages except those properly authorized for removal. When a 100 percent inspection is impractical, conduct frequent unannounced spot checks. A good package-control system assists in the movement of authorized packages, material, and property.

7-46. Property controls are not limited to packages carried openly, but they include the control of anything that could be used to conceal property or material. Personnel should not be routinely searched except in unusual situations. Searches must be performed according to the local SOP.

7-47. All POVs on the installation should be registered with the PM or the installation's physical-security office. Security personnel should assign a temporary decal or other temporary ID tag to visitors' vehicles to permit ready recognition. The decal or the tag should be distinctly different from that of permanent-party personnel.

7-48. When authorized vehicles enter or exit a restricted area, they undergo a systematic search, including (but not limited to) the—

- Vehicle's interior.
- Engine compartment.
- External air breathers.
- Top of the vehicle.
- Battery compartment.
- Cargo compartment.
- Undercarriage.

7-49. The movement of trucks and railroad cars into and out of restricted areas should be supervised and inspected. Truck and railroad entrances are controlled by locked gates when not in use and are manned by security personnel when unlocked. The ID cards/badges are issued to operators to ensure proper ID and registration for access to specific loading and unloading areas.

7-50. All conveyances entering or leaving a protected area are required to pass through a service gate manned by security forces. Drivers, helpers, passengers, and vehicle contents must be carefully examined. The examination may include—

- Appropriate entries in the security log (including the date, operator's name, load description, and time entered and departed).
- A check of the operator's license.
- Verification of the seal number with the shipping document and examination of the seal for tampering.

7-51. Incoming trucks and railroad cars must be assigned escorts before they are permitted to enter designated limited or exclusion areas. Commanders should establish published procedures to control the movement of trucks and railroad cars that enter designated restricted areas to discharge or pick up cargo (escorts will be provided when necessary).

7-52. The best control is provided when all of these elements are incorporated into access-control procedures. Simple, understandable, and workable access-control procedures are used to achieve security objectives without impeding operations. When properly organized and administered, access-control procedures provide a method of positively identifying personnel who have the need to enter or leave an area.

TACTICAL-ENVIRONMENT CONSIDERATIONS

7-53. Access-control procedures during tactical operations may establish additional challenges for the commander. In some instances, the commander cannot provide a perimeter barrier (such as a fence) based on METT-TC. Commanders are still required to provide security measures for restricted areas, although they may not always have the necessary assets. Early-warning systems and the use of guards become crucial. A restricted area may become a requirement without prior notice during an operation. Figure 7-3 and Figure 7-4, page 7-14, are examples of temporary tactical restricted and exclusion areas.

7-54. Commanders must plan for these considerations when developing their budget. Funding must be requested and set aside to support physical-security requirements during tactical operations. Resources will not always be available; therefore, commanders must implement procedures that support access-control measures. Improvising will become common practice to overcome shortfalls concerning access-control equipment in the field.

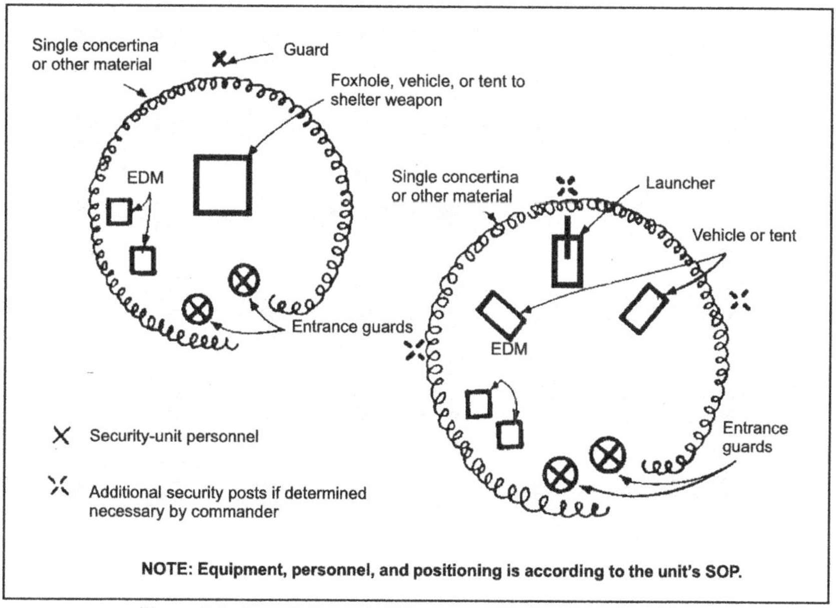

Figure 7-3. Sample Layout of Temporary Tactical Restricted Areas

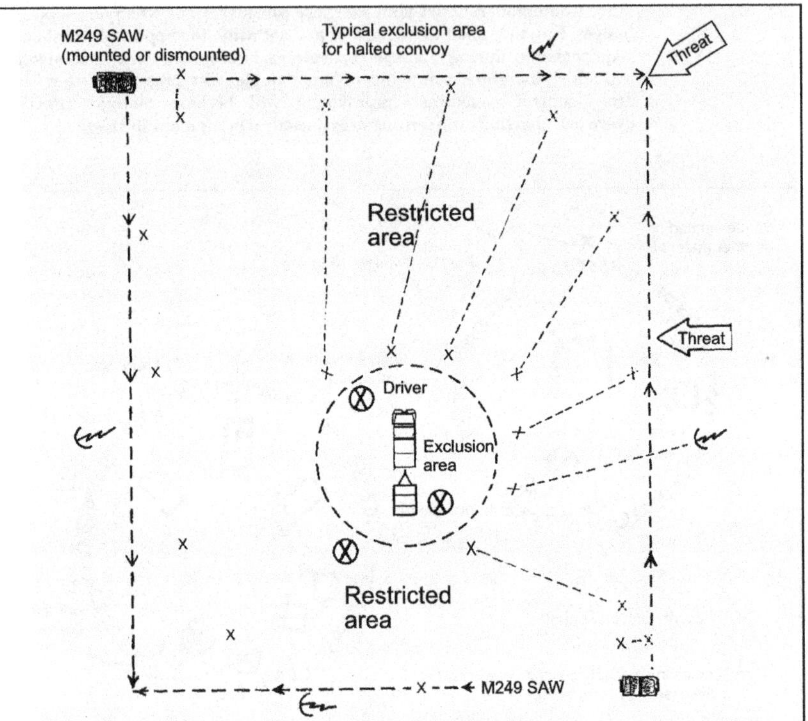

NOTE: Other vehicles and personnel, to include the LPs, form a defensive perimeter inside the primary directions of fire of the M60 machine guns. During attack, alternate machine-gun crews should be placed in the perimeter away from primary machine guns and be spaced so that they can be used as replacements and so that large gaps do not appear in the perimeter.

Legend:

X MP rifleman

Temporary exclusion area (size depends on terrain and threat force)

Principal direction of fire

Listening post

Figure 7-4. Sample Layout for a Temporary Tactical Exclusion Area

Chapter 8

Lock and Key Systems

Locks are the most acceptable and widely used security devices for protecting facilities, classified materials, and property. All containers, rooms, and facilities must be locked when not in actual use. Regardless of their quality or cost, locks are considered delay devices only. Some locks require considerable time and expert manipulation to open, but all locks can be defeated by force and with the proper tools. Locks must never be considered as a stand-alone method of security.

INSTALLATION AND MAINTENANCE

8-1. The USACE is responsible for installing locking devices in newly constructed facilities. Installation-level engineers are responsible for maintaining the locking devices. Physical-security personnel must work closely with engineer personnel to ensure that locks meet the standards and are installed according to applicable regulations. One source of assistance and information is the DOD Lock Program Technical Support Hotline at the Naval Facilities Engineering Services Center, Port Hueneme, California.

TYPES OF LOCKING DEVICES

8-2. The degree of protection afforded by a vault, a safe, or a filing cabinet may be measured in terms of the lock's resistance. Locking devices are listed in TM 5-805-8. Types of locking devices include key and combination locks.

8-3. ARs 190-11, 190-51, 50-5, and 50-6 prescribe specific types of locks for specific types of facilities. AR 380-5 prescribes standard facilities for storing classified material and contains guidance for different storage requirements.

KEY LOCKS

8-4. Key locks consist of, but are not limited to, the following:

- Cylindrical locksets are often called key-in-knob or key-in-lever locks. They are normally used to secure offices and storerooms. The locking cylinder located in the center of the doorknob distinguishes these locks. Some cylindrical locksets have keyways in each of the opposing knobs that require a key on either side to lock and unlock them. Others unlock with a key, but may be locked by pushing or rotating a button on the inside knob. These locks are suitable only for very low-security applications. Using these locks may require compensatory measures in the form of additional locks on containers within the room.
- Dead-bolt locks are sometimes called tubular dead bolts. They are mounted on the door in a manner similar to cylindrical locksets. The primary difference is in the bolt. When the bolt is extended (locked),

the dead bolt projects into the doorframe at least one inch, and it cannot be forced back (unlocked) by applying pressure to the end of the bolt. The dead-bolt lock has the potential for providing acceptable levels of protection for storerooms and other areas where more security is desired. It is recommended for use in military housing as an effective security measure in the installation's crime-prevention program. In situations where there is a window in or adjacent to the door, a double-cylinder dead-bolt lock (one that requires a key to open from either side) should be used.

- Mortise locks are so named because the lock case is mortised or recessed into the edge of the door. The most common variety of mortise locks has a doorknob on each side of the door. Entrance doors often have an exterior thumb latch rather than a doorknob. The mortise lock can be locked from inside by means of a thumb turn or by a button. Mortise locks are considered low-security devices since they weaken the door in the mortised area.
- Drop-bolt locks (often referred to as jimmy-proof locks) are normally used as auxiliary locks similar to dead bolts. Both the drop-bolt lock body and the strike have interlocking leaves similar to a door hinge. When closed, locking pins in the lock body drop down into the holes provided in the strike and secure the locking system. Since the lock body and the strike are interconnected with locking pins when closed, the lock essentially becomes a single unit and is extremely difficult to separate.
- Rim-cylinder locks are mounted to the door's inside surface and are secured by screws in the door face. These locks are generally used with drop-bolt and other surface-mounted locks and latches. They consist of an outer barrel, a cylinder and ring, a tailpiece, a back mounting plate, and two mounting screws. The tailpiece screws are usually scored so that the lock can be tailored to fit varying door thicknesses.
- Unit locks are ideal in heavily traveled facilities (such as hospitals or institutional buildings). These locks are a complete, one-piece unit that slides into a notch cut into the door's latch edge. The one-size cutout of the door edge simplifies the door preparation for the lock.
- Mechanical, push-button combination locks are digital (push buttons numbered 1 through 9) combination door-locking devices used to deny area access to any individual not authorized or cleared for a specific area. These locks are normally used for access control and should be backed up by door locking devices when the facility is unoccupied.
- Padlocks are detachable locks that are typically used with a hasp. Low-security padlocks, sometimes called secondary padlocks, are used to deter unauthorized access, and they provide only minimal resistance to force. Low-security locks are made with hardened steel shackles. Precautions must be taken to avoid confusing these locks with similar brass or bronze locks. The brass or bronze locks are commonly used but do not meet the security requirements of the hardened shackled locks. High-security padlocks may be used to secure AA&E. They provide the maximum resistance to unauthorized entry when used with a high-security hasp.

8-5. Some locks have interchangeable cores, which allow the same key system to include a variety of locks. Padlocks, door locks, cabinet locks, and electrical key switches can all be operated by the same key system. Because these cores are removable by a special key, this system allows for rapid rekeying of locks in the event that the key is compromised.

8-6. Locks are keyed in several different ways. When several locks are keyed differently, each is operated by its own key. When they are keyed alike, one key will open them all. Locks that are master-keyed are keyed differently, yet have one key that will open them all. Master-keying is done for convenience and represents the controlled loss of security. Master-keying is not used unless permitted by regulation.

COMBINATION LOCKS

8-7. Combination locks are available as padlocks or as mounted locks. They are low-security padlocks with combinations that are either fixed or changeable. Combination locks may be either mechanical or electronic. They are operated by entering a particular sequence of numbers. When the correct combination is entered, the lock's bolt is retracted. Combination locks used for securing classified material must meet Federal Specification FF-L-2740.

8-8. Although the lock is the most accepted and widely used security device, it is only a delay device and should never be considered as a positive bar to entry. A lock can (and will) be defeated. The best defense for locking devices is a good key-control program. Refer to AR 190-51, Appendix D, for standard key and lock procedures. Additional key and lock procedures for AA&E can be found in AR 190-11, Chapter 3.

Chapter 9

Security Forces

The security force for an installation or a facility provides the enforcement element in the physical-security program. This force consists of personnel specifically organized, trained, and equipped to protect the command's physical-security interests. It is a commander's most effective tool in a comprehensive, integrated, physical-security program. Vulnerability tests are periodically conducted to determine and ensure the state of readiness of security forces (see Appendix K).

TYPES OF SECURITY FORCES

9-1. On installations, security forces may be MP forces, security police, DOD civil-service security guards, or contract guards. Interior guard duties are performed by installation unit troops on a roster basis. MP forces normally perform security duties that require higher degrees of training and experience. These include—

- Security of restricted areas.
- Security of specific sensitive gates.
- Supervisory or coordinated roles with other military or DOD civil-service security guards.
- Responsibility for monitoring and responding to intrusion alarms.

9-2. An MP unit may perform the entire physical-security function alone based on METT-TC, the area, and the facilities. When an MP unit cannot assume responsibilities for all of the physical-security requirements in the command, other forces may be required. Additional forces may consist of the following:

- Personnel furnished by units of the installation's command on a daily or weekly basis. While this method has the single advantage of providing additional manpower, it has the disadvantages of rapid turnover and the lack of training. If this manpower is used, personnel should be assigned the least sensitive posts or patrols. For extended augmentation, units may be attached to MP units. The MP unit may also be augmented by reserve units or units in rotation.
- The combat-arms branches (especially the infantry) may attach their forces to MP units and may be designated as security guards assisting in the required operations.
- Military or paramilitary units of the host country may also be attached to or operate in coordination with MP forces. They may also be supplemented with national police of their own country.

- The installation's band may be a source of military force during wartime. (The band is assigned enemy-prisoner-of-war [EPW] duty as a wartime duty.) The band is doctrinally capable of providing security at the division tactical operations center (DTOC) and ASPs, assisting in the perimeter defense of the command post (CP), and operating the dismount point for the CP. It is capable of providing access control at the DTOC and the ASPs and augmenting or relieving security personnel on the defensive perimeter.

9-3. Civil-service security guards are uniformed civilian employees from a government agency. They are customarily trained and organized along military lines. The organization may be completely civil service or may be composed of civil-service personnel under military supervision. In either case, they are under the operational control of the PM or the security officer.

9-4. Labor-service personnel (local civilian personnel) have been organized and used successfully in theaters of operation. These types of units were organized after World War II and since that time have established enviable records in the physical-security field. They are distinctively uniformed, organized, and equipped. They have set and maintained the highest security standards, resulting in a minimal loss of property. While not military organizations, these units have successfully developed a high sense of duty and esprit de corps that has been reflected in their outstanding contributions to the physical security of installations in overseas commands.

AUTHORITY AND JURISDICTION

9-5. It is most important that the PM or the security officer determine (and instruct his security force in) the extent and limitations of the commander's jurisdiction in the field of law enforcement and investigations. Those jurisdictions include—

- Jurisdiction of place.
  - Military installations and facilities. Whether state or federal law or both are applicable on a military installation or facility depends largely on the nature of jurisdiction over the land involved. The amount of federal jurisdiction may vary between different areas of the installation or facility. The legal formalities of acquiring jurisdiction over land under the control of the Secretary of the Army are accomplished at DA level and according to the provisions of AR 405-20. Information and advice relating to jurisdictional questions should be referred to the local SJA.
  - Areas outside of military installations. Areas outside of military installations are generally subject to state and local laws; however, there are exceptions. Information and advice in this regard should be obtained through the local SJA.
  - Overseas areas. In overseas areas, jurisdiction varies according to the military situation and existing international treaties, contracts, and agreements. Guidance should be obtained in each instance from the commander and the SJA and set forth in appropriate command directives.

- Jurisdiction of personnel.
  - Jurisdiction of personnel generally follows the limitations of jurisdiction of the installation.
  - MP forces have jurisdiction and authority over personnel as described in AR 190-14 and related publications.
  - Authority for federal civilian employees assigned to security, police, and guard duties is derived from the installation's commanding officer. These personnel can have no more authority than he possesses and are subject to any limitations imposed thereon.
  - Security-force personnel may enforce all offenses under the Uniform Code of Military Justice (UCMJ), military regulations, federal laws and regulations, and state laws where applicable.
  - Security-force personnel may be given the same authority as MP forces over all personnel subject to military jurisdiction, including apprehension, detention, and search.
  - Civilian security-force personnel have no specific grant of authority over civilians other than the right of citizen's arrest.
  - The commander is the source of jurisdiction and authority for all other personnel assigned to security-force duties.

PERSONNEL SELECTION

9-6. Regardless of the use of structural, mechanical, or electronic equipment, the human element in security operations makes the difference between success or failure. Commanders and supervisors have a responsibility to ensure that security personnel who control access to restricted areas and classified activities are qualified based on criteria in AR 380-67. Personnel who perform physical-security duties must be disciplined and alert, have sound judgment, be confident and physically fit, and possess good interpersonal communication skills.

SECURITY CLEARANCE

9-7. Security-clearance criteria for security positions must be based on the security classifications of the information to which access will be granted. Security positions are normally designated as sensitive and require a secret security clearance. ARs 381-20 and 380-67 describe criteria and procedures governing security clearances. Appropriate civilian-personnel regulations should also be consulted when civilians are involved.

9-8. Positive evaluation of the reliability of all personnel must be made before they are entrusted with classified or sensitive information. (The Individual Reliability Program is prescribed in AR 190-56.) Follow-up action must be made on all personnel who are granted a security clearance to ensure that they continue to meet the criteria for their clearance. Personnel not meeting or adhering to the prescribed standards must have their security clearances revoked and thereby lose their access to areas containing classified information or material (see AR 380-67).

ORGANIZATION AND EMPLOYMENT OF FORCES

9-9. The organization of a security force will vary, depending on circumstances and the forces available. Forces consist of—

- Mobile patrols. A mobile detachment of ground, sea, or air forces dispatched to gather information or carry out a security mission.
- The response force. A mobile force with appropriate fire support (usually designated by the area commander) to deal with Level II threats in the rear area (Army). This is normally an MP function.
- Reserves. That portion of a force withheld from action or uncommitted to a specific course of action so as to be available for commitment at the decisive moment. Its primary purpose is to retain flexibility throughout an offensive action.
- Any combination of these three.

9-10. Instructions to the security force should be issued in writing. These instructions are normally in the form of general, special, or temporary orders. They should be carefully and clearly worded and include all phases of each assignment. They should be reviewed at least monthly to ensure that they are current. Categories of instructions of each are as follows:

- General orders are those orders that concern the security force as a whole and are applicable at all posts and patrols.
- Special orders pertain to a permanent post or patrol. Each permanent post or patrol should have special orders issued concerning the location, duties, hours manned, arms, ammunition, and other equipment required and the instructions for using force in enforcement and apprehension activities.
- Temporary orders are issued for a short period and cover a special or temporary situation. If it can be predetermined, such orders should indicate the period of time for which they are valid.

9-11. A security-force SOP that outlines policies, organization, authority, functions, and other required information should be prepared for required reading. Each security-force member should be held responsible for full knowledge and understanding of the contents of the SOP. Each installation PM, physical-security officer, or chief of a guard force should conduct periodic inspections and tests to determine each individual's degree of understanding of these instructions. Instructions should be provided in writing regarding the safeguarding and control of the SOP. Its contents may not be classified; however, the information could assist an intruder in breaching security.

HEADQUARTERS AND SHELTERS

9-12. The location of the security force's headquarters will depend on the size and layout of the installation or facility. The objectives are the efficient control of the security force and the adequate security of vital activities. On a small installation, there is frequently only one full-time entrance that may be supplemented by several part-time entrances. At these installations, the logical location of the headquarters would be at or near the main entrance. On larger installations, it might be better to locate the headquarters near the center of the cantonment area.

9-13. The security force's headquarters should be the control point for all physical-security matters for the installation and the monitoring point for protective alarm and communication systems. This office should have a reliable and independent means to contact nearby civil authorities. A list of key telephone numbers should be available for use in emergency operations.

9-14. Personnel shelters should be available to protect the guards from the elements. The design can be temporary or hardened and include adequate space for guard-force personnel only. The facility should have heat, ventilation, storage space for essential accessories, lighting that will not expose the occupant, and good visibility in all directions.

EXECUTION OF SECURITY ACTIVITIES

9-15. Security personnel must exercise good interpersonal communication skills when carrying out their duties with other employees. Bad employee relations can result if security personnel become impertinent and assume powers not rightfully theirs. Security personnel must understand the methods and techniques that will detect security hazards and assist in identifying violators and intruders.

9-16. Written reports or journals are recommended for security activities. These should be prepared by either the security force's supervisor or the personnel at the security post. These reports should record all activities, actions, and visits at the security post.

9-17. It must be strongly emphasized that security personnel will be used for security duties only and should not be given other routine functions except as directed by the commander or his representative. Security personnel should have no fire-fighting or similar duties regularly assigned. Such emergencies offer an excellent diversion to cover an intruder's entrance. Consequently, during such times, security personnel must be exceptionally alert when performing their duties. However, the security force may be cross-trained in other areas (such as fire fighting) so that they may be used when required and when circumstances permit (such as when they are off duty).

9-18. Personnel who are assigned to fixed posts should have a designated method of relief. The security force's shift supervisor should establish a relief schedule (about every two hours) according to local policies and the SOP. A simple but effective plan of operation should be worked out for the security force to meet every foreseeable emergency. Practice alarms should be conducted frequently to test the plan's effectiveness. Such plans should be designed to prevent a diversion at one point on the installation, drawing off the guards or distracting their attention from another section of the installation where unauthorized entry may be made. Routes and times for security patrols should also be varied at frequent intervals to preclude establishing a routine that may be observed by potential intruders.

TRAINING REQUIREMENTS

9-19. The extent and type of training required for security forces will vary according to the importance, vulnerability, size, and other factors affecting a particular installation or facility. The training program's objective is to ensure that all personnel are able to perform routine and emergency duties competently and efficiently.

BENEFITS OF PROPER TRAINING

9-20. Efficient and continuing training is the most effective means of obtaining and maintaining maximum proficiency of security-force personnel. Regardless of the selection process, new personnel seldom have all of the qualifications and experience necessary to do the job. In addition, new or revised job requirements frequently mean that personnel must be retrained. Training can bridge the void between ability and job requirement.

9-21. Supervisors need to remember that all personnel do not have the same training needs. It is a waste of valuable time to train an individual in a subject that he has already mastered. Past experience, training, acquired skills, and duty assignments should be evaluated for each person as an aid in planning an effective training program.

9-22. A good training program benefits both the installation and the security force. The task of supervising the security force is made easier, there is much less wasted time, fewer mistakes are made, and there is less friction with other agencies. A good training program helps to instill confidence through developing increased skill proficiency. The training program provides for more flexibility and better physical protection, fewer required personnel, and less time to learn duties. Training establishes systematic and uniform work habits.

BASIC TRAINING

9-23. As a minimum, personnel (including civil-service security personnel) who have not had security training should receive training in their security duties. This training includes—

- The care and use of weapons, if required. No person should be placed on security duty unless weapons training has occurred within the past 12 months. Weapons training must be according to AR 190-14.
- Areas of responsibility and authority of security personnel, particularly on apprehension, search and seizure, and the use of force.
- The location and use of first aid and fire-control equipment and electrical switches.
- Duties in case of emergencies such as alerts, fires, explosions, and civil disturbances.
- Common forms of sabotage and espionage activity.
- The location of hazardous and vulnerable equipment and material.

IN-SERVICE TRAINING

9-24. All newly assigned individuals are given special instructions for each post. When possible, their first assignment should be with an experienced person. Additional in-service training and periodic retraining to review basic material and procedures are continuous requirements.

9-25. Scheduling in-service training and classes to enable all of the security force or a complete shift to participate is often difficult. Therefore, the supervisor must exercise good judgment when scheduling training to ensure that each person has the opportunity to receive the training.

EVALUATION OF TRAINING

9-26. Testing designed to evaluate performance is a necessary step in the training program. These tests may be oral or written or may be a type of performance test. They should be administered annually to ensure that the entire force maintains high standards of proficiency. A testing program also helps to improve training by—

- Discovering gaps in learning.
- Emphasizing main points.
- Evaluating instructional methods.

9-27. Security training received by personnel at their units must be entered in unit training charts or records. The record serves to—

- Indicate individual degrees of skill.
- Establish priorities of instruction.
- Present a consolidated picture of the security force's training status.
- Help certify guard personnel.

SUPERVISION

9-28. A security supervisor is tasked with overseeing and directing the work and behavior of other members of the security force. To obtain maximum performance from each member of his force, the supervisor must have a complete understanding of leadership principles and be capable of applying them.

9-29. The supervisor is responsible for understanding the operations of all posts. Additionally, he is often responsible for selecting, inducting, training, and ensuring the productivity, safety, morale, and advancement of guard-force members.

9-30. To ensure an alert, presentable, and efficient security force, the leadership must provide consistent and intelligent supervision. To earn the respect and cooperation of the guard force, supervisors must be professional in their conduct. The security force's morale and efficiency is a direct reflection of the quality of its supervision.

9-31. The ratio of supervisory personnel to security personnel should be determined by the individual characteristics of each installation. At small installations, the ratio may be higher than at large installations.

9-32. There must be sufficient supervision to enable the inspection of each post and patrol. It is also essential that supervisors be in contact with security headquarters to control emergencies that may arise. Specific duties of a supervisor include the inspection and briefing of the relief shift and the inspection of posts, vehicles, and equipment during visits to posts and patrols.

SUPPLEMENTS TO SUPERVISION

9-33. Various means and devices may be used as supplements to personnel supervision. These include the following:

- **Recorded tour systems.** Personnel record their presence at strategic points throughout an installation by using portable watch clocks or similar devices. These are effective means of ensuring that such points are regularly covered. This system provides an after-the-fact type of supervision.
- **Supervisory tour systems.** A signal is transmitted to a manned central headquarters at the time the post is visited. These systems provide instantaneous supervision and a means of detecting interference with normal security activities and initiating an investigation or other appropriate action.

9-34. All personnel on security duty should be required to report periodically to headquarters by the usual means of communication. The frequency of such reports will vary, depending on a number of factors. Regularity should be avoided to preclude setting a pattern by which an intruder can gauge an appropriate time for entrance.

MANAGEMENT

9-35. The physical-security supervisor is responsible for managing and developing the security organization. A physical-security program is greatly enhanced by a well-developed educational program.

9-36. The physical-security supervisor acts as an advisor and assists in formulating policies for the installation's physical-security measures. The goal should be the best security within the restrictions of the commander's budget guidance. Physical-security planners must remember that anyone can provide adequate security with unlimited funds; however, this is not a realistic approach. There must be a constant endeavor to effect justifiable economy where possible without jeopardizing the physical-security program.

UNIFORMS

9-37. All security-force personnel are required to wear the complete prescribed uniform as outlined in their special orders. Deviations from the prescribed uniform should not be made except for items to protect the guard force's health, comfort, and safety. The duty uniform will be worn during all tours of duty and may be worn during off-duty hours only between the place of residence and the place of duty. Each member of the security force is required to maintain high standards of appearance.

VEHICLES

9-38. The security force should be furnished with sufficient and reliable vehicles to maintain patrol standards established by the installation commander. Vehicles assigned to the force should be equipped with two-way radios to obtain the greatest possible use of all personnel and vehicles.

FIREARMS

9-39. Before issuing weapons, the security force will be briefed on the use of force. Security-force personnel will be issued weapons as prescribed by AR 190-11 and the unit's SOP. The commander may prescribe other weapons for the security force based on needs and requirements. Weapons normally are loaded with live ammunition, except where prohibited for safety reasons. The use of privately owned weapons while on duty is not authorized. Weapons and ammunition issued to security-force personnel will not be removed from the installation except in the course of official duty. When not in use, weapons are secured in arm racks in storage rooms as prescribed by AR 190-11.

9-40. Weapons are inspected as necessary to ensure proper maintenance. A written report is prepared and filed on the discharge of any weapon except for authorized and supervised training. The patrol supervisor or an MP investigator prepares the report (DA Form 3975).

9-41. Ammunition supplies for the security force's use must be maintained in secured storage containers according to AR 190-11. Ammunition must be issued only under proper supervision for authorized purposes. Ammunition issued to members of the security force must be accounted for by individual members immediately upon completion of duty. Any ammunition unaccounted for will be the subject of a report of its disposition by the individual.

COMMUNICATIONS

9-42. The security force should be equipped with two-way radios. These may be vehicle-mounted and portable, or they may be telephones. A secure-voice capability should be used where possible. This equipment is considered essential for the efficient operation of the security force and the accomplishment of its assigned mission. Proper use and care by security personnel will enhance the equipment's usefulness and capability.

MISCELLANEOUS EQUIPMENT

9-43. Security managers or supervisors should obtain other equipment necessary to accomplish their security mission. Items in this category may include (but are not limited to) warning lights; sirens; spotlights; portable lights; flashlights; first aid kits; traffic-control devices; and items of wear for the health, comfort, and safety of security personnel. Some of this equipment may require local purchase.

MILITARY WORKING DOGS

9-44. The requirements for physical protection of installations or facilities within the US and overseas theaters of operation continue to increase. Manpower available for this purpose has always been (and probably will continue to be) limited. The MWD, properly trained and properly used, can enhance a physical-security program. See AR 190-12 and DA Pam 190-12 for information regarding the use of MWDs.

SUMMARY

9-45. A security force is the critical element of a successful physical-security program. It is as strong as its weakest member. A comprehensive training program is essential to a knowledgeable, disciplined, and alert security force. A well-trained security force will be prepared to respond to a security breach.

Chapter 10

In-Transit Security

In-transit security subjects the movement of cargo to different, and frequently, more demanding aspects of physical security. Cargoes may be moved via port, rail, pipeline, or convoy. Regardless of the mode of movement, commanders must aggressively apply the principles of physical security to their protection. Security forces must be provided at the most vulnerable areas of each cargo movement.

IN-PORT CARGO

10-1. Ports and harbors are prime targets for enemy and criminal activities. Perimeter areas of these facilities are more vulnerable because of the extensive distance and exposed beach or pier areas. Terminal areas may include fully developed piers and warehouses or may be an unimproved beach where logistics-over-the-shore (LOTS) or roll-on/roll-off (RORO) operations are conducted.

10-2. If a Theater Army Area Command (TAACOM) MP unit must provide security for cargo in a port, the main effort is to provide security from the perimeter of the port outward. Security measures focus on aggressive patrolling to detect, report and, if need be, combat enemy threats. Measures may include—

- Conducting route and area reconnaissance patrols.
- Developing police intelligence in the area of operations (AO).
- Controlling traffic in the area surrounding the port.
- Conducting mounted or dismounted patrols (with MWDs, if available) around the port's perimeter.
- Establishing an access-control/ID section.
- Watching for diversions of supplies out of the port.
- Providing a response force to react to incidents inside the port's perimeter.
- Providing observation and early warning of threat ground and air attacks.

10-3. When providing security for cargo, the focus is on providing a security overwatch for the cargo as it moves from the port to the combat area. Inside a port's perimeter, access to cargo is limited by—

- Operating random mounted or dismounted patrols (with MWDs, if available).
- Using combined patrols as a response force for incidents inside the perimeter.

- Controlling access to the most restricted areas.

10-4. On occasion, the MP may have to safeguard highly critical cargo inside a port's perimeter. The type and degree of security provided is based on logistical security information. Some examples are the—

- Types and values of the cargo stored.
- Vulnerability of the cargo to a land threat.
- Likelihood of theft, diversion, pilferage, or sabotage by military personnel, local workers, black marketers, or enemy agents.
- Location and nature of the port facilities.
- HN agreements.
- Degree of entrance and exit controls.

10-5. Safeguarding the most critical cargo waiting to be transferred to land transport is the priority. The following measures help to safeguard stored cargo:

- Establishing access-control procedures.
- Searching bundles and packages being taken from the area.
- Examining trip tickets and documentation of cargo vehicles.

10-6. If the restricted area is a pier or other maritime environment, access from the water must be controlled as well as from the land. Entry on the landward side of a pier can be limited with fencing, pass control, and aggressive patrolling; but the part of the pier that protrudes over the water is accessible from the sides and from below. Methods for securing the pier along its water boundaries include—

- Patrols (both walking on the pier and in small boats).
- Protective lighting.
- Log booms.
- Nets.
- Buoys or floats.
- Anchored or pile-mounted navigational aids and signaling devices.
- Barges.

10-7. While most of the barriers described above will stop or impede access to facilities from boats or swimmers, nets are among the most effective. Well-marked, partially submerged objects are also effective; however, there may be legal prohibitions against placing barriers that may constitute a hazard to navigation. These barriers should be placed only after coordination with and approval by the appropriate legal and HN authorities. Sometimes it is best to close off the waterside of a pier. A floating boom will keep small boats out. Suspending a cable or a chain-link net from the bottom of the boom will deny access underwater.

10-8. At least two security zones must be established on a facility's waterside—the reaction zone and the keep-out (exclusion) zone. Security forces in these zones notify vessels, craft, and swimmers that they are entering restricted waters and should alter their course. Security forces may stop and search intruders if necessary. Security zones should extend at least 1,000 meters from the nearest protected asset; however, in some port areas this large security zone is not possible. In such cases, other measures (such as boat patrols) must be increased to mitigate the possibility of attack.

10-9. A reaction zone extends from the high-water mark to a distance beyond the maximum range of anticipated waterborne threats. Security forces will stop and challenge intruders inside the reaction zone.

10-10. The keep-out zone is the zone closest to the protected assets. It extends from the asset to the maximum range of anticipated threat weapons. Security forces should prevent the entry of all unauthorized craft or vessels into this zone. The tactical response force (in this case, a boat) may be used. In addition to organic security, forces may be provided by HN or contracted personnel.

10-11. To keep the cargo secured while transferring from one transport method to another, the traffic moving in and out of cargo-handling areas must be controlled. MP forces can—

- Set up a single access-control point.
- Erect field-expedient barriers. Truck trailers or other large vehicles can be used to constrict the traffic flow if permanent barriers are not in place.
- Limit entry to mission-essential personnel, vehicles, and equipment (as designated by the port authority).

10-12. A holding area should be provided if gates are used by vehicles other than cargo vehicles. Cargo vehicles can pull into the holding area while they are being checked. The holding area should be large enough to handle the volume and size of traffic. A wooden deck or platform at, or slightly higher than, the level of the truck bed can be used to facilitate checking. The platform must be at least as long as the vehicle (such as an empty flatbed trailer). Such a platform makes it quicker and easier to observe and check cargo.

10-13. Cargo is less likely to be diverted if a close watch is kept on cargo documentation and container safety. Containerized cargo is less likely to be stolen or sabotaged. However, containers must be watched closely as they are filled and sealed. Cargo can be pilfered before the seal is applied. An unsealed container can be moved to a stacking area; or someone may apply a false seal, break the seal later, remove the cargo, and then apply a legitimate seal.

10-14. At access-control points—

- Inbound and outbound containers should be inspected. Signs of damage or unserviceability should be observed.
- Containers must be inspected for the presence of seals or locks and hinges. Their serviceability should also be checked.
- The document's transport number, container number, and seal number should be checked to ensure that they match those numbers on the transportation control-and-movement document. (Check the seals by handling them, not simply by a visual check.)
- Containers with valid documents only should be allowed to pass inbound or outbound through the control point.

RAIL CARGO

10-15. Because a train's movement is determined directly by the condition of the tracks, cargo moving by rail is particularly vulnerable to attack. The destruction of switches, signals, or the track may be a delaying harassment; or it could trigger a major catastrophe. Since railroads can be such high-value targets, the commander may task MP or other US forces to provide on-board security for critical cargo.

10-16. Most train crews consist of four or five people who control the train—the engineer, a conductor, a fireman, a senior brakeman, and a brakeman or a flagman. The conductor is the train commander unless a transportation railway service officer is assigned to the train. The train commander is responsible for the train's operation and security. He makes all decisions affecting the train. The security force's commander is responsible for the cargo's security. The train crew and the security force watch for and report any discrepancies or interruptions to normal procedures at any time during the movement. Information about the movement is usually sent along the movement route by the chief dispatcher through a telephone circuit.

10-17. A four- to six-person security force is usually enough to secure railway shipments of sensitive freight, but additional security forces may be needed for moving critical cargo. In addition to a military security force, the shipper or loading agency may send specially trained personnel with highly sensitive cargo. The number of MP in a train security force depends on the—

- Sensitivity of the freight.
- Priority of need for the freight.
- Terrain over which the train will pass.
- Length of the train.
- Duration of the trip.
- Degree of threat.

10-18. Security forces prepare and maintain a record (by car number) of guarded cars in the train. Security forces can ride in—

- A specific car that requires protection.
- The caboose.
- A security-force car. (If only one security car is used, it should be near the center of the train; if more than one is used, cars should be spaced to provide the best protection for the train.)

10-19. The security force on a train must keep a constant check on car doors, seals, wires, and locks to detect tampering. The following instances must be noted and reported immediately:

- Irregularities in procedures.
- The presence or actions of unauthorized persons.
- Deficiencies or incidents that occur.

10-20. When planning rail-cargo security, the time schedule for the rail movement must be obtained. A map reconnaissance of the route should be provided, detailing bridges and tunnels that are especially vulnerable.

10-21. Security-force actions should be planned at scheduled stops or relief points, and forces should be deployed according to these plans. Locations of MP units and other friendly forces should be plotted along the route, and their radio frequencies and call signs should be noted. An intelligence report covering the route should also be obtained. This report should indicate sites where sabotage may occur, attacks may be expected, or thefts and pilferage are likely.

10-22. The shipper is responsible for the security of all carload freight until it is turned over to the Transportation Railway Service and the loaded cars are coupled to a locomotive for movement. The shipper or field transportation officer should complete the freight waybill or the government bill of lading. This report shows the car number, a brief description of contents, the weight of the load, the consignor, the consignee, the origin, and the destination. In addition, it may show special instructions for the movement or security of the car and its contents. Careful documentation is essential for—

- Securing the shipment.
- Locating cars with critical cargo.
- Ensuring that priority movement is authorized.

10-23. Transportation officers are responsible for the completeness, correctness, and proper handling of waybills. Each car must have a waybill; this allows cars to be detached or left behind should they become defective en route. If this occurs, a team from the security force must remain with the cargo until they are relieved.

10-24. Railway cars are sealed after loading. A seal shows that a car has been inventoried and inspected. The standard method of sealing a railway boxcar door (in addition to padlocks or wires) is with a soft metal strap or a cable seal that contains a serial number. Maintaining rigid accountability of all seals is necessary to prevent the undetected replacement of an original seal with another. While sealing does not prevent pilferage, a broken seal is a good indicator that the car and its contents have been tampered with. Train security forces or operating crews can easily check the seals on cars when the train stops. Broken seals should be reported immediately to help pinpoint the time and place of a possible theft. When vehicles are shipped by railcar, sensitive and high-value items must not be secured in the vehicles. Container-express (CONEX) and military-van (MILVAN) containers are ideal for shipping these and other small items on flatcars since they greatly reduce the chance of pilferage. These containers must be locked and sealed and, if possible, placed door to door for additional security.

10-25. When operations permit, cars containing highly pilferable freight, high-priority cargo, or special shipments are grouped in the train to permit the most economical use of security forces. When flatcars or gondolas are used to transport sensitive or easily pilfered freight, security forces should be placed where they can continuously observe and protect these cars.

10-26. When the train is stopped, security forces should dismount and check both sides of the train, verifying that seals, locks, and wires are intact. They must report a broken seal immediately to help pinpoint the time and place of the theft.

10-27. If the security force is relieved by another security force while en route, a joint inspection of the cars is conducted. The relief force signs the record being kept on the guarded cars. Consignees assume responsibility for the security of loaded freight cars at the time they arrive at their destination. When the trip is complete, the receiver or his agent will inspect the cars. The security force obtains a receipt for the cars, which is then attached to the trip report. The trip report should include—

- Dates and times the trip started and ended.
- Any additional information required by the local SOP or command directive.
- Recommendations for correcting deficiencies or for improving future security on trains.

10-28. Because unloading points are highly vulnerable to pilferage and sabotage, cars should be unloaded as soon as possible to reduce the opportunity for loss. MP forces are normally not available for the security of freight in railway yards. For more information regarding rail cargo, see FM 55-20.

PIPELINE CARGO

10-29. Pipeline systems are widely used in a theater of operation to transport bulk petroleum products or other liquids. Such systems are open to a number of security threats from the point of entry to the point of final delivery. Pipeline systems are composed of storage and dispersing facilities, pump stations, and extended pipelines. They also include discharging facilities for tankers at ports or other water terminals.

10-30. The type and extent of risk to a pipeline varies with the level of conflict in the AO. In a communications zone, the chief hazard is likely to be pilferage. Pipelines can be tapped by loosening the flange bolts that join sections of pipe or by cutting holes in the hose line. The risk rises if gasoline is scarce and expensive on the civilian market. Sabotage is a security hazard during all levels of conflict. It is committed by any method such as simply opening pipe flanges, cutting a hose line, or setting fires and causing explosions to destroy portions of the line.

10-31. In areas of conflict, the likelihood of sabotage and interdiction increases. Pipeline systems are vulnerable to air attacks, especially aboveground sections of the pipeline, pump stations, and storage facilities.

10-32. Security forces should be deployed in the best manner to provide coverage to the most vulnerable portions of the pipeline that are at the greatest risk to enemy, terrorist, partisan, and ground attack. Patrols should be set up to screen isolated areas and remote pumping stations. Sensors should also be considered, along with aerial security. Security patrols will—

- Detect, report, and respond to attacks on or sabotages of the pipeline.
- Monitor critical parts of the pipeline on a routine but random basis.
- Monitor ground sensors and other intrusion-detection devices. These are often used at pump stations and elsewhere along the pipeline to detect and identify threats to the system.

- Check line-pressure devices in pipeline and pumping facilities. These devices monitor the flow and detect breaks in the line, which may indicate pilferage of gasoline (or other petroleum products).

10-33. Dedicated security forces are rarely sufficient in number for the surveillance of an entire pipeline system. All available supporting forces (in the course of their normal duties) should observe and report items of intelligence for further investigation. Examples of suspicious activities in the pipeline area might include the unusual presence of commercial tanker trucks, the appearance of gasoline drums or cans, or an increased use of motor vehicles in fuel-scarce areas. Other resources available to the commander for coordination and support include HN and MP elements responsible for the AO, as well as the security officer of the petroleum group or battalion.

CONVOY MOVEMENT

10-34. As convoy movements are tactical in nature and are discussed in detail in FM 19-4, they will be briefly discussed here. When moving by convoy, consideration should be made for the following:

- Congested traffic areas.
- Travel during night hours when traffic is reduced rather than travel during daylight hours when traffic congestion is heaviest.
- National holidays. Traffic may be three times heavier than on a normal day. Also, if you are moving a convoy overseas on a national holiday, the HN people may not be receptive to your action and the result may be unwanted reactions on their part.
- The use of a marked HN police vehicle in conjunction with the convoy. The HN people are more receptive to an activity when it is represented by one of their own. Additionally, the HN police may be able to diffuse a potential crisis.
- Security of the convoy. Security of the convoy is foremost important both during movement and stops. During extended or overnight stops, special consideration must be given to securing the loaded vehicles.

Chapter 11

Inspections and Surveys

Inspections and surveys are valuable tools to a commander's physical-security program. These tools collectively measure and identify the readiness of a commander's physical-security program. The survey provides the installation commander with an overall security posture of the installation.

INSPECTIONS

11-1. Physical-security inspections are conducted at DA installations, activities, and facilities by trained physical-security inspectors. Some facilities on an installation may be exempt from inspection due to their mission. These facilities are inspected under the guidance of regulations and directives unique to those activities. Inspection personnel will be trained and will conduct inspections according to AR 190-13.

11-2. A physical-security inspection is a recorded assessment of physical-security procedures and measures implemented by a unit or an activity to protect its assets. The inspection is recorded on DA Form 2806-1-R (see AR 190-13 for using the form).

COORDINATION

11-3. Liaison and coordination should be established with other agencies on the installation before an inspection. The director of facility engineers can provide information to benefit the overall security program. Other agencies, such as MI (threat analysis) and local law-enforcement agencies, may have input essential to the security program.

SECURITY LIBRARY

11-4. A security library is necessary to help personnel prepare for and conduct an inspection. This library may include—

- The mission and history of each activity to be inspected.
- Previous inspection reports.
- A copy of the most current risk analysis.
- The SOPs and ARs specific to physical security.

ENTRANCE INTERVIEWS

11-5. Entrance interviews are usually required before the actual inspection. During the interview, the inspector establishes a rapport with the unit representative. The inspector identifies the following during the interview:

- All members of the inspection team.

- An overview of the last inspection.
- Areas to be inspected and the order of inspection.
- A review of waivers, work orders, and exceptions.
- Changes to the unit's mission (if any).

CONDUCTING INSPECTIONS

11-6. The inspection should be conducted from the outside to the inside of the facility, activity, or area with regard to the following:

- Observation of the facility will be conducted during all hours of the day.
- Interviews of managerial and operational personnel will be performed.
- Security forces should be inspected so as not to disrupt the mission (if possible).
- An assessment should be made of security-force training, especially if security-force knowledge proves inadequate.
- Inspection of entry and movement control by the guard force should not hinder operations.
- All communications systems used by the guard force should be thoroughly inspected. The guard force should have two reliable and efficient means of communication, one of which is a radio.
- Inspections should be conducted according to regulations appropriate for the facility.

EXIT INTERVIEWS

11-7. Exit interviews should be conducted as soon as possible after the inspection. The commander should be informed of any deficiencies or compliments noted. A rating on the inspection's results will not be provided during the exit interview. The approving authority, not the inspector, will determine the inspection rating. The rating will be forwarded to the unit along with the final report.

11-8. Recommendations will be made according to regulations. Written reports should be forwarded through channels in a timely manner according to the PM's SOP. The commander's report of actions taken will be required and reviewed by the PM's staff.

SURVEYS

11-9. A physical-security survey differs from an inspection in that a survey covers a formal assessment of an installation's physical-security program. Each survey includes a complete reconnaissance, study, and analysis of installation property and operations. The survey provides the commander with an assessment of the installation's overall security posture. It consists of the threat and the mission, and it advises the commander on the installation's physical-security program's strengths and weaknesses.

PHYSICAL-SECURITY SURVEY

11-10. The physical-security survey is a formal recorded assessment of an installation's physical-security program. See AR 190-13 for further information on this type of survey.

SECURITY-ENGINEERING SURVEYS

11-11. While a security-engineering survey is largely an engineer function, it must be coordinated with physical-security personnel to be successful. A security-engineering survey is the process of identifying (by means of an on-site survey) engineering requirements associated with facility enhancements for physical security and antiterrorism, including an IDS installation. This type of survey should be conducted when planning new construction, renovations, or upgrades to existing facilities where there are likely to be physical-security requirements. A security-engineering survey may also be requested by the PM or an equivalent security officer to evaluate existing security. This survey—

- Identifies assets to be protected.
- Identifies threats to these assets and the level of protection required to protect them.
- Identifies the protective measures.
- Determines the cost of the protective measures.

Appendix A

Metric Conversion Chart

This appendix complies with current Army directives which state that the metric system will be incorporated into all new publications. Table A-1 is a conversion chart.

Table A-1. Metric Conversion Chart

Metric to English			English to Metric		
Multiply	**By**	**To Obtain**	**Multiply**	**By**	**To Obtain**
Length					
Centimeters	0.0394	Inches	Inches	2.54	Centimeters
Meters	3.28	Feet	Feet	0.0305	Meters
Meters	1.094	Yards	Yards	0.9144	Meters
Kilometers	0.621	Miles (stat)	Miles (stat)	1.5609	Kilometers
Kilometers	0.540	Miles (naut)	Miles (naut)	1.853	Kilometers
Millimeters	0.039	Inches	Inches	25.40	Millimeters
Area					
Square centimeters	0.1550	Square inches	Square inches	6.45	Square centimeters
Square meters	10.76	Square feet	Square feet	0.0929	Square meters
Square meters	1.196	Square yards	Square yards	0.836	Square meters
Volume					
Cubic centimeters	0.610	Cubic inches	Cubic inches	16.39	Cubic centimeters
Cubic meters	35.3	Cubic feet	Cubic feet	0.0283	Cubic meters
Cubic meters	1.308	Cubic yards	Cubic yards	0.765	Cubic meters
Milliliters	0.0338	US liq ounces	US liq ounces	29.6	Milliliters
Liters	1.057	US liq quarts	US liq quarts	0.946	Liters
Liters	0.264	US liq gallons	US liq gallons	3.79	Liters
Weight					
Grams	0.0353	Ounces	Ounces	28.4	Grams
Kilograms	2.20	Pounds	Pounds	0.454	Kilograms
Metric tons	1.102	Short tons	Short tons	0.907	Metric tons
Metric tons	0.984	Long tons	Long tons	1.016	Metric tons

Appendix B

Sample Installation Crime-Prevention Handbook

This appendix provides guidance on planning, organizing, directing, and controlling installation crime-prevention programs. It provides guidance on developing an installation program, criminal analyses to identify crimes, guidance on which crimes to address, command and individual countermeasures for particular crimes, and program-evaluation procedures.

SECTION I — INSTALLATION CRIME-PREVENTION PROGRAMS

B-1. In the past few years, the Army has shifted an increasingly larger percentage of its manpower from combat-service-support activities to combat organizations. This change means that fewer MP personnel are available to support a larger number of units. To meet this challenge, it is necessary to reevaluate the way we do business and to emphasize those programs or procedures that have the greatest impact on our installation crime rates. Crime prevention is one program that can have a major impact on installation crime rates at a relatively minor cost in both dollars and manpower. It takes less effort to discourage a criminal from perpetrating a crime or to teach a soldier to avoid becoming a victim than it does to investigate a crime, identify the offender, prosecute him, and punish him. In addition, a proactive approach to law enforcement can help maintain the high quality of service life that can improve the retention of first-term soldiers.

B-2. The Army is a large organization that performs a variety of activities in many different environments. Crimes that are major problems on one installation may be totally absent from others. For example, most military installations have a significant number of robberies while most depots have none. Because of this, any rigid, centrally controlled program—no matter how carefully thought out—is bound to be inappropriate in many locations. Therefore, DA has elected to provide only the most general guidance and to allow commanders to develop crime-prevention programs that address their local problems.

CRIME-PREVENTION WORKING GROUPS

B-3. The installation is the smallest practical level for implementing crime-prevention programs. If these programs are developed and implemented at a lower level, then crime is often not eliminated but is merely displaced from units with good programs to units with less effective programs. Also, crime does not affect personnel only when they are in their place of duty. In many cases, a company commander's troops are victimized in areas over which he has little control. Unit commanders are responsible for implementing many anticrime measures; however, the selection of overall program goals, the ID of appropriate countermeasures, and quality control should be done at the installation level.

B-4. Crime prevention must always be recognized as a commander's program rather than as an MP program. MP personnel have the expertise to analyze data, identify major problems, and develop lists of possible countermeasures. They should perform these functions in support of an installation crime-prevention council appointed by the installation commander and composed of representatives of all of the installation's major organizations and activities. The advantages of using this type of system are—

- It provides representatives of all major segments of the post population with a forum where they can identify criminal problems that are of the greatest concern to them.
- It allows the representatives of all major commands to review the available options to counter a crime and to select the level of resource commitment that is compatible with their missions and internal priorities.
- It helps ensure that the resources of the entire community, rather than only those of the MP force, are mobilized to attack the problem.
- It is easier to obtain the support of the whole population if its representatives are instrumental in the development of the program.

CRIME-PREVENTION OFFICERS

B-5. The installation's crime-prevention officer is normally a senior NCO or an officer who has a solid background as an MP investigator or a physical-security inspector (PSI). He supports the installation council by performing a crime-data analysis to identify problem areas, drafting programs for the council's consideration, inspecting the implementation of council-mandated measures, and coordinating the efforts of unit crime-prevention officers in the implementation of the crime-prevention program.

B-6. As a member of the PM's staff, the crime-prevention officer develops the law-enforcement section of the crime-prevention program, develops and maintains the written crime-prevention plan, and coordinates crime-prevention programs with civilian police agencies and community groups.

B-7. Crime-prevention officers are also appointed in each organization down to the company level. At this level, written crime-prevention plans are not required; however, SOPs are established. The crime-prevention officers serve as their organizations' focal points for coordinating installation crime-prevention plans; they supervise the implementation of the installation's program within their organizations.

CRIME-PREVENTION PROGRAM DEVELOPMENT

B-8. The starting point for developing a crime-prevention program must be a thorough analysis of criminal activity on the installation. This identifies significant criminal problems that are susceptible to crime-prevention efforts. Crimes that are most susceptible to crime-prevention measures are those for which a high probability of reoccurrence exists. Crimes such as murder normally are not repetitive and are poor candidates for inclusion in the crime-prevention program. Since it is seldom practical to attack all criminal problems simultaneously, they should be prioritized based on their impact on the command's ability to perform its mission and their impact on installation personnel. Next, the whole range of countermeasures that can be used to combat each problem must be identified (see Figure B-1). Table B-1, page B-4, identifies (by offense) programs that have been successful in countering specific criminal problems. Sections III and V of this appendix contain discussions of the strengths, weaknesses, and applicability of the countermeasures listed in Table B-1. Once developed and prioritized, the list of criminal problems and possible countermeasures must be presented to the installation crime-prevention council for action. The council should decide which crimes will be addressed and which countermeasures will be used for each crime. The council must then identify specific objectives for its anticrime campaigns.

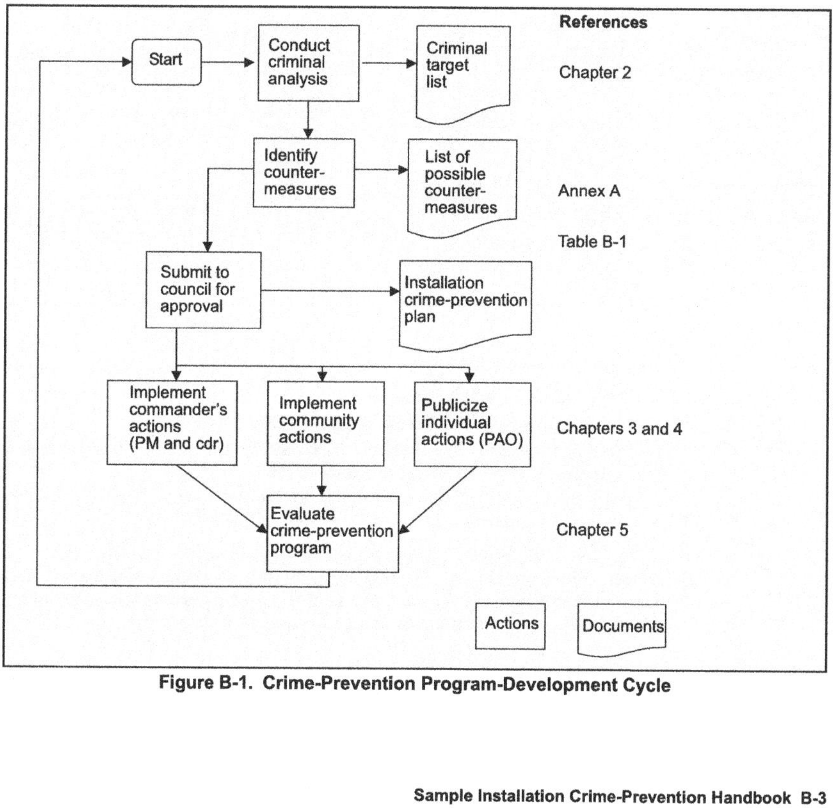

Figure B-1. Crime-Prevention Program-Development Cycle

Table B-1. Offenses Countermeasure Matrix

Crime	Command/Law-Enforcement Countermeasures												Community Programs					
	Crime Hot line	Lighting	Environmental Changes	Patrol/Surveillance	Publicity Campaigns	Residential Security Survey	Juvenile Programs	Fraud Programs	Employee Theft	Arson Programs	CCTV	Warning Signs	Neighborhood Watch	Operation ID	Neighborhood Walks/Escorts	Mobile Patrols	Project Lock	Education
Arson				X	X					X			X			X		X
Auto theft	X	X			X								X		X	X	X	X
Burglary/ housebreaking	X	X	X	X	X	X							X	X	X	X		X
Employee theft	X				X				X		X	X		X				X
Fraud	X				X			X				X						X
Larceny	X		X			X			X					X				X
Rape	X	X	X	X	X	X							X		X	X		X
Robbery	X	X	X	X	X						X		X		X	X		X
Juvenile delinquency	X			X	X		X						X		X	X		X
Vandalism	X	X		X	X		X						X		X	X		X

B-9. Objectives must identify—

- What crime will be reduced.
- What target population will be addressed.
- What specific changes and behaviors on the part of the victims or perpetrators will be encouraged.
- What actions the command must take to reduce the opportunity for the crime to occur.

B-10. Once objectives have been clearly defined, specific areas of responsibility should be assigned to each council member (based on their organization's primary area of responsibility) and major milestones should be identified for developing the campaign against each targeted crime.

TRAINING

B-11. The prerequisite skills for successful performance as an installation crime-prevention officer are best developed through on-the-job experience as the supervisor of MP investigations or physical-security inspections. More important than any technical skill is the cultivation of a frame of mind that instinctively examines each case to determine not only what occurred, but also how the crime could have been prevented. Technical skills (such as criminal-data analysis) that may not have been developed as an MP investigator or physical-security supervisor are presented in courses taught by several civilian agencies. These classes should be used to the fullest extent possible.

CIVILIAN CRIME-PREVENTION ORGANIZATIONS

B-12. There are many civilian crime-prevention organizations at the national, state, and local levels. Many of these organizations have produced crime-prevention material (including posters, radio spots, and leaflets). Material and programs sponsored by civilian agencies should be used to support Army crime-prevention efforts. However, when material from a source outside of DOD is used, a copyright release must be obtained. Normally, it is necessary to get a release for each separate item that is used. If there is any doubt as to the necessity of securing a copyright release, the crime-prevention officer should refer the matter to the local SJA.

SECTION II — CRIMINAL ANALYSIS

B-13. Criminal analysis is a system for identifying trends and patterns where they may exist. It is a routine, ongoing function for the PM and battalion- and brigade-level staffs. Criminal analysis is the foundation upon which the installation force-protection program is based. Moreover, criminal analysis is an integral component of the police intelligence-operations function and is applicable across the operational continuum. An effective criminal analysis establishes the following:

- Crimes having a significant impact on the installation.
- The segments of the population being victimized.
- The ID of criminals/perpetrators.
- The most common time of occurrence.
- The areas that experience the highest number of incidents.
- Offense information (such as types of weapons or victims' actions that contribute to the offense).
- Information critical to an installation's VA.
- Information essential in formulating a successful patrol-distribution plan.
- Police information and criminal intelligence fused with tactical intelligence.

B-14. With this type of information, specific countermeasures are developed to reduce the opportunity for a crime to occur or to remove the incentives for perpetrators. Without an effective criminal analysis, the overall security effort is unfocused. Moreover, the installation/base patrol-distribution plan may be skewed. When this occurs, broad countermeasures are implemented and meaningful results are lost.

B-15. A professional analysis of criminal data is essential for protecting soldiers, units, and installations. The Military Police Management Information System (MPMIS) provided the initial software capable of assisting in the development of criminal-analysis information. As this system is modernized and replaced by the Military Police Automated Control System (MPACS), the MP Corps and the CID will have interface with the Army Battle Command System (ABCS). The analysis of police, criminal, and tactical information results in the development of police intelligence (PI). The PI, which is coordinated with the Assistant Chief of Staff, G2 (Intelligence) (G2) or the Intelligence Directorate (Joint Command) (J2), provides relevant information and intelligence (RII) that greatly contribute to the commander's critical information requirements (CCIR). The RII provided vertically and horizontally ensures that the common operational picture is properly shared.

SOURCES OF INFORMATION

B-16. The basic source for identifying crimes that warrant examination is the installation's Law Enforcement and Discipline Report (DA Form 2819). The information in this report can be used to plot the amount of and seasonal variations for each major type of crime. This will tell little about conditions that produce the crime, but it is useful in identifying crimes that can be eliminated from detailed analyses due to low frequency rates.

B-17. The Law Enforcement and Discipline Report identifies which major category of crime should be targeted, but it does not tell which type of crime is causing the most problems. For example, It may indicate that robbery is a problem, but it will not discriminate between robbery of commercial establishments (such as the post exchange [PX], the Class VI store, or banks) and muggings of individuals. Since countermeasures for these two types of robbery are different, it is necessary to collect additional information. The best sources for this information are the military-police report (MPR) and the CID report of investigation (ROI). If there are fewer than 200 cases in the past year for a particular type of crime, they should all be examined. With a larger annual caseload, a random sample that is large enough to give results of plus or minus 5 percent accuracy should be examined.

B-18. General factors must be identified for any type of crime. These are the types of victims, the perpetrators, geographical data, and chronological data.

VICTIMS

B-19. It is essential to determine (as precisely as possible) the segment of a post population that is being victimized. Junior soldiers live in different areas on post than officers and senior NCOs. They patronize different clubs and, for the most part, work in different areas. Information programs must use multiple sources to ensure that the same message reaches the different portions of the population. If the specific population segment that is being victimized is not identified, it is possible to spend a large amount of resources and have little impact on the targeted crime because the message is not getting to the people who need it.

PERPETRATORS

B-20. As with the victim profile, data on perpetrators is essential to ensure that countermeasures are targeted against the correct population. When on-post auto thefts are committed by civilians with legitimate reasons for entering the installation, it may not make sense to increase security (as a countermeasure) at post entrances. Countermeasures are developed based on specific profiles identified by case type and the information described above. Successful countermeasures depend on correct criminal analyses. An analysis and profile of perpetrators will help to focus MP/CID countermeasures.

GEOGRAPHICAL DATA

B-21. The MPR and the ROI should contain a street address or a description of the incident's location (such as "in a parking lot on the east side of building 1409"). This type of description fulfills requirements for identifying the location of the incident in court. It is further enhanced by identifying the normal duty hours and the type of activity that takes place in the location (for example, "in a parking lot on the east side of the NCO club (building 1401), normal operating hours 1800-0400" or "in a parking lot on the east side of building 1408, a troop billets for Company A, 15th Cav"). This type of information may help to develop a list of specific types of areas where a particular crime is occurring, making it easier for the PM to provide increased MP patrols in these areas. For example, if there have been several incidents in which vehicles have been broken into and stereos removed, it would help to know that 90 percent of them occurred in the parking lot by troop billets.

CHRONOLOGICAL DATA

B-22. As with geographical data, the more specific the time-of-occurrence pattern is for a crime, the easier it is to apply sufficient resources to affect the crime rate. For each crime having a significant impact, determine the following:

- Major seasonal variations.
- Monthly variations. Is there a concentration of crimes immediately before or after payday?
- Weekend occurrences. Is there a concentration of incidents on weekends? Each day represents about 14.25 percent of the week. Concentrations of crime higher than that for any particular day may be significant.
- Time of day. Is there a particular period that accounts for a disproportionate share of the incidents?

CRIME-SPECIFIC FACTORS

B-23. In addition to the factors that should be examined for all crimes, the following crime-specific factors are useful in analyzing specific offenses:

- Housebreaking and burglary.

    - The type of building that was attacked (family-housing unit, troop barracks, PX, and so forth).
    - Whether the facility was occupied or unoccupied.
    - The point of entry (door, window, and so forth).
    - The method of entry (unsecured door, forced door, forced window, and so forth).
    - The property that was stolen. (Was it marked?)
- Robbery.
    - The number of perpetrators.
    - The perpetrator's method of operation.
    - The type of weapon used.
    - The type of robbery (street mugging, residential robbery, or commercial robbery).
    - The victim's injuries.
    - Actions by the victim that contributed to his being targeted.
- Larceny.
    - The type of property taken.
    - Whether the property was secured or unsecured.
    - The perpetrator's method of operation.
- Auto theft.
    - The type of vehicle stolen (POV, motorcycle, and so forth).
    - Whether the vehicle was secured or unsecured.
    - Whether the vehicle was recovered and where it was recovered.
    - Whether the vehicle was stripped of parts.
    - The perpetrator's method of operation.
- Forgery.
    - The type of document that was forged.
    - How the document was obtained.
    - The type of ID used in passing the forged document.
    - The con games or techniques used.
- Rape and sex offenses.
    - The perpetrator's method of operation.
    - The relationship between the victim and the perpetrator (blood relatives, acquaintances, or strangers).
    - The degree of force used.
- Aggravated assault or murder.
    - The relationship between the victim and the perpetrator.
    - The motivation.
    - The weapon used.

B-24. Personnel who have tried to use MPRs and ROIs as a sole source of data know that in many cases the required information is not contained in sufficient detail to be useful. To correct this situation, investigators and inspectors must be trained to recognize conditions that contributed to the crime as well as the information needed to identify and prosecute the offender. As a minimum, the PM ensures that all law-enforcement personnel know general and specific crime factors that are required to analyze each type of crime. He must also provide feedback through the law-enforcement operations staff when incomplete reports are received. If this is done, high-quality data can be collected without generating additional reports to collect criminal information for analysis.

B-25. The MPRs and ROIs are not the only sources of information on which to base criminal analyses. Physical-security inspection results, summaries of common deficiencies noted by the inspector general, data from the SJA claims section, and information summaries from reports of survey on lost government material can all produce worthwhile information on conditions that lead to the commission of crimes. Additionally, a review of civilian criminal statistics and civilian police information may contribute to the pattern analysis. The reviewing authority must understand which information is relevant. The MP commanders ensure that the CCIR are published in operations orders (OPORDs), operation plans (OPLANs), and SOPs. This effort ensures that MP forces at every level recognize relevant information.

INDIVIDUAL CRIMINAL ANALYSIS

B-26. A criminal analysis is most effective when it is applied to the class of criminal offenses with a high probability of recurrence. Single-incident crimes do not lend themselves to analyses. Most crimes against persons do not usually benefit from analyses, with the notable exceptions of rape, robbery, and related combinations of offenses (such as kidnapping and rape; robbery and attempted murder; burglary and rape; and burglary, robbery, and kidnapping). Analyzing isolated criminal offenses has some value (such as gaining knowledge of where these offenses are most likely to reoccur). However, this knowledge is usually difficult to use effectively for preventive purposes.

B-27. There are universal factors available for analysis for most crimes. The availability of these factors varies greatly between criminal types and specific reported offenses.

B-28. In addition to the universal factors, there are an almost infinite number of factors that may be considered specific to a particular criminal class or type. These crime-specific factors are data elements that are usually recorded during the reporting of a particular type of offense and are used for analysis purposes.

B-29. Crime-specific factors provide information that can be used by the analyst to connect crimes with similar characteristics. Information regarding physical evidence may have considerable value in the analysis of several criminal types (such as burglary and auto theft). Therefore, the suitability of different crimes to analyses depends on the general factors, specific modus operandi (MO) factors, and physical evidence. With these considerations in mind, different criminal types and classifications are examined to determine their applicability.

RESIDENTIAL HOUSEBREAKING AND BURGLARY

B-30. Although housebreaking and burglary are two of the most difficult crimes to prevent, they are the most suitable for analyses. The typical housebreaker or burglar establishes an MO pattern based on successful past offenses. Usually, a burglar will continue to commit similar crimes until he is apprehended.

B-31. The available information for evaluation must consider the specificity, accuracy, and value of the information received. Informational factors are presented in the order of relative importance. It should be emphasized that the factors listed in Figure B-2 are for analysis purposes; the order for investigative solutions may vary.

COMMERCIAL HOUSEBREAKING

B-32. The analysis of commercial housebreaking is in many respects easier than that of residential housebreaking or burglaries. More pertinent information is available to the analyst for analyzing commercial housebreaking (which is more specialized and will exhibit more specific MO characteristics). The analyst normally has specific information regarding the point and method of entry, the victim/target description, and the property-loss description. The time factors for commercial housebreaking may be of less value in a commercial-housebreaking analysis than in a residential-burglary analysis since most commercial housebreakings occur at night or during weekends.

B-33. The commercial housebreaker is generally more mobile than the residential burglar. An analysis of commercial housebreaking is not as

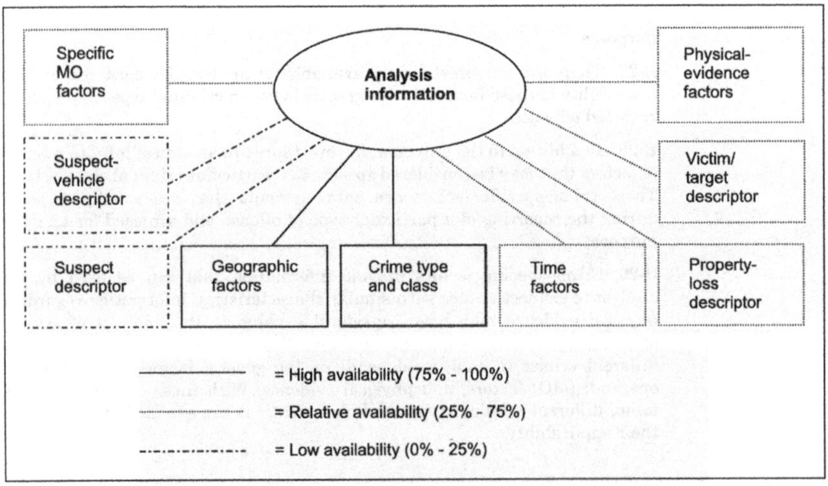

Figure B-2. Residential Housebreaking and Burglary

restricted in geographical area as an analysis of residential burglaries. The commercial burglar may travel a considerable distance to attack a particular type of business. The residential burglar is less discriminating because he has a greater number of potential targets.

B-34. The analysis of commercial housebreaking should be systematically directed to examining the information factors shown in Figure B-3. A commercial housebreaker may be a juvenile criminal, an addict, a professional thief, a soldier, or an enemy agent. Each of these types provides a different set of factors to analyze.

ROBBERY

B-35. Robberies are well suited for analysis. This class of criminal offender usually operates in a given geographical area. The commercial robber generally seeks a sizeable amount of cash; while targets of the street robber frequently include credit cards, checks, and other valuables in addition to cash.

B-36. The presence of physical evidence is more probable in street robberies than in commercial robberies because the criminal offender frequently discards evidence (such as purses and wallets) after completing the offense. The mugger frequently uses surprise as part of his MO in order to reduce his chance of apprehension by physical ID.

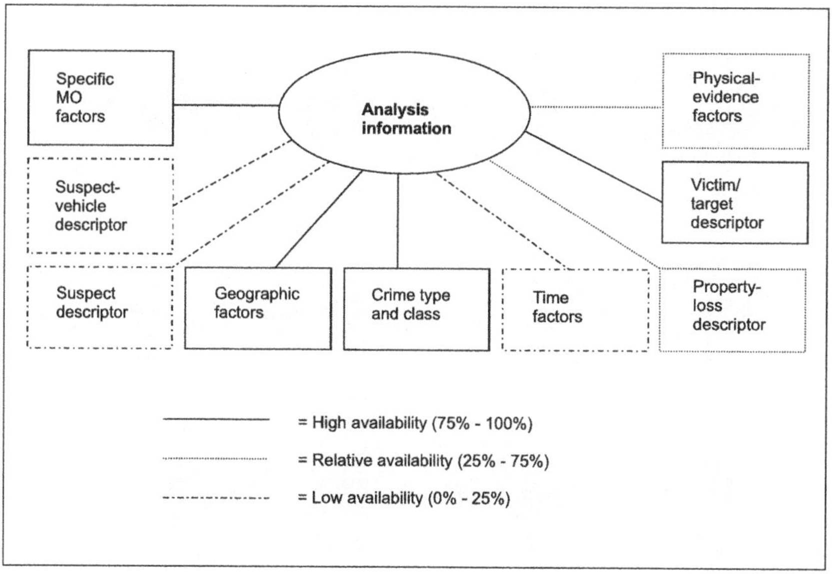

Figure B-3. Commercial Housebreaking

B-37. The probability of universal and crime-specific informational factors for a typical robbery is a systematic method for examining the case. See Figure B-4 for the analysis of a robbery case.

AUTO THEFT

B-38. Auto theft is probably the best-suited crime for analysis. The analyst frequently has more information available on any given auto theft than any other crime. Suspect descriptive information is usually lacking in auto larceny cases. However, the greater availability of crime-specific information makes auto theft extremely receptive to an analysis.

B-39. Auto-theft offenders include joyriders, professional car dismantlers, and wholesalers of stolen vehicles. In many cases, these offenders will establish and maintain a particular MO until apprehended.

B-40. Helpful elements in analyzing auto-theft cases are the availability of information, the presence of physical evidence, and the ability of the stolen property to be traced. Of special importance are the geographic factors, suspect-vehicle descriptors, and property-loss descriptors. In analyzing an auto theft, the analyst usually has information concerning two geographic locations for analysis—where the vehicle was stolen and where it was recovered. The suspect-vehicle descriptors, property-loss descriptors, and victim/target descriptors are the same (see Figure B-5). This greatly augments the analysis of auto thefts. In an examination of specific MO factors in an auto

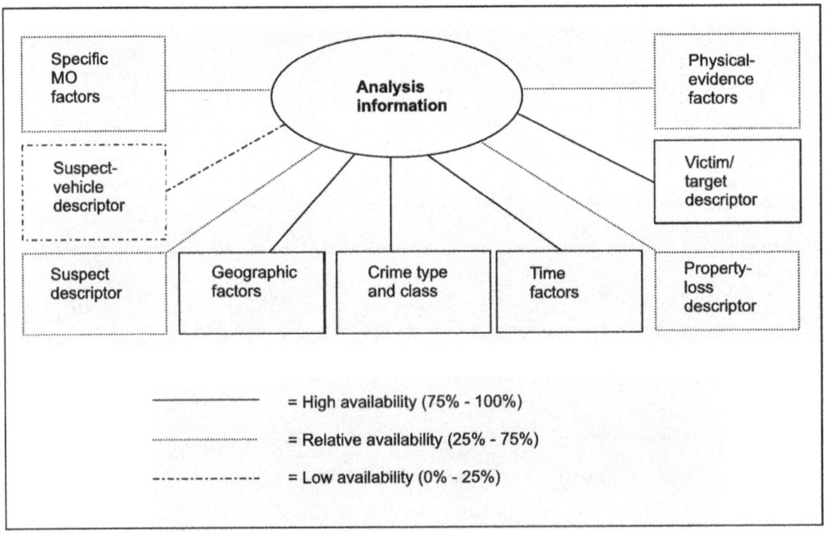

Figure B-4. Robbery

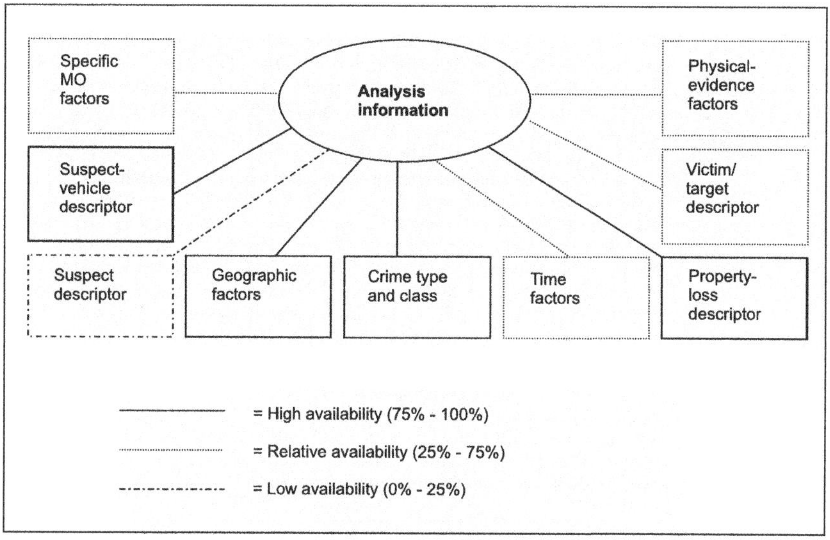

Figure B-5. Auto Theft

theft, the analyst will place emphasis on the vehicle's condition when recovered. Factors of primary importance in analyzing auto-theft cases are the vehicle's recovery location, the vehicle's make and model, and the degree of stripping. Coordination with local civilian police or HN police may reveal potential buyers of stolen automobiles.

B-41. Larceny and theft cases are not usually well suited for analysis. However, benefits are derived from analyzing larceny offenses when specific factors are selected, such as auto accession thefts.

B-42. The problem with trying to analyze general theft cases results from the large volume of reported offenses, the large number of possible crime types, and the many possible MO patterns. To analyze theft cases, the analysis operation must first restrict the number of cases and crime classification to a workable level. This may be done by classifying general theft cases into special categories and analyzing them by considering only special MO factors. Logical classifications include thefts of autos, auto accessories, bicycles, items on shipping docks, and tools and equipment.

B-43. Many of the general theft cases are helpful in analyzing other types of crimes. For example, an increasing trend in stolen automobile parts may have common perpetrators involved in auto theft and stripping. When particular patterns appear, a specific in-depth analysis can be conducted. It is essential that information developed during the analysis process be shared with the G2/ J2, the SJA, the CID, and others (as authorized).

RAPE AND SEX CRIMES

B-44. Rape and sex crimes fall into two distinct categories—those in which the offender is known to the victim and those in which he is unknown (stranger to stranger). Those cases in which the victim knows the offender are of limited informational value to the analyst. The in-depth analysis of rape and sex crimes is restricted to those cases in which there is no apparent relationship between the suspect and the victim. Rape and sex offenses in which the victim knows the suspect are often crimes of opportunity. Usually, these offenders do not establish a particular MO. On the other hand, the rapist or the sex offender committing stranger-to-stranger crimes will usually provide definite MO patterns, making this crime well suited for analysis. For example, the burglary/rape motive is particularly well-suited for analysis, as is the kidnapper/child molester. The advantages to analyzing stranger-to-stranger rape and sex offenses lies in the seriousness of the crime, its relative rareness, and the availability of information for a particular offense or group of offenses.

B-45. As with robbery, the suspect descriptors are important to the analyst for examining rape and sex offenses (see Figure B-6). The victim descriptors are also important in analyzing rape and sex offenses. In many cases, the perpetrator will restrict his attacks to victims of a certain age, a particular race, or a particular occupation grouping.

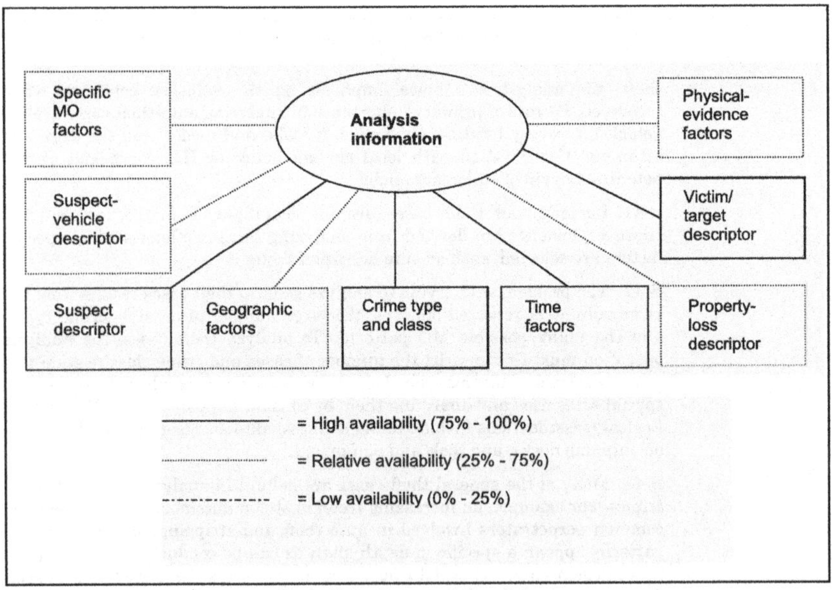

Figure B-6. Rape and Sex Crimes

FORGERY AND FRAUD

B-46. Forgery and fraud cases can provide correlative information for other reported criminal offenses such as burglary and strong-armed robbery. The analysis of forgeries is also aided by the fact that most forgers are repeaters who have established definite MO patterns.

B-47. An analysis of forgery and fraud cases can generate information that is disseminated to area merchants in the form of check-warning bulletins. The use of these bulletins can enhance community relations.

B-48. The primary problem in analyzing forgery and fraud cases is the immediate availability of information. In some cases, a delay of between three or four days to several months occurs between the time the offense was committed and the time it was reported.

ASSAULT AND MURDER

B-49. Most criminal offenses constituting assault and murder do not lend themselves to analysis. The comparative rarity of these crimes involving complete strangers makes it difficult to plot and predict assaults and murders. A tremendous amount of time and research is necessary to predict assaults and murders effectively. This data may identify specific areas with high rates of violent crime. This allows the commander to identify unsafe areas and warn soldiers to stay away from them.

CRIMINAL-ANALYSIS PROCEDURES

B-50. To organize the flow of information within a PMO and to facilitate analyses in support of crime-prevention efforts, the staff reviews a list of offenses. This list contains offenses that lend to analyses. The analyses provide the azimuth for crime suppression and countermeasures (for example, focused investigation/surveillance and patrol distribution). When final action on an MPR or an ROI covering one of these offenses is completed, it is forwarded to the staff for review.

B-51. The staff uses preprinted work sheets containing the factors of particular interest for an offense. They annotate and review the specific criminal factors. If some of the information is not available, the MP or CID investigator continues the investigation. The period covered by each work sheet will depend on the volume of cases at the particular installation. For a low-incidence-rate crime (such as robbery), one work sheet will suffice for the entire calendar year. For a more frequent crime (such as larceny), it may be necessary to use a different work sheet for each month or quarter. A summary sheet can be used to keep track of the data from all of the reporting periods in the year.

B-52. The data-collection sheets give specific information on individual types of crime. This information, together with general crime trends that are identified by comparing the data from DA Form 2819 for the current period to earlier periods and the geographic data from the crime-occurrence map, gives most of the information that is required to develop tightly targeted crime-prevention programs. While this system requires additional forms (the data-collection sheets and the crime-occurrence map) to be developed for use by the

MP Intelligence Officer (US Army) (S2), it does not increase the patrol's workload.

B-53. In addition to the information recorded in the forms, handwritten notes should be kept on items or trends that become apparent to the analyst during the review. For example, as a result of reading the cases on robbery to extract information for the collection form, the analyst may notice that a high proportion of the victims were attacked while crossing an unlighted athletic field near a troop billet area. This should be recorded so that it may be used in publicity campaigns and so that it may be passed along to patrols who routinely patrol the troop billet areas.

B-54. A geographic analysis is performed to determine information not available in the original data-element source (for example, a crime report). When a pin is used to indicate a crime location on a map, the relationship to other reported crimes of the same type becomes apparent to the analyst. Mapping is usually the source for a geographic analysis.

B-55. Mapping-analysis techniques involve the use of a map to depict the actual geographical relationships between particular criminal events according to prescribed data elements. A pin map displaying the actual locations of burglaries over a period of time is an example of such a technique.

B-56. In selecting a mapping technique, several things must be considered. These include the number of data elements that are to be recorded in the mapping, the retrievability of stored data, and the number of maps to be maintained.

B-57. The number of data elements to be recorded on a map can dictate the technique to be used. For example, if the case number is to be recorded in conjunction with the location of the offense, a dot (color-coded, self-adhesive paper disk) map or a flag-pin map can be used, whereas a typical color-coded pin map cannot. If particular MO factors are recorded, various color-coded and marked pins can be applied to identify specific MO patterns.

B-58. In addition, the length of time a particular map or set of maps is maintained should be considered. No set rules are established regarding a map's maintenance time; however, a number of agencies develop annual statistics based upon maps. Generally, maps are maintained for short-term (quarterly) and long-term (annually) periods for each type of crime. This decision should be based on the volume of criminal activity, physical-space limitations, the specific mapping technique adopted, and the number of maps to be maintained. As geospatial-terrain data becomes available, mapping techniques can be fully integrated into the MPACS.

B-59. Stored map data can be easily retrieved for comparing data. Pin maps can be photographically recorded and digitized for software applications. Computers and scanning equipment can be used to store the data for later retrieval. Color-coded paper dots placed on acetate overlays covering area maps facilitate the retrieval of recorded data. Digitized photographs of the maps are excellent sources of evidence for trials and court-martials. Moreover, the developed maps may contribute to the CCIR.

B-60. The number of maps to be maintained must be based on physical-space limitations, staffing, the types of crimes selected for analysis, and the maintenance period for each map. There are five types of crime suited to geographical analysis (burglary, robbery, housebreaking, auto theft, and sex crimes). Other crimes (such as theft of government property) may be depicted as a geographical analysis and contribute to RII. Housebreaking and burglaries can be placed on the same map by using different colored pins for each type of crime. The incorporation of numbered pins of different colors can be used to designate the type of premise attacked. Auto-theft and recovery maps may be kept on the same map with recorded information on thefts of vehicle parts.

B-61. The number of maps maintained should reflect the needs of the agency according to the volume of reported crimes and the different data elements or informational factors to be recorded. Therefore, the more detailed the analysis, the more maps are required. The actual number of maps maintained must be left to the discretion of the PM or the commander.

B-62. Periodically, a summary of the information that is collected, countermeasures, and operational/tactical procedures are briefed to the installation commander. Whether this is done monthly, quarterly, or semiannually will depend on the volume of reported crime and the commander's preference.

CRIMINAL-ANALYSIS SUMMARY

B-63. Criminal analysis is an important element to the MP function of PI. The techniques mentioned in this appendix are not all-inclusive. The PM/CID commander establishes formal procedures for integrating data and developing PI. It may require the formation of an ad hoc team, or it may be accomplished with computers and the MPACS. Criminal analysis requires close coordination with the G2/J2, the public affairs office (PAO), civilian police, and others based on METT-TC.

SECTION III — COMMAND AND LAW-ENFORCEMENT COUNTERMEASURES

B-64. This section discusses actions that commanders or PMs can take to reduce crime on Army installations. In many cases, the actions described have applications other than strictly crime prevention. However, only the crime-prevention aspects are discussed.

B-65. Most of the data available on the effectiveness of law-enforcement measures in reducing crime comes from studies conducted by civilians. While most of the findings of these studies are applicable to crime-prevention programs on Army installations, differences in population, the degree of control that can be exercised by authorities, and other environmental factors may dictate that the civilian recommendation be modified before implementation on military installations.

CRIME HOT LINES

B-66. A crime hot line is a dedicated crime-reporting telephone number located at the MP or security desk. This hot line allows anyone in the community to make an immediate report of an observed crime or suspicious activity. An effective crime-reporting program is a deterrent to crime and enhances law-enforcement responses to such incidents. Considerations in implementing a reporting program include the following:

- It should be publicized that personnel reporting incidents are allowed to remain anonymous if they desire. Some individuals will report their observations only if they know they can remain anonymous. If deemed feasible, neighborhood-watch blocks could be provided designated block numbers that could be used when reporting crimes or other suspicious activities. This system would allow neighborhoods to receive feedback on the disposition of the reported incident.
- The hot line's phone number should be easy to remember. This could include a number where extension digits are all the same, are in ascending or descending order, or spell out a word (such as 4357 [HELP]). Sticker labels listing the number could be placed on the phone with other emergency numbers for quick reference.
- The program should be well-publicized.
- Members on the installation should be educated on the desired procedures for reporting incidents.
- Each call should be documented. Records should be maintained concerning the results of these calls to evaluate the program's effectiveness.

CRIME PREVENTION THROUGH ENVIRONMENTAL DESIGN

B-67. Crime-prevention practitioners are recognizing the importance of considering design and physical planning in crime reduction. The crime-prevention officer has an opportunity to influence the design of facilities through the installation planning board. However, to be effective, he must understand a number of concepts about the relationship between the physical design of buildings and crime occurrences. These include the concepts of territoriality, natural surveillance, and defensible space.

TERRITORIALITY

B-68. Historically, a single-family home on its own piece of land and somewhat isolated from its neighbors (but often by as little as a few feet) has been considered to be the family's territory. The single-family home sits on a piece of land buffered from neighbors and the public street by intervening grounds. At times, symbolic shrubs or fences reinforce a boundary. The positioning of lights in windows that look out on the grounds also act to reinforce the claim.

B-69. Unfortunately, as the population has grown and the need for housing has increased, the trend toward developing single-family units has been paralleled, if not surpassed, by the development of row houses, apartment buildings, and various high-rise structures. Architects, planners, and designers involved in developing structures have not paid a great deal of attention to crime control or the need for an individual or a family group to identify with its home in a manner that might affect crime. Therefore, most families living in apartment buildings consider the space outside their apartment door to be distinctly public. In effect, they relegate responsibility for all activity outside the immediate confines of their apartment to public authorities. A question is whether environmental design can be used to extend the boundaries of these private realms, subdividing public space outside quarters so that more of the common space comes under the resident's influence and responsibility.

B-70. Through extensive research of efficiently functioning housing developments, a number of mechanisms have been identified that may be used in the design process (or may be added after construction). These mechanisms encourage the residents of multifamily dwellings to identify more with the ground or area around their immediate home site and to assume responsibility for its protection. Presented below is a brief discussion of a number of the mechanisms.

- **Site design.** If the grounds around a set of quarters can be directly identified with a particular building and the residents of that building take a personal interest in the use or upkeep of that area, they will play a role in protecting it. Through proper site design, a recreational area adjoining a building may be used as a buffer zone by providing play equipment for young children and seating areas for adults. The fact that children play and adults sit in these areas serves to increase the residents' concerns with the activities taking place there. Strangers are usually recognized and their activities come under observation and immediate questioning.
- **Street design.** Research has shown that by the placement, enclosure, or rerouting of streets and traffic the nature of a particular area can be changed and the crime rate reduced. For example, a particular portion of a street might be closed to vehicular traffic, and play equipment and seats may be added. In a number of areas where this technique has been used, it has been found that most residents know or at least recognize people up and down the block and strangers on the street are identified. Similar approaches that involve rerouting traffic, using one-way streets, or blocking off streets has lowered the crime rate in some areas.
- **Symbolic barriers.** The types of barriers that planners may use in laying out an area include open gateways, light standards, low walls, and plantings. Both physical and symbolic barriers serve the same purpose—to inform an individual that he is passing from a public to a private space. Symbolic barriers identified by residents as boundary lines serve as defining areas of comparative safety. Many places warrant the use of symbolic barriers, including transition points between a public street and the semipublic grounds of a building; an area between a building's lobby and its corridors; or hallways on particular floors of a building.
- **Internal design.** Although economics may sometimes enter the picture, a building's interior may be designed for specific groupings of apartment units and shared entrances. These factors may cause the residents of these apartments to develop a concern for the space immediately adjacent to their dwelling. For example, on each floor of an apartment building, two to four families might be required to share

a common corridor area. The apartment doors would be grouped around that common corridor, and access to elevators or stairs might be screened by a glazed partition. The net effect would be that the floor's residents would adopt the corridor as a collective extension of their dwelling unit and would take an increased interest in its maintenance and use.

- **Facilities and amenities.** The location of particular facilities (such as play and sitting areas and laundry facilities) will tend to give an area a high intensity of use and support the idea of territoriality. The presence of residents involved in various activities (children at play and people chatting or engaged in other types of activities) allows for casual surveillance by concerned members of the family and screens out possible intruders.

B-71. Reducing the number of apartment units grouped together to share a collectively defined area and limiting the number of buildings that comprise a housing project are important factors for creating an environment that residents will help to protect. Research has documented the fact that housing projects comprised of fewer high-rise buildings (two to four) have lower crime rates than projects containing a larger number of buildings. Based on this finding, it is argued that there appears to be much less freedom of movement in the public spaces of the smaller high-rise projects. Unlike buildings and large developments, every building of a small grouping usually has an entrance directly off a public street. These dwellings more closely resemble middle-income, high-rise developments and look more private.

B-72. As a crime-prevention officer, you may not be in a position to directly use these techniques. However, your familiarity with these approaches and the value of their use in the crime-prevention process are important elements in your arsenal of tools to create public involvement in reducing crime. In particular, the purpose of outlining these tools is not to equip you to be a designer, but rather to equip you to communicate with those who are involved in that profession. The discussion that follows will further enhance your ability to converse with designers.

NATURAL SURVEILLANCE

B-73. Experience has shown that the ability to observe criminal activity may not be adequate to stimulate an observer to respond with assistance to the person or property being victimized. The decision to act depends on the presence of motivational conditions, including—

- The degree to which the observer has developed a sense of personal and property rights that are being violated by the criminal act.
- The degree to which the observer feels that the event is within his area of influence.
- The observer's ability to clearly identify whether the act is unusual for the particular area.
- The observer's identification with either the victim or the property being vandalized.
- The degree to which the observer believes he can effectively alter the course of events he is observing.

B-74. Based on these conditions, a number of mechanisms have been identified that can be used to design the grounds and internal areas of apartment units, housing developments, and other residential areas to facilitate natural monitoring of activities taking place. By providing opportunities for surveillance through the positioning of windows in relation to stairs, corridors, or outside areas, continual natural observation will be maintained and crime will be deterred. If such steps are taken, the security of observed areas will be understood by the potential criminal, making him think twice before committing a crime.

B-75. The first of these natural surveillance mechanisms involves the positioning of service areas and access paths leading to apartment buildings to facilitate surveillance by residents and authorities. For example, buildings might be designed so that their entries face and are within 50 feet of a street, so that well-lit paths lead to the front door or the lobby, and so that the lobby is arranged to afford good visibility from the street. Other related steps focus on the strategic placement of windows, fire stairwells, lobby lights, and mailboxes so that they can be easily viewed from the street. Elevator waiting areas on each floor can also be designed so that they can be seen from the street level. Research has proven that if steps such as these are taken, residents will be more likely to become involved with protecting the facility, MP patrols will be in a better position to observe what is going on, and criminals will be discouraged from vandalizing the site.

B-76. A second technique that might be used to increase surveillance is to design facilities so that people within them will naturally view commonly used paths, entries, and play and seating areas during their normal household activities. This concept also focuses on the strategic placement of windows, lighting, and open areas so that natural surveillance by residents is improved.

B-77. Another mechanism involves the subdivision of housing areas into small, recognizable, and identifiable groupings that improve visual surveillance possibilities. Research has shown that in housing developments where the surveillance of a neighbor's outside activities was possible, residents were found to be very familiar with everyone's comings and goings. The overall effect was to cement collective identity and responsibility through social pressure.

DEFENSIBLE SPACE

B-78. Defensible space is a term for a range of combined security measures that bring an environment more under the control of its residents. A defensible space is a residential environment that can be used by inhabitants for the enhancement of their lives while providing security for their families, neighbors, and friends. The physical mechanisms suggested to create safety and improve upkeep (as part of the defensible-space concept) are self-help tools wherein design catalyzes the natural impulses of residents rather than forcing them to surrender their shared social responsibilities to any formal authority.

B-79. Research has revealed investigative techniques that might be used to modify existing housing areas to make them more secure. The following methods may require alteration or adaptation to the particular situation on your installation:

- Widening major pathways and using colored decorative paving.
- Differentiating small private areas (front lawns) outside each dwelling unit from the public path with low, symbolic walls.
- Adding public-seating areas in the center of public paths far enough from private-dwelling units to eliminate conflicts of use but close enough to be under constant surveillance by residents.
- Designing play areas as an integral part of open space.
- Adding new and decorative lighting to highlight various paths and recreation areas at night and extending the residents' surveillance potential and feeling of security.
- Adding seats and path networks to recreational facilities where large, central court areas exist. This increases the interest and usability of the areas.
- Redesigning parking and play areas around buildings to create the illusion that the buildings are grouped where natural opportunities exist.
- Modernizing building entrances to create breezeways into building courts and to accommodate a telephone intercom for opening entry doors to the lobby.
- Providing video surveillance of public grounds and central paths by security of public monitors.
- Installing audio surveillance capabilities in elevators and at the doors of residences.

CRIME-PREVENTION MODEL

B-80. The model for crime prevention through environmental design is based on the theory that action must be taken to counter crime before it occurs. The critical element in this model is the environmental-engineering component. It provides both direct and indirect controls against criminal activity by reducing the opportunity for crime through science and technology and the use of various urban planning and design techniques. The model explains what environmental engineering is and how it supports crime prevention. With this information, you may be in a better position to understand and respond to questions and discussions on how urban design and planning can have an impact on the installation's criminal element.

THE ENVIRONMENTAL INFLUENCE ON CRIMINAL BEHAVIOR

B-81. The basic theory that supports crime prevention through environmental design is that urban environments can influence criminal behavior in two ways. First, the physical surroundings in which people live have an effect on each individual. These physical characteristics include noise, pollution, overcrowding, and the existence and unmonitored spreading of refuse and other unsightly waste. The second element that must be dealt with in the environmental-engineering formula concerns the social characteristics of the community that provide individuals with social relationships to which they must respond. Characteristics such as alienation, loneliness, anxiety, and dehumanization are seen as keys to criminal behavior.

B-82. In terms of these environmental characteristics, buildings are all too often constructed to be dangerous, with corridors and passageways hidden from public view. Elevators, basements, and storage and washroom areas are also laden with danger due to their design. Various large-scale housing developments are not secure in that they are often isolated from the main flow of traffic (both human and automobile) and are closed to public use and public view.

B-83. With regard to altering the social characteristics of the community and their relationship to criminal behavior, it should be recognized that behavior is future-oriented, not past-oriented. A man steals so that he can have a car or money in the future, not because in the past he experienced psychic trauma, a broken home, poverty, or delinquent associates. Criminal behavior can be explained directly in terms of the consequences of behavior and in terms of noncriminal variables such as poverty, race, or social class. Criminal behavior is viewed as a problem to be dealt with and not symptomatic of other problems (such as poverty, mental conflict, class conflict, unemployment, or undereducation). To change criminal behavior, it must be dealt with directly by removing the environmental reinforcement that maintains the behavior. The approach advocated is to change the environment to which the individual responds.

ACTION APPROACHES TO CRIME PREVENTION THROUGH PHYSICAL PLANNING

B-84. The primary focus of crime prevention has been on what architects, planners, and other nonpolice professionals can do in terms of various physical-planning strategies to reduce criminal opportunity. Experienced MP personnel have long recognized that certain physical conditions can contribute to the rate and nature of crime. They have also developed a capability to identify high crime-risk locations by noting such factors as poor lighting and weak points of entry as potential targets. The critical job is to identify specific areas concerning physical planning and design that can be responded to and actions can be taken against on the installation.

B-85. Attempting to reduce crime or the fear of crime by regulating physical environments is easier said than done. In fact, although crime prevention can be built into almost every aspect of community planning, it is often ignored for a number of reasons. For example, fragmentation of responsible agencies is a key problem. In addition, crime has historically been looked upon as the exclusive responsibility of MP forces; not of those in charge of education, housing, or health and welfare. Yet, with an understanding of crime prevention combined with the knowledge that design techniques can change the opportunity for criminal behavior, PSIs will be able to talk the language of the planner and the designer and to be able to advise them from a police perspective.

B-86. It is notable that a number of civilian police agencies have become involved in the physical-planning process and have achieved notable results from their work. For example, the Fremont, California, Police Department has been involved in a planning process and maintains that law enforcement should become an integral part of the master- or comprehensive-plan review to screen all redevelopment plans for safety and crime hazards. Working with other units of municipal government as well as architects and designers, the department drew up a set of model guidelines for the evaluation of projects. The model included evaluation criteria dealing with such subjects as the accessibility of buildings to patrol units; traffic flow and off-street parking provisions; and the location and regulation of cul-de-sacs, playgrounds, common greens, fences, and security entrances. In addition, working with agencies such as the American Institute of Architects, the National Public Works Association, the Association of Public Utilities, and others, the department identified a number of subjects that are of specific concern to police officers and that should be considered in the design and planning stage. As a result of these efforts, the following list of design concerns was developed by the department:

- Building setbacks (front, side, and rear).
- Wall construction, interior and exterior (industrial, commercial, and residential).
- Door construction, setbacks and security (industrial, commercial, and residential) (including carports, garages, and sliding-glass doors).
- Windows and skylights, setbacks, heights (from ground), show-window displays, and the type of frame or pane.
- Stairs (stairwells and staircases).
- Balconies.
- Utility boxes.
- Fences, walls, hedges, screens, setbacks, heights, and louvers.
- Parking (public and private).
- Lighting (industrial, commercial, and residential).
- Streets, sidewalks, and walkways (locations, slopes, curvature, grades, and the length of a block).
- Alleys (blind and through alleys).
- Visibility of valuables (people, safes, cash registers, and personal property).
- Signs (street signs and signals, traffic signs and signals, and advertising signs).
- Accessibility; approach, entrance, and exit (pedestrian, vehicular, services, residential, commercial, and industrial).
- Public utilities and easements (gas, water, telephone, and electrical).
- Public areas and facilities (public restrooms, parks, bus stops and shelters, playgrounds, recreation halls, and so forth).
- Street trees and shrubbery (types, heights, and locations).

B-87. With this information, you can improve the security aspects of the community's physical-planning process. There is a probability that the work and recommendations of the National Advisory Commission on Criminal Justice Goals and Standards may help you in your efforts to get involved in the design process. More specifically, the Commission noted that "every police agency should participate with local planning agencies and organizations, public and private, in community physical planning that affects the rate or nature of crime or the fear of crime."

B-88. The future role of MP forces in crime prevention through physical planning will depend on the PSI's initiative. The perspectives and knowledge of what is happening in this field, combined with a working knowledge of the language, should equip PSIs to sell this approach as part of the overall program. It is important to point out that others who represent professions other than crime prevention are aware of the relationships between urban planning and crime. Remember, they deal in concepts, approaches, and ideas that have not involved the realities that law-enforcement personnel face.

SPECIALIZED PATROL TACTICS AND SURVEILLANCE

B-89. Specialized patrol operations use a variety of tactics in attempting to control identified crime problems. The most common tactics include uniformed tactical patrols and suspect and area surveillances. The following paragraphs discuss these tactics in terms of their target crimes, operation requirements and characteristics, and established or perceived levels of effectiveness.

B-90. The appropriateness of a given tactic depends on the characteristics of a particular crime problem. The selection of specialized patrol tactics should be made on the basis of a careful and continuous analysis of crimes. Most crimes can be addressed by more than one tactic. Several tactics might be tried in an effort to find the best one, and it is quite possible that the most effective approach to a given crime problem will include the combination of several tactics.

UNIFORMED TACTICAL PATROLS

B-91. A uniformed tactical patrol is the most traditional and widely used form of specialized patrol. It is a simple, straightforward approach to specialized patrol that involves the same procedures and techniques used by MP officers on routine patrol. These include constant visible movement throughout an area to generate a sense of police presence, careful observation of street activity, vehicle and pedestrian stops, and citizen contacts. The difference between uniformed tactical patrols and routine patrols are that uniformed tactical patrols use these tactics in an intense, concentrated fashion. MP officers are relieved of the responsibility for responding to routine calls for services so that they can devote their full time and attention to patrol, thus intensifying its impact. In addition, uniformed tactical operations typically deploy a number of MP officers in target areas, thereby increasing the level of patrol in these areas.

B-92. Uniformed tactical patrols can be used to control virtually any type of suppressible crime (for example, crimes that can be viewed from locations where the police have a legitimate right to be and those that can be potentially affected by police operations). These suppressible crimes include street robberies, purse snatches, vehicle thefts, burglaries, and housebreakings. Uniformed tactical patrols can also have an impact on other types of crime as officers use observation, field interrogation, and citizen contacts to develop information on the locations, activities, vehicles, and associates of suspects.

B-93. The primary purpose of uniformed tactical patrol is deterrence. This tactic is based on the assumption that highly visible, active patrols will deter potential offenders. By increasing the perceived probability of apprehension, conspicuous patrol is thought to reduce the likelihood that crimes will occur. If the deterrence should fail, heightened patrol coverage is believed to increase the probability of the immediate apprehension of the suspects.

B-94. Uniformed tactical patrols are often used to saturate an area that is experiencing a particularly serious crime problem. Although it has been widely used for years, saturation patrol has never been clearly and adequately defined. Exactly what level and intensity of patrol constitutes saturation has never been determined, nor have the effects of different levels of patrol been clearly established. It is difficult to prescribe the level of uniformed tactical patrols that should be used to disrupt a crime pattern in a particular area. This should be determined through an analysis of the size and characteristics of the area of concentration of each potential target crime pattern coupled with an assessment of manpower availability.

B-95. Some patterns can be effectively handled by a very small increase in the patrol level. For example, in Portland, Oregon, Operation Crime Reduction Involving Many People (CRIMP) had a significant impact on street robberies and assaults in the city's skid-row area by initiating two-officer, uniformed foot beats in the area. This was sufficient to saturate the primary locations of the target crimes during the high-crime hours, and it led to a substantial reduction in these crimes with little apparent spillover into adjacent areas.

B-96. Other departments have used saturation patrols on a much larger scale. For example, in the mid-1950s, the New York City Police Department attempted to saturate an entire precinct by assigning over 200 additional officers to the precinct's patrol force. Foot beats covering extremely small areas were arranged in straight lines so that an officer could keep the entire street area of his beat in view at all times. The four-month experiment led to a dramatic reduction in crime in the precinct. Compared with the same period in the previous year, street muggings fell by 89.9 percent, burglaries where entry was made from the front of the building dropped by 78 percent, and so on. The only crime category that was not affected was the relatively private crime of murder. Since crime displacement was not examined in the experiment, its true impact remains unknown. The experiment strongly suggests that massive additions of police manpower can have a substantial effect on crime. The problem is that most departments do not have the ability to conduct even a limited version of this experiment.

B-97. The amount of resources required for saturation patrols can vary tremendously, and there is no definite way of determining how much additional patrol is needed. This can best be decided on a problem-by-problem basis, using experience and evaluations of past efforts as a guide. As a rule, patrol augmentation should be sufficient to affect rather quickly the perceptions of would-be offenders concerning the level of police activity in a particular area. It appears that for such a strategy to be effective, sufficient resources should be applied in a limited area to ensure a true saturation effect.

B-98. Uniformed tactical patrols can use various modes of transportation. Patrol cars are most often used; however, foot patrols can be effective in congested service districts and bikes have been used to good advantage in high-density residential areas. The mode of transportation should be selected based on visibility, mobility, and access to citizens.

B-99. Some specialized units deploy MP officers in unmarked police cars. This is done in an effort to effect a balance between overt and covert operations, hopefully gaining many of the advantages of both. Unmarked cars may also be readily available since in many departments investigators work primarily during the day, which leaves their vehicles free for specialized patrol in the evening and early morning watches. This approach has serious drawbacks. First, unmarked MP cars are somewhat less visible than marked cars, yet they are still easily recognizable as police vehicles to large segments of the public (especially when the officers in them are in uniform). Second, the use of these cars in uniformed tactical patrols could lead to the sacrifice of some of the deterrent effects of high visibility without realizing the apprehension benefits of truly covert patrols.

B-100. Several patrol techniques have been tried to increase the visibility of uniformed patrol and to enhance the sense of police presence. A tandem patrol uses two marked patrol cars that follow each other at intervals of one-half to several blocks. Two units can also patrol on parallel streets one block over or in an alley. Another approach combines foot and vehicle patrols in an effort to increase visibility. Officers park their marked cars in conspicuous locations in high-crime areas and then are transported to other high-crime areas where they patrol on foot. The frequent repetition of this procedure is seen as a way of multiplying patrol visibility.

B-101. Unless a target area is more or less completely saturated (as in the New York experiment), MP patrols should move in a random, unpredictable pattern. This will make it difficult for potential offenders to plan their crimes on the basis of observation of patrol activities. One department found that its uniformed-tactical-patrol operation was actually helping burglars to commit break-ins. Interviews with confessed burglars indicated that they were aware of a visible patrol passing through the neighborhoods at regular intervals and they planned their crimes accordingly.

B-102. In addition to increasing the level and visibility of patrol in high-crime areas, uniformed tactical patrols often use aggressive patrol tactics involving frequent vehicle and pedestrian stops. MP patrols stop, question, and perhaps search citizens when there are reasonable grounds for suspecting that the citizens may have committed, may be committing, or may be about to commit a crime. Since the concept of reasonable suspicion is vague, MP officers have a considerable amount of discretion in conducting field interviews. Field interviews that do not result in either immediate arrest of the citizen or in alleviation of the officer's suspicions are usually documented in field-interview cards or spot reports. Field interviews serve to generate information about the activities of probable suspects; more importantly, they make the suspects aware that the police know of their presence in a given area, regard them as suspicious, and are watching them closely. This is expected to reduce the likelihood that they will commit crimes, at least in the area in which the tactical force is working.

B-103. The extensive use of field interviews is often highly controversial. Tactical units that emphasize field interviews have been accused of being storm troopers who constantly harass citizens and abuse their rights. There is sometimes a certain amount of truth to the allegations of harassment, especially when the activities of a particular type or group of suspects are being closely monitored. Not surprisingly, individuals who are stopped and questioned frequently are likely to complain, particularly if they have reason to be concerned about close police scrutiny. A recent study of field interviews in San Diego found that when conducted at moderate levels, field-interview activities do not have a major effect on police and community relations. The majority of citizens in all three study areas accept field interviews as legitimate and properly conducted police activity. The majority of all citizens who were the subjects of contacts felt that the contact was justified and properly conducted. The study found that the suspension of field interviews in a particular area was associated with a significant increase in the level of suppressible crime and when reinstated in the area, these crimes declined. The results of the San Diego study provide confirmation for the widely held belief that field interviews can have an important deterring effect on suppressible crime without doing irreparable damage to police and community relations.

B-104. The potentially negative impact of field interviews on certain segments of the community can be held to a minimum if the interviews are conducted in a professional manner. Citizens should be informed of the reasons why they are being stopped. They should be detained for as little time as possible and should not be subjected to verbal or physical abuse. There is also no need to stop everything that moves.

B-105. While experience and a limited amount of research indicate that uniformed tactical patrols can have a positive impact on the level of suppressible crime in areas, the overall effectiveness of this tactic is a controversial and much-debated issue. The principal concern is that rather than reducing crime, uniformed tactical patrols may simply lead to its displacement from one area to another or from one period to another.

B-106. When conducting uniformed tactical patrols, a PM activity should carefully monitor crime trends for indications of possible displacement effects. This is an important aspect of evaluating the impact of tactical operations, and it will provide information to guide future deployments and tactical decisions.

AREA SURVEILLANCE

B-107. Covert patrol and surveillance of high-crime areas can be used to make apprehensions for crime problems. These problems include those for which there are no suspects who warrant personal surveillance, the suspects are too numerous to permit personal surveillance, and there are too many potential targets to conduct either physical or electronic stakeouts. Examples of these types of problems would be a rash of residential burglaries or auto thefts in a particular area.

B-108. This tactic simply involves the covert patrol of a particular area and the observation of suspicious or unusual activities and occurrences that might indicate the likelihood of a crime. Suspicious individuals are not stopped but are watched until they either commit an offense or the officers' suspicions are removed.

B-109. The list of various techniques that can be used in an area surveillance is virtually endless. The following are some techniques that have been effectively used by specialized patrol units:

- Mingling with citizens at the crime scene to pick up information on possible suspects.
- Maintaining rooftop surveillance of a shopping center's parking lot to locate larcenies from vehicles.
- Surveilling housing areas by posing as maintenance workers.
- Following likely crime victims such as elderly citizens leaving a bank.
- Surveilling rooftops for unusual activity from aircraft or higher buildings. Binoculars are used to facilitate surveillance, and rooftops are marked so that street units can be dispatched to check out suspicious circumstances.

B-110. As in all types of plain or uniformed patrol, care should be taken to ensure that area surveillance is truly covert. Rental vehicles that can be changed frequently provide an excellent, though expensive, means of covert transportation.

B-111. Police on covert patrol should be dressed to blend in with the environment in which they are working, and they should have apparently legitimate, nonpolice-related reasons for being where they are. Several specialized units have found that surveillance teams composed of one male and one female officer can work in many situations without arousing suspicion. An apparently married or romantically involved couple lingering in a park, meandering slowly down the street, or sitting together in a parked car would generally appear less suspicious than two male MP officers doing the same things. Finally, it should be noted that in some small neighborhoods where residents know each other well, covert surveillance may be difficult, if not impossible, since the presence of any stranger arouses immediate curiosity and suspicion.

SUMMARY

B-112. These tactics represent the basic approaches that specialized patrol operations normally take in trying to control suppressible crime. Some of the tactics (such as uniformed tactical patrols) are directed primarily at crime deterrence, while others (such as suspect surveillance) are used to achieve apprehensions for target crimes. The tactics are most commonly used independently of one another. However, there are some indications that the combined use of several tactics in an integrated operation might be an effective way of coping with particular types of crime. Especially promising is the coordinated use of highly visible and covert patrols. A visible patrol force could be deployed to a particular area to deter crime there and direct it toward other areas in which MP forces using covert tactics are working. To date, efforts to direct criminal activity to areas or targets where MP forces are set up to make apprehensions have only been tried on a sporadic basis. However, this appears to be a promising approach to crime control and warrants greater attention in the future. It can be viewed as the creative use of crime displacement.

B-113. Many of the tactics mentioned previously could not be effectively used by MP patrols that are responsible for handling routine calls for service. It is obvious that the effectiveness of decoy operations, physical stakeouts, and suspect surveillance would be destroyed if the MP officers had to interrupt these activities to handle calls for service. It is also generally unwise to have MP officers in civilian clothes respond to calls for service.

B-114. The importance of crime analyses in identifying crime problems and suggesting appropriate tactics deserves repetition. Specialized patrol units should be deployed to address clearly identified crime patterns, and they should choose their tactics on the basis of an analysis of these patterns and a comprehensive knowledge of the area of occurrence. The nature and characteristics of a particular crime pattern should be the driving force behind the selection of tactics. None of the tactics of specialized patrol can be effectively used unless crime patterns have been identified and analyzed.

PUBLICITY CAMPAIGNS

B-115. Public-information campaigns are an essential part of every crime-prevention effort. The installation PAO can provide assistance in—

- Informing the public of the magnitude of the local crime problem.
- Disseminating information on crime circumstances.
- Generating the interest and enthusiasm necessary to initiate and sustain community crime-prevention programs.

B-116. The PAO should be a member of the installation's crime-prevention council and must be involved from the start in planning crime-prevention efforts. There are many information vehicles at the installation level available to carry the message to the public, including—

- Digital marquees.
- Radio.
- CCTV.
- Installation and unit newspapers.
- Posters and leaflets.
- Commanders calls and similar meetings.
- Town-hall meetings.

B-117. Not all media are equally effective in reaching a particular segment of the post population. To ensure that the media campaign is effective, identify the segment of the population requiring the information and the exact message to be communicated. The more specific the target audience, the more effective the media campaign will be.

B-118. Whatever media is used, it is important to provide coordinated input to the PAO well in advance of the desired publicity campaign date to allow for writing, rewriting, and publishing. For material to be published in a magazine, a minimum lead time of six months is normally required. For installation newspapers, a lead time of a week may be sufficient. Check with the PAO to determine the correct lead times for local publications.

B-119. When addressing civic groups or school audiences, films are often more effective than straight lectures because of the animation and other special techniques used to illustrate critical points. Several Army films are available as are a large number of commercial films.

RESIDENTIAL-SECURITY SURVEYS

B-120. A security survey is an in-depth, on-site examination of a physical facility and its surrounding property. The survey is conducted to determine a residence's security status, to identify deficiencies or security risks, to define the protection needed, and to make recommendations to minimize criminal opportunity. Because of several common characteristics, one expert has likened the security survey to the traditional criminal investigation. This comparison hinges primarily on the facts that both techniques are systematic in nature; are aimed at identifying the method of a criminal act; and are, in effect, more an art than a science. It should be recognized, however, that the survey has two other advantages. First, it can be conducted before the commission of a crime; second, it can offer protection against, rather than just remedial action after, criminal victimization.

THE RESIDENTIAL-SECURITY SURVEY AS A CRIME-PREVENTION TOOL

B-121. A residential survey is not a substitute for an aggressive neighborhood watch. It supplements these efforts and should be established after Neighborhood Watch and Operation ID programs are established.

B-122. The security survey is the primary tool used in crime prevention to recognize, appraise, and anticipate potential loss in residential areas. It is often defined as the backbone of a local crime-prevention program. In practice, it combines the security experience, training, and education of the local crime-prevention officer and focuses on a single element—the analysis of a residential facility.

B-123. The inherent value of surveys has been proven by nearly 300 civilian police departments across the country that have established crime-prevention bureaus or units. An even broader endorsement of the survey technique, however, was provided by the National Advisory Commission on Criminal Justice Standards and Goals, which stated that, "Every police agency should conduct, upon request, security inspections of businesses and residences and recommend measures to avoid being victimized by crime."

B-124. In short, the security survey is a tool that informs a homeowner of the particular areas in which his home is susceptible to criminal victimization along with steps that can be taken to reduce and minimize that potential. Further, the survey is a tangible document that reflects the efforts of the MP force not only to be responsive to community needs, but to get the community more directly involved in the criminal-justice process.

THE CRIME-PREVENTION SURVEY—A PROCESS OF INVESTIGATION

B-125. An important factor in understanding the residential-security survey is that it must be considered as an ongoing process. While a particular survey will result in specific recommendations, each survey will provide a foundation for future action. In combination, these surveys will provide a database that can be used in the analysis of the community's crime problems and guide action toward the resolution or reduction of the problem on a community-wide basis.

B-126. As a starting point, five steps must be used in carrying out the actual survey, while four additional steps must be remembered afterward. These steps are—

- Analyzing the overall environment (neighborhood, block, and so forth).
- Assessing the general vulnerability of the premises.
- Defining the specific points of vulnerability.
- Recommending specific security procedures.
- Including specific remedial hardware recommendations.
- Urging the implementation of the recommendations.
- Conducting a follow-up to ensure that recommendations have been implemented.
- Keeping crime statistics to evaluate the survey's effect and the implementation of recommendations.
- Conducting a second survey of the premise's statistical analysis to determine the alteration in criminal activity in the areas surveyed.

B-127. This process is the continued involvement and participation of the police with the community. Once the survey is completed, the job is not finished. In fact, if this posture is assumed, it may later be learned that the recommendations were not implemented and that the work was done in vain. This can easily lead to a loss of community-wide public confidence in the programs. Similarly, even in those cases when recommendations are implemented, additional crime might be experienced. Attention to this fact, coupled with immediate follow-up, will be essential to avoid losing the confidence gained from the original survey. Through prompt actions, the proposal of additional tactics may become the final step needed to substantially reduce criminal opportunity. In short, the security-survey process is not a one-shot operation and it is not to be looked upon as a simple, straightforward task. It is a continuing, difficult, rigorous, yet effective approach to reducing crime that has not been systematically applied by police agencies in this country.

THE ROLE OF POLICE IN IMPLEMENTING CRIME-REDUCTION PROGRAMS THROUGH SECURITY SURVEYS

B-128. At a minimum, there are two ways to encourage people to improve their personal security. First, you can organize, conduct, and participate in broad-based public-information and -education programs that make use of such media as radio, television, and the press. Second, you can organize, undertake, and follow up on a series of person-to-person security surveys. Clearly, both of these techniques have their advantages. The security survey has a unique quality that does not exist in the public-education program. It provides an added incentive on the part of citizens to implement recommendations because of the personal relationships and respect established during the actual survey by a crime-prevention officer.

B-129. For example, consider a homeowner listening to someone on television recommending improvements of security by installing better locks or alarm systems. You might ask yourself these questions: How secure am I? Are the locks that I have on my doors adequate? If they are inadequate, what kind of locks should I install? Do I really need an alarm system? What can the MP officers do for me in terms of making recommendations about security? In short, you would be aware of the possible need for improved security, but you would not know enough of the specifics to warrant real action. Because home security is a complicated matter requiring careful analysis and the installation of carefully tailored systems, implementation must be approached in a personal way. Certainly, public-education programs can assist in some aspects of crime prevention (such as prevention against schemes, auto theft, and personal protection), but they are not as effective in causing improved security as is personal contact.

B-130. Moreover, for the first time, police are placed in a position where they can actually provide a different kind of advice or service that can be offered in an environment where a crisis has not yet occurred. This is a truly unique opportunity for crime-prevention officers.

CONDUCTING THE SURVEY

B-131. To develop a proper perspective of the types of crimes that a PSI will most frequently be trying to reduce, a review of the cases is necessary to get a broader feel for the actual conditions in your area. During this review process, pay specific attention to photographs in the files. Study crime scenes in an effort to identify the type of security device that was defeated. In particular, if a door was used as a point of entry, note whether it was of adequate construction; if the door frame was broken or separated; if hardware, such as strike plates and door trim was inadequate; and so forth. In addition, review photos to determine if lock cylinders were pulled or if door chains fastened to trim moldings were simply pulled away to permit easy entrance. If photographs of the crime scenes are not available, visit as many crime scenes as possible. While doing so, photograph security risks you can study later and use as examples in future presentations to community groups.

B-132. By becoming familiar with the MO of persons committing such crimes as burglary and larceny, you will be better-equipped to understand potential risk situations and to point them out to potential victims. Quite surprisingly, many of the cases investigated were invited by some obvious crime-risk hazard that was overlooked by a resident. Also, many additional crime risks existed at a particular crime scene that a burglar could have exploited. Such vulnerability might be an indication of other crime targets within the community to which you should pay particular attention in your survey work.

B-133. To be an effective security surveyor, it is necessary to develop an intimate knowledge of the crime factors in your community. You do not have to become a statistician; however, the more you know and understand about crime problems, the methods used in your community, and security devices or systems that were defeated, the better you will be equipped to analyze the potential crime when surveying a home.

B-134. In addition to the general types of crimes that occur in your community, it is necessary to develop an understanding of the details of particular types of offenses. For example, with regard to residential housebreaking, you should be familiar with the types of burglaries and approaches used in particular sections of the community.

B-135. On the surface, it might appear as if this would be a monumental task. However, in terms of housebreaking, you might pull the files for the last quarter and carry out the following steps:

- Tally the number of times entry was made through the front door, rear door, or through a window.
- Identify the approach used for entry (kicking the lock, throwing a shoulder through the door, jimmying the lock, and so forth).
- Determine how often a window was broken or removed or a mechanism was used to force the latch (when entry was made through a window).
- Attempt to determine whether security devices were used in residences (such as alarms, special lighting, or other systems).
- Identify the burglar's general escape route (down a back alley, through a schoolyard, and so forth).

B-136. In developing an understanding of the MO of crimes, statistics that illustrate exactly what is happening in your community will be valuable tools. Not only will you be able to use this information in explaining crime risks while you are surveying the site, but it will be invaluable in making public presentations.

B-137. Only when you have developed the ability to visualize the potential for criminal activity will you have become an effective crime-scene surveyor. It is important that when you arrive on a survey site, you are prepared to give a property owner sound advice on the type of security precautions he should consider.

B-138. In summary, to be a good crime-prevention surveyor, you have to be a good investigator. You must understand the criminal's method of operation and the limitations of standard security devices. In addition, you must be knowledgeable about the type of security hardware necessary to provide various degrees of protection. (Specific information on conducting surveys is contained in Chapter 11).

JUVENILE CRIME PREVENTION

B-139. Law enforcement has long recognized the importance of establishing personal contact with youth. This need was first met with the "Officer Friendly" concept. The goal of this concept was to enhance the image of police among younger children and attempt to negate the unfavorable image other segments of society provided children. This concept has evolved into today's Drug Abuse Resistance and Education (DARE) program which has spread to most law-enforcement agencies.

B-140. Crime-prevention officers can have an impact on juvenile crime by establishing a positive interaction with juveniles. Many areas suffer a high crime rate of which over 50 percent is attributable to juveniles. Nationally, it is known that youth 17 years of age and under (who make up just 16 percent of the population) commit 42 percent of the crimes that cause injury or loss of property. As the punitive aspects of the juvenile justice system today are of questionable value, it is logical that crime-prevention officers should develop programs aimed at juvenile crime.

B-141. The basic problem in establishing any juvenile crime-prevention program is the destruction of preconceived ideas by both youths and law-enforcement personnel. This obstacle can only be overcome by the building of an honest rapport with individual youths. Just as a crime-prevention officer must sell crime-prevention procedures to the public, he must also sell himself and his program to the youth.

B-142. One method of establishing the needed rapport with youth reverts to a similar system used in the Officer Friendly concept. Classroom presentations in high schools and junior high schools can be used to destroy the preconceived ideas held by youth regarding police and law enforcement in general. Once an MP officer enters the classroom, he discovers that the majority of students obtain their ideas of police one of two ways—

- From television shows that portray every officer as a slow-witted moron (such as the old "Car 54" show).
- From their peers at school who relate every negative aspect of law enforcement.

B-143. However the students gain their information, it seldom resembles fact and must be changed to effectively establish a juvenile crime-prevention program.

B-144. In making classroom presentations for crime-prevention purposes, the crime-prevention officer must display honesty at all times. When confronted by uncomfortable questions, the MP officer should render the law-enforcement's point of view and make a truthful explanation. Obviously, some answers will not always be well received, but they will establish the credibility of his presentation. Further, when making these presentations, many areas dealing with specific areas of crime prevention will arise (such as rape prevention, property engravement, residential and apartment security, vehicle theft, and burglary prevention).

B-145. In dealing with schools, the crime-prevention officer can establish many positive contacts with individuals inside the school system that will prove to be of great value. Individuals such as administrators, counselors, and faculty members offer the crime-prevention officer assistance in many different ways. Perhaps the most valuable assistance is in the area of establishing a juvenile crime-prevention program that is based on peer influence.

B-146. A crime-prevention program aimed at peer influence is essential to a crime-prevention unit. Peer influence is directly related to the Determination Theory of Delinquency Causation. Simply stated, the theory holds that delinquency occurs as a result of external influences on youth. While attempting to control all external influences on youths is an impossible task, the control of peer influence is possible and perhaps the most important external influence to be controlled.

PEER-INFLUENCE CRIME-PREVENTION PROGRAM

B-147. A peer-influence crime-prevention program can be established after the crime-prevention officer establishes himself in the junior/senior high schools as described earlier. The basic concept of the program is the negation of undesirable peer influence on predelinquent youths and the establishment of a positive peer influence.

B-148. The peer-influence program functions within the school district. A problem youth is detected at the junior high school level by his teacher, counselor, or the crime-prevention officer. When it is believed that the peer-influence program could serve a youth, his counselor arranges a conference with him. During the conference, the counselor attempts to determine the basic problem experienced by the youth and has him complete an application for admittance into the program.

B-149. When the application is complete, the counselor contacts the youth's parent or guardian and explains the program. If the parent agrees to allow participation, the application is sent to the parent for his signature.

B-150. When the completed application is returned to the counselor, it is forwarded to a senior high school counselor for review. If necessary, the senior high school counselor may contact the junior high school counselor personally to discuss the youth in more detail. The high school counselor then selects a suitable senior high school student who has been accepted for the peer-influence program and matches the student with the junior high school youth.

B-151. Once the match has been made, the senior high school student is given a briefing on the junior high school student by the crime-prevention officer. The senior youth is instructed to contact the junior youth (preferably at the junior youth's home) and introduce himself to the youth and to the youth's parent or guardian. After the initial contact has been made, the senior youth is instructed to plan activities with the junior youth and spend as much time as possible with him.

B-152. Each senior youth that volunteers for the program is required to complete an application. This application gives personal background information and references that are thoroughly checked by the crime-prevention officer. The youth's academic record is examined, but prime consideration is given to citizenship factors rather than grades. The crime-prevention officer personally interviews each applicant.

B-153. When a senior youth is selected for the program, he is given a basic outline to follow. He is warned not to preach to the junior youth but to lead him by allowing him to observe the senior youth's actions at different activities. The senior youth must devote the necessary amount of time to limit the junior youth's opportunity to continue the negative associations he had in the past.

B-154. While the peer-influence program is simple in nature, the results can be remarkable. During a three-year period in Irvin, Texas, approximately 600 students participated in the program. From this number, seven arrests were made from the group with one juvenile accounting for four of the seven arrests. As about 30 percent of the juveniles participating were legally judged delinquent, the relapse rate was officially established as 0.1 percent.

B-155. While juveniles judged as delinquent can participate in the program, the crime-prevention officer should concentrate his efforts on those juveniles identified as predelinquent by counselors or faculty members. Again, the personal contact of the crime-prevention officer and the school faculty is essential to the program's success. A personal rapport should be cultivated and maintained.

B-156. Operationally, the peer-influence program can be administered by the collective efforts of school personnel and MP officers who are committed to the program. Once the program is established, it will probably not require the services of an officer on a full-time basis (depending on the number of students involved). However, constant contact is necessary; the more time that can be devoted, the more success will be apt to follow.

CRIME-PREVENTION PRESENTATIONS FOR ELEMENTARY AND JUNIOR HIGH SCHOOLS

B-157. As the traditional American family structure has experienced a significant decline over recent years, more and more emphasis is being placed on schools to provide guidance for children. As a crime-prevention officer who is frequently invited to speak to students, you can play an important role in assisting the school to provide this guidance. To stress the importance of the above statement, many experts in child behavior believe that the school is second only to the family in molding a child's socialization and behavior. Many people feel that the school is increasingly replacing much of the family socialization process. For this reason, schools are the most logical environment to begin a juvenile crime-prevention program.

B-158. Crime-prevention officers can have an impact on the most common types of juvenile crime by preparing specific lectures for particular grade levels in the elementary and junior high schools. Classes discussing shoplifting, vandalism, and bicycle theft are well received by students and have a wide-reaching effect on the community as a whole due to the children relaying the information to their parents.

B-159. A number of films are currently on the market that assist in making crime-prevention presentations in the classroom. In addition to specifically addressing juvenile crimes, the films directly relate to the peer-influence causation of crime.

OVERVIEW OF SELECTED JUVENILE CRIME-PREVENTION PROGRAMS

B-160. In addition to the programs described above, there are a variety of other juvenile programs. You will notice that some of the programs are designed to offer activities for the juvenile while at the same time attempting to provide one or more of the following:

- Positive peer influence.
- Education.

- Police/juvenile cooperation.
- An understanding of law-enforcement and MP duties.

B-161. One program deals with the enforcement aspect, with the goal being truancy reduction. Also mentioned are programs designed for child safety and victimization reduction.

DARE Program

B-162. DARE is a drug-abuse educational program that gives children and parents the skills to recognize and resist the subtle and overt pressures that cause them to experiment with drugs, alcohol, and violence. This unique program uses uniformed law-enforcement officers to teach a formal curriculum to students ranging from kindergarten to twelfth grade. DARE gives special attention to fifth through ninth graders to prepare them for entry into middle/junior high and high school where they are most likely to encounter pressures to use drugs or alcohol.

Police Explorer Programs

B-163. Explorer programs are run in conjunction with the Boy Scouts of America and a sponsoring police agency. The programs offer a wide range of flexibility in that they may be structured along the lines of a police cadet program with all activities centered around law-enforcement training. This type of explorer program usually seeks juveniles who have an active interest in pursuing a law-enforcement career. Being an explorer in this type of program will offer the student the opportunity to view the duties of a police officer and to take part in minor police activities (such as traffic direction and observation riding in patrol cars). The selection and training of the students for this type of program strictly depend on the sponsoring agency.

B-164. Other police-sponsored explorer programs may be designed to provide the members with other nonpolice activities (such as hiking, camping, and canoeing). At the same time, this type of program will offer the student the chance to meet with police officers and learn more about police functions, though this is a secondary purpose. Selection for this program would not be as stringent as the above-mentioned program. A local council office of the Boy Scouts can provide additional information on this program.

Alateen

B-165. The Alateen program is run in cooperation with the Alcoholics Anonymous organization. This program's intended purpose is to provide assistance for a child's emotional needs if he or she is faced with family problems caused by alcoholism of one or both parents. Your local chapter of Alcoholics Anonymous can provide more information on this program.

Truancy-Enforcement Program

B-166. Various cities have successfully adopted truancy-enforcement programs that focus attention on identifying and apprehending youths that are on unexcused absence from school. The primary goal behind this program is to prevent or reduce the incidence of daytime residential burglary. The San Angelo, Texas, Truancy Enforcement Program selected a target area in an effort to see how effective the program would be in reducing daytime burglaries. The initial evaluation of the program showed positive results. The San Angelo program uses uniformed crime-prevention officers in marked police cars patrolling around the school and known hangouts. All street personnel in the patrol division are also encouraged to check school-age individuals. Those students under 17 years of age who are suspected of truancy are taken to the school office while those over 17 years of age are detained only long enough to obtain their name and other identifying information, which is then turned into the school office. The assistant principal in charge of attendance completes the investigation and takes the appropriate action. The action varies (depending on the circumstances) but can include counseling with the student, the teacher, and the parents or, in extreme cases, suspension for a short period.

B-167. The emphasis behind these programs is to inform youths of the facts of the law. It was shown that many of the problems with the youthful offenders would be eliminated if they were presented the facts of the law to which they are subject. The courses can be adapted to fit the needs and desires of the individual school system, and the extent of the areas covered will depend on the commitment and availability of the local MP officers. Various civilian school districts have incorporated these law-enforcement courses as part of their regular curriculum.

Criminal Victimization Reduction

B-168. Programs designed toward this goal normally are structured to reduce the possibility of a child becoming a victim of a child molester. Most commonly, officers give a verbal presentation to groups of children using visual aids that help to act as a positive reinforcement to learning. Some effective methods have been the use of "talking traffic lights," "talking bicycles," and "talking motorcycles." These items are usually equipped with a tape-recorded message that can be controlled by the officer conducting the program. The talking traffic light uses a full-size traffic light that flashes red or green according to the type of person that is being described.

VANDALISM

B-169. Today's vandals often attack their own territory. School vandalism—the illegal and deliberate destruction of school property—is committed by students themselves. They break so many windows that in large districts the funds spent annually on replacing broken windows could pay for a new school. Vandals destroy about $3 million worth of school-bus seats annually, and they commit enough arson to account for 40 percent of all vandalism costs.

B-170. School vandalism outranks all other assaults on private and public property. At the end of the 1973 school year, the average cost of damages from vandalism was estimated at $63,031 per school district. That figure could have paid the salaries of eight reading specialists or could have financed a school breakfast program for 133 children for one year. A typical school's chance of being vandalized in a month are greater than one in four, and the average cost of each act of vandalism is $81. Yet, these figures do not include the hidden costs of school vandalism—increased expenses for fencing, intrusion and fire detectors, special lighting, emergency communications equipment, and vandal-resistant windows. Every dollar spent in replacing a window or installing an alarm cannot be spent on education.

B-171. School vandalism can also have enormous social cost. The impact of an 89-cent can of spray paint used to cover a wall with racial epithets far exceeds the monetary cost of removing the paint. An abusive word scrawled across a hallway wall can destroy student morale, disrupt intergroup relations, undermine the authority of an administration, or even close the school. Incidents with high social costs damage the educational process as much as those with high monetary costs. Today's vandal is not a hardened, war-scarred veteran. Instead of grizzled whiskers, he sports peach fuzz. He is almost literally the boy next door. In fact, the typical vandal differs quite dramatically from the typical juvenile delinquent.

B-172. It is significant that vandals fall into a well-defined and relatively narrow age group. They are usually early adolescent males who are highly subject to group pressures and transitory impulses. It is not at all unusual for adolescents to act out whatever is controlling them at the moment—rage, boredom, pent-up energy, or the sheer joy of "wreckreation." While there are conditions that may predispose or provoke a youth toward vandalism, the problem seems to be almost human nature. Few among us have never written on a sidewalk or scrawled initials on a school desk. Vandalism cuts across all strata of society, all geographic regions, and all racial lines.

B-173. The causes of vandalism remain obscure. Though research addressing the "why" of vandalism is growing, it has yet to yield clear-cut answers. Among the motivating factors often cited are anger, frustration, hostility, bitterness, alienation, futility, inequality, restricted opportunity, emotional pain, failure, prejudice, revenge, and the need for attention. Although much of the research is convincing, the fact remains that many vandals do not appear to be among the most angry, frustrated, hostile, alienated, or needy youth. Only a small fraction of the youngsters who fall into that category actually commit acts of vandalism. So, while most experts agree that vandalism is not totally senseless, they do not claim to fully understand its causes. In fact, vandalism is often not understood by vandals themselves. Many vandals report that they do not know why they did it. Many others, according to case reports, offer the unsolicited observation that destruction is fun. Still others express satisfaction and exhilaration. Few consider themselves criminals. For the time being, we can conclude only that motives for vandalism are diverse. But the whys notwithstanding, the vandal profile suggests that our task is, in large part, to anticipate and redirect the impulses of young teenagers.

B-174. Schools are by no means the helpless victims of early adolescence. Many school factors, most of which are amenable to change, influence the amount of vandalism that schools experience. The following characteristics are typical of schools that suffer high property damage or loss:

- Vandalism is higher when there is poor communication between the faculty and the administration (such as when the principal fails to define policy or makes policy decisions unilaterally).
- Hostility and authoritarian attitudes on the part of teachers toward students often result in students "taking it out" on the school.

- Limited contact between teachers and students reduces student involvement with the school and increases the likelihood of vandalism.
- Schools characterized by intense competition for leadership positions suffer greater property damage and loss.
- The chances for vandalism increase when the students do not value their teachers' opinions of them.
- Schools at which students strive to get good grades experience more vandalism.
- Parents of students in high-damage schools express less favorable attitudes toward their schools than do other parents.
- The school is a convenient target for vandalism when it is close to students' homes.
- The chances for vandalism are increased when grades are used as a disciplinary tool.
- Damage is greater in larger schools where there is more property to destroy. This correlation between school size and vandalism prevails regardless of whether the school is located in an urban, suburban, or rural setting.
- Fewer offenses occur when rules are well understood by students and are consistently and firmly enforced by teachers and administrators.

Vandalism Prevention

B-175. If the special problems of early adolescence, often intensified by social or personal pressures, interact with school conditions to produce vandalism, then preventive measures must address the nature of both the child and the school. Furthermore, prevention must include both physical and human responses. At present, most vandalism-prevention or -reduction programs rely on physical security—bigger and better electronic alarm systems, patrol guards, dogs, tamperproof locks, and window grilles. These techniques help, but they address only 20 percent of the problem—those incidents involving breakage. These incidents usually occur when school is not in session and in the absence of witnesses. The techniques have little effect on the day-to-day trashing of the school or on the disruptive acts aimed at the school's routine (bomb threats, the setting of fires, and false fire alarms) that are committed during school hours. The most sophisticated physical and electronic barriers are not sufficient to keep vandals from what they consider an attractive target. In fact, it has been argued that alarms and armed guards, besides lowering student and staff morale, often themselves become a challenge, inviting rather than deterring vandals. Vandalism prevention requires not a narrow or piecemeal approach, but a varied and comprehensive effort that includes both physical and human components geared to the school's specific problems. Furthermore, an effective long-term program must involve the community, parents, neighbors, police, and civic groups as well as students, teachers, and school administrators.

B-176. Schools are an easy target for vandals. Most are public, secular, and often unoccupied. Most will remain public and secular; but they need not remain unoccupied, unprotected, or unobserved. The following are techniques that have made some schools less vulnerable to vandals. These are especially effective against problems occurring during nonschool hours.

B-177. **Occupy the School.** Employ a custodial force around the clock. In most schools, the entire custodial force works at one time, leaving the school at night. As an alternative, custodians can be assigned staggered shifts so that the school is always occupied. Twenty-four-hour custodians are particularly appropriate in schools suffering sporadic property damage that demand more than a roving patrol but less than permanent security guards.

B-178. Invite police to use the school buildings at night. Police can be issued keys to the schools in their patrol areas so that they can use school offices to write their reports. This places a police officer in the school when it might otherwise be unoccupied, and it places a police car in front of the school.

B-179. Bring the community into the school. The school is an excellent place for recreational programs; health clinics; adult-education classes; counseling centers; community gatherings; and Boy Scout, Girl Scout, and Parent-Teacher Organization (PTO) meetings. The presence of people in the school building not only reduces the opportunity for vandalism, but also stimulates community and student interest in the school.

B-180. **Watch the School.** Use school neighbors as eyes and ears. Ask nearby homeowners to watch the school and report suspicious activities. Emphasize careful observation and rapid reporting, but discourage direct involvement in any situation observed. Such programs work best if they are organized but based on informal involvement, if they are accompanied by overall involvement of parents and community with the school, and if they offer some sort of prestige to participants.

B-181. **Employ Roving Patrols.** A uniformed patrol used instead of or in conjunction with an alarm system can deter vandalism. The individuals who patrol should establish rapport with neighborhood youths and open communication with community leaders. They should also vary their patrol patterns.

B-182. Hire student patrols during the summer and on weekends. The school district or community can provide its youth with part-time or summer employment and, at the same time, curb vandalism by paying students to patrol the school grounds during weekends, holidays, and summer vacations. These students should be paid an adequate wage and considered an integral part of the school's security force.

B-183. **Control Access to the School by Using an Alarm System.** Alarms are the most expensive vandalism control measure a school can use. While they can detect vandals, they cannot apprehend them; they can merely signal the alarm-system monitor, which may be miles away. They can, however, be an efficient deterrent and should be considered as part of any comprehensive plan to control vandalism. If alarm systems are linked with a surveillance camera, the chances of apprehending intruders are greatly increased.

B-184. **Design the School With Vandalism Prevention in Mind.** The following designs for preventing vandalism can be implemented when building a school:

- Limit ground-to-roof access.
- Eliminate low, overhanging roofs.
- Avoid unnecessary exterior fixtures.
- Plant trees that cannot be climbed near buildings.
- Consider raising as much of the school building as possible from ground level.
- Build the school at some distance from residential areas. While it should be located near the homes of most of those it serves, it will suffer less property damage if there is a buffer zone between it and surrounding residential areas.
- Design the school with plenty of defensible space so that the normal flow of school traffic allows continuing, casual surveillance of the premises.
- Use vandal-resistant surfaces. Use harder surfaces in damage-prone areas. For walls, use epoxy paint or glazed tiles that are easily and inexpensively replaced or repaired; use small wall panels and keep replacement panels in stock; and place permanent signs, building names, and decorative hardware at a level that cannot be reached from the ground. Replacing damaged areas immediately shows a sense of pride in the appearance and helps to eliminate copycat acts of vandalism.
- Plan windows carefully. Avoid windows that are vulnerably placed. Use small panes of glass to simplify replacement; use thick, tempered glass, thick acrylic, or Plexiglas® for windows in heavily traveled or hangout areas. Avoid useless windows in student stores, administrative offices, and industrial-arts storage areas.
- Plan entries with multiple uses in mind. Install flexible internal gates to block off specific areas or corridors when necessary. Provide separate exterior entries for community and student use. Inside the building, create areas for informal gatherings near entrances and exits by installing soft-drink machines and benches.
- Locate or relocate the playground's access roads to provide better surveillance by roving patrols.
- Consider outdoor lighting. Opinions on this issue are divided. Many schools report a decline in vandalism after installing hardened exterior night lighting. Others report that elimination of all night lighting reduces vandalism, presumably because young adolescents are afraid of the dark. If lighting is used, it should be directed away from windows to keep vandals from seeing the process of destruction or its outcome.
- Channel graffiti. Graffiti artists will usually select light, smooth surfaces rather than dark, rough surfaces. Therefore, school officials can channel graffiti onto one or two walls designed to withstand such treatment. Students or maintenance staff can paint most walls at regular but not too frequent intervals. One wall can be officially designated a "legitimate" graffiti wall, though this approach removes some of the challenge inherent in informal graffiti.

B-185. **Make the Target Less Attractive.** The school is not only an easy target but also an attractive one. It represents enforced learning, discipline, and mandatory attendance to students who are, simply by virtue of their age, rejecting adult standards and striving to achieve independence from adult control. Students are additionally provoked if the school functions in an impersonal, undemocratic, repressive, or manipulative manner, further increasing the likelihood of vandalism.

B-186. **Revise the Curriculum.** The list of characteristics associated with vandals and vandalized schools indicates that property damage and loss are higher when competition for rewards is intense, the availability of rewards is limited, or the distribution of rewards is unfair. All of this suggests that school policy and atmosphere have a direct bearing on school vandalism. Below are changes in school governance that can help remove the features that make a school an attractive target for vandals. These procedures, especially in combination, have proven effective against all forms of vandalism, including those that most commonly occur while school is in session.

- **Alternative schools.** Though originally designed to perform educational functions, alternative schools have proven effective in reducing school violence and vandalism. They provide an option to students who are not benefiting from the regular academic classroom. These schools operate within or alongside the traditional school. They are characterized by low student-to-teacher ratios, intense individual counseling, and extensive use of the community as a learning resource. They offer an alternative to suspension, personalize the learning environment, and provide successful experiences to students who have performed poorly in the past.
- **Law-related education and police-school liaison programs.** In many communities, the police department has assigned school resource or liaison officers to local junior and senior high schools. These officers may on occasion assume policing duties, but their primary function is to counsel students and teach law-related courses. These activities acquaint students with the positive role that law plays in our society and personalize the relationship between the "cop on the beat" and "the kid on the corner."
- **Action learning.** This term refers to apprenticeship programs as well as training in practical aspects of adult life. The former allows students academic credit for community work (such as tutoring and assisting physicians, lawyers, or other professionals). The latter provides instruction in skills such as checkbook balancing, comparative shopping, and applying for a job. Both address the boredom and frustration that are linked to truancy, violence, and vandalism.
- **Reward distribution.** The school's reward structure is related to school crime. Although the school may offer attractive incentives, most students receive very little in the way of rewards. Many of those who lose out still care about the competition. They become frustrated, and they vent their anger on the apparent source of their problems, the school. It should be possible to spread the rewards around without compromising performance standards, perhaps by recognizing improvement as well as achievement or by acknowledging forms of achievement other than scholastic, athletic, and social.

B-187. **Change Administrative Policy and Organizational Structure.** According to the National Institute of Education's Safe-School Study, the principal's leadership is a critical factor in reducing or preventing school crime. Strong principals who are visible, available, and responsive to students and staff appear to be most successful in eliminating violence and vandalism. Those who are less successful are often described as unavailable and ineffective.

B-188. The Safe-School Study also found that the exercise of discipline through clear enunciation and even-handed enforcement of rules is associated with a low incidence of school crime. To increase the likelihood that students will find school a fulfilling experience, many districts are establishing minischools—schools small enough to allow the individual student to feel significant. Similarly, several large schools are currently functioning on a house basis—the school is divided programmatically into several smaller units with which students can more easily identify.

B-189. **Involve the Students.** Establishing a vandalism fund uses peer pressure to the school's advantage. The community or the school district puts aside a certain amount of money and announces that the funds will be used to cover the costs of vandalism. Any money left over reverts to the students to be used as they choose. This plan works because it educates students about the costs of vandalism, allows them to see the positive results of curbing property damage and, most important, gives them full responsibility for the problem.

B-190. Several school districts have established voluntary student security advisory councils that conduct workshops and small group discussions focusing on vandalism and violence prevention. These committees increase awareness of the school's problems, generate recommendations for action, and give students opportunities to participate in school decision making.

B-191. Older students can be helpful in influencing younger students. In several communities, junior and senior high school students visit fourth- and sixth-grade classrooms where they show a film about vandalism and then lead a discussion on the causes and consequences of vandalism.

B-192. School beautification projects involve students in the care of the school building and grounds in an attempt to increase their pride in the responsibility for the school. The more effective projects are selected by the students themselves and continue throughout the year. The projects focus on marginal students rather than school leaders.

B-193. **Involve the Parents.** An open-door policy allows parents to go directly to their child's classroom whenever they wish without securing a special visitor's permit from the office. This simple policy offers parents concrete evidence that they are indeed welcome at school.

B-194. In some schools, parents serve on antitruancy committees along with teachers and students. They visit youngsters and their families in an effort to resolve truancy problems. In other schools, parents serve as hall monitors, supervise extracurricular activities, and help in classrooms. Parents are also the school's best source of guest speakers and contacts for work-study or apprenticeship programs. Some school districts have initiated faculty men's clubs to acquaint fathers with the male teaching staff and to encourage them to assume more responsibility for their children's progress in school.

B-195. **Involve the Community.** In some communities, students and law-enforcement, school-district, and city personnel sponsor daylong forums on vandalism. The forums introduce citizens to the causes and costs of vandalism and give them an opportunity to voice their concerns and initiate preventive programs.

B-196. Police departments can initiate public-relations programs within schools and youth-service agencies. In addition, they can enlist the help of youth in preventing vandalism through police-sponsored groups, such as the Police-Youth Service Corps, which recruits adolescents from high-crime areas to work as public-safety aides. Similarly, law students from neighboring universities can be brought into the schools at minimal or no cost to discuss the legal implications of vandalism.

Picking up the Pieces

B-197. The preventive measures listed on the preceding pages can function as part of a long-range, proactive response plan. However, they do not address the question of immediate response. What should the school do, right away, about 20 broken windows, a cherry bomb in the toilet, or recurring racial graffiti on the wall? The answer is to repair the damage immediately. Quick repair keeps perpetrators from admiring their handiwork, lessens the epidemic effect of vandalism, and minimizes any social impact the act may have.

B-198. Initiate formal record-keeping procedures, and ensure that they are followed. Schools faced with serious problems should begin recording all acts of vandalism. They should also consult law-enforcement personnel about when police should and should not be called. When a school begins to have problems, it should work with the juvenile justice system so that the two institutions can coordinate their efforts with regard to school-age offenders.

B-199. Careful record keeping allows a school to plot the incidence of vandalism to find out precisely where and when each type of offense is occurring. For example, using incident analyses, the National Institute for Education's Safe-School Study found that—

- Fire and bomb threats most often occur on Tuesdays.
- School-property offenses tend to occur with greater frequency toward the end of each semester, especially in November and December.
- Break-ins and school-property offenses occur most often on weekends and Mondays.

B-200. This type of information is invaluable in planning a vandalism reduction and prevention plan.

Restitution

B-201. Restitution is a set of legal and administrative procedures through which the school receives payment from vandals for damages they cause. While it seems reasonable to require payment for damages, restitution does not appear to be worth the effort. In the first place, most vandals are not caught. In the Los Angeles School District (which has an aggressive restitution program), only 30 percent of the offenders are ever identified.

From this 30 percent, most restitution is paid before matters get to court. Going through lengthy legal processes to obtain the rest is simply not cost-effective. However, a parent faced with the possibility of a court case may make a greater effort to keep his or her child out of trouble.

CONCLUSION

B-202. The goal of crime prevention is to reduce crime through public awareness and education. The skills that crime-prevention officers acquire can eventually benefit all segments of our communities. It is logical then for crime-prevention officers to attack one of the major crime problems in our society—juvenile delinquency. Efforts of crime-prevention units will certainly be lacking if the juvenile-crime problem is not given a high priority and if juvenile-crime programs are not established.

FRAUD

B-203. Two types of fraud can be affected by crime-prevention efforts—fraud against the government and fraud against individuals in the Army. Since the countermeasures for each of these categories of fraud are different, they will be discussed separately.

FRAUD AGAINST THE GOVERNMENT

B-204. Fraud of this type is a loss to the Army due to manipulation of systems from within the government with criminal intent. Typical examples of fraud are—

- The diversion or theft of government property by falsifying documents such as purchase orders, shipping documents, and so forth.
- False claims for temporary-duty (TDY) pay, for reimbursement for losses due to the movement of household goods, or for reimbursement for material reported as stolen.
- Overcharges or underproduction on contracts with the Army.
- Bribery.
- Kickbacks to secure purchase orders.
- Use of one's official position for personal gain.

B-205. This list is not all-inclusive. There are many more ways to defraud the government. Some of the factors that make it easy to perpetrate a fraud are—

- The lack of independent verification of records, transactions, and reports.
- The lack of adequate supervision.
- Unrealistic budgeting and acquisition requirements.
- Failure to correct deficiencies identified by existing systems.
- Concentration, at the operational level, of responsibility and authority for an entire process in one individual.

B-206. There are built-in measures to discourage fraud for most Army material-control systems. These programs are routinely examined by the inspector general (IG) along with other command inspectors. A crime-prevention officer's main function is to encourage the reporting of fraud by ensuring that the community understands what type of activities should be reported and by ensuring that the numbers are posted in work areas where personnel are in a position to detect fraud.

FRAUD AGAINST INDIVIDUALS

B-207. The second type of fraud is con games or consumer fraud. This crime is perpetrated by professional con men, unscrupulous businesses, and amateurs who see a chance to make a fast buck. Most of the time, the operator offers a "something for nothing" deal. By the time the victim realizes that he has been duped, the con man is long gone. Frequently, the victim is so embarrassed that he has been duped that he does not report the crime. The only way that a crime-prevention officer can help prevent this crime is to publicize common con games and illegal business practices and to encourage victims to report them. When a case of fraud occurs, the specific con technique that was used should be advertised widely to prevent others from being fooled by the same ploy.

Repair Fraud

B-208. Repair frauds are simple to execute but difficult to detect. Some unscrupulous repairmen do not fix the problem, but they charge you anyway. Some use inferior parts; others charge you for work that you did not expect or need. Some also do "insurance" work; they will repair one thing but ensure that something else will soon go wrong. The following are methods that you can use to protect yourself from repair fraud:

- Shop around. Ask friends, neighbors, or coworkers for references. When you find repairmen you trust, stick with them.
- Do not try to diagnose the problem yourself unless you are an expert. The mechanic may take your advice, even if it is wrong. If you know exactly what the problem is, do not tell the mechanics; wait and see if their recommendations agree with your diagnosis. That way you will know whether needless repairs are suggested.
- Try to get several detailed written estimates before any work is done. Compare job descriptions and materials to be used. Be sure to ask if there is a charge for an estimate.
- Ask for the old parts to ensure that replacements were really installed.
- Ensure that you get a guarantee on any work done.
- Make sure that the work was done before you pay. Take your car for a test drive. Plug in the refrigerator. Test the television.
- Ask your local consumer affairs office about the laws in your state and what specific protection it gives regarding professional services to be licensed or certified.

Home-Improvement Fraud

B-209. Home repairs and improvements can be costly. Be cautious if somebody offers to do an expensive job for an unusually low price, if a firm offers to make a free inspection, or if the workers "just happened to be in the neighborhood." These are the favorite tricks of dishonest home-repair firms.

Some firms offer a price you just cannot resist. Once you sign the contract, you learn why—they never deliver the service. Others send door-to-door inspectors to do a free roof, termite, or furnace inspection. These free inspections will turn up plenty of expensive repairs. Some fly-by-night companies will offer to do the work on the spot. When they leave, you may be left with a large bill and a faulty repair job. To avoid home-improvement fraud, try the following:

- Get several estimates for every repair job. Compare prices and terms. Check to see if there is a charge for estimates.
- Ask your friends for recommendations, or ask the firm for references and check them.
- Check the ID of all inspectors.
- Call the local consumer affairs office or the Better Business Bureau to check the company's reputation before you authorize any work.
- Be suspicious of high-pressure sales tactics.
- Pay by check, never with cash. Arrange to make payments in installments—one-third at the beginning of the job, one third when the work is nearly completed, and one-third after the job is done.

Land Fraud

B-210. Real estate can be a great investment. The enterprising real estate salesperson knows how anxious you are to find just the right property, especially for an investment or a retirement home—a nice warm climate, not too crowded, or a new development. Some dishonest agents will promise you anything—a swimming pool, country club, or private lake—to get your name on the contract. Even if the sales agent promises you luxury, they may not guarantee the basics such as water, energy sources, and sewage disposal.

B-211. Most land developers offering 50 or more lots (of less than 5 acres each) for sale or lease through the mail or by interstate commerce are required by law to file a Statement of Record with the Housing and Urban Development Administration (HUD). This document tells you almost everything you need to know about your future home—legal title, facilities available in the area (such as schools and transportation), availability of utilities and water, plans for sewage disposal, and local regulations and development plans. All of this information must be given to you in a property report prepared by the developer. Always ask to see this report before you sign anything.

B-212. If the developer does not give you a copy of the property report for the lot you are considering, you can obtain it from HUD for a fee. Send requests to: Department of Housing and Urban Development, Office of Interstate Land Sales Registration, 451 Seventh Street South West, Washington, DC 20410.

Door-to-Door Sales

B-213. Beware of the following door-to-door sales tactics:

- "Nothing like it in the stores!" This is a true statement. The vacuum cleaners in the stores are probably of better quality and come with a better warranty.

- "Won't find this price anywhere." This is also a true statement. The prices in the stores are probably lower.
- "Easy credit!" This is another true statement. They do not care what your credit rating looks like. Once you sign for the purchase, paying for it is your problem. Be wary of low monthly payments. Find out the total amount you will pay over the life of the loan, then subtract the actual cost of the item itself. The difference is what you will pay in interest. Your bank, your credit union, or a local legal-aid society can tell you if the interest rate is fair.

B-214. Be cautious of these words, and be firm if the salesperson pressures you to buy. If you do get trapped, you are protected by a Federal Trade Commission regulation. When you make a purchase in your home totaling $25 or more, the salesperson must give you a written contract and two Notice of Cancellation forms. You have three days to change your mind and use one of these forms to cancel your contract.

Charity Fraud

B-215. Charity fraud does a lot of harm. The swindler takes advantage of a person's good will and takes his cash—money that was meant for people in need. You can ensure that any money you give gets into the right hands. Remember these pointers when somebody asks you for a donation:

- Ask for ID, either the organization's or the solicitor's. Find out what the charity's purpose is and how the funds are used. Ask if contributions are tax deductible. If you are not satisfied with the answers, do not give money.
- Give to charities that you know. Check out those you have never heard of or others whose names are similar to a well-known charity.
- Do not fall for high-pressure tactics. If solicitors will not take no for an answer, tell them no anyway, but do not give them your money.
- Be suspicious of charities that only accept cash. Always send a check made out to the organization, not to an individual.
- If a solicitor reaches you by telephone, offer to mail your donation. Shady solicitors usually want to collect quickly.

Self-Improvement Fraud

B-216. Con artists know that everyone wants to look better, feel better, and be a better person. Selling worthless plans and cures is one of the easiest ways for them to make a "quick buck." The following ads can look tempting:

- "Miracle reducing plan."
- "Look like a model in only five days."
- "Learn to speak Spanish while you sleep."
- "You can have a new, dynamic personality."

B-217. What can you do? Be careful! Read the small print. Know what the product contains. You should check with your doctor before you embark on any diet or exercise program.

Medical and Health Fraud

B-218. Most of us do not know much about medicine; that is why we go to doctors. It is also why we fall for miracle cures and other phony health products and services. Patent medicines, health spas, and mail-in lab tests should be warning signs for the potential consumer. A laboratory in Texas advertised nationally that it had perfected a fail-safe urine test for cancer. More than 15,000 tests were made at $10 each before authorities stopped this fraudulent company.

Unsolicited Merchandise

B-219. Cagey con artists will send you a gift in the mail—a tie, a good luck charm, or a key chain. What do you do with it if you did not order it? If you are the kind of person they are looking for, you will feel guilty and pay for it, but you are not obligated to. If you—

- Have not opened the package, mark it "return to sender." The post office will send it back at no charge to you.
- Open the package and do not like what you find, throw it away.
- Open the package and like what you find, keep it, free of charge. This is a rare instance where the rule of "finders, keepers" applies unconditionally.

B-220. Whatever you do, do not pay for it. Look at your gift as an honest-to-goodness way of getting something for nothing. Do not get conned if the giver follows up with a phone call or a visit—by law the gift is yours to keep.

Mail Fraud

B-221. The following are examples of mail fraud:

- **The contest winner.** "You've won! This beautiful brand-name sewing machine is yours for a song. To claim your prize, come to our store and select one of these attractive cabinets for your new machine. Bring this letter with you and go home with a new sewing machine for next to nothing." Treat an offer like this carefully. Shop around before you claim your prize. Chances are, the cost of the cabinet will be more than the machine and cabinet are worth.
- **The missing heir.** You have just received a very official looking document. The sender is looking for the rightful heirs to the estate of someone with your last name. It could be you. To find out, just send $10 for more information. There may be thousands of people with your last name, and letters like these are often mailed nationwide. Even if there really was claimed estate, it is highly unlikely that you would be an heir. Save your money; why help a swindler get rich?

B-222. These are just two examples of mail fraud. Many of the other frauds described in this section can be handled through the mail. When they are, the US Postal Service can launch a full-scale investigation. If you think you have been cheated in a mail-fraud scheme—

- Save all letters, including envelopes.

- See if your neighbors or business associates received the same material.
- Contact your local postmaster who can direct you to your regional Postal Inspector's Office.

SUMMARY

B-223. The following are recommendations to avoid being duped:

- Do not believe something-for-nothing offers. You get what you pay for.
- Be suspicious of high-pressure sales efforts.
- Take your time. Think about the deal before you part with your money.
- Get all agreements in writing. Insist that agreements are made in "plain English," not "legalese."
- Read all contracts and agreements before signing. Have a lawyer examine all major contracts.
- Compare services, prices, and credit offers before agreeing to a deal. Ask friends what their experiences have been with the firm or service in question.
- Check the firm's reputation with your local consumer affairs office or the Better Business Bureau.

INTERNAL THEFT

B-224. Employee theft is a major problem in most large organizations. It is estimated that one-third of all business bankruptcies are a result of theft. Of course, in the Army there is no danger of the organization going out of business; however, waste and internal theft can divert significant amounts of critical resources from mission-essential activities. Unlike many businesses, the Army has recognized that internal losses can be a problem, and most areas have adequate controls mandated by regulation. As crime-prevention officers, we should monitor reports of survey, physical-security inspection reports, and results of crime-prevention inspections to ensure that control measures are being followed. When it is apparent that the correct measures are not being followed, the problem must be identified for the senior commander's action.

B-225. Research has shown that an organization's atmosphere is just as important as management controls and physical security in preventing employee theft. To control this problem, the following items are essential:

- Employees, both civilians and soldiers, must believe they are part of a professional organization that expects superior performance from all of its members. When second-rate work is accepted, the lax attitude carries over into property-control procedures and losses increase. Standards must be high, but fair.
- Leadership must set the example. If leaders, supervisors, and managers take advantage of their positions to use government material or services for their own benefit (even in minor ways), their employees will also feel justified in diverting Army resources to their own use.

- The organization must show a genuine concern for the problems of its personnel. If an employee feels he has been treated unfairly, it is easy for him to justify stealing from the organization. He believes that he is only taking what he would be getting if the organization was fair.
- The organization must take appropriate disciplinary or administrative actions in cases of theft. There can be no acceptable level of internal theft. If employees believe that an activity considers a certain level of loss to be acceptable, then material theft will grow rapidly. Everyone will consider the material he takes to be within the 2 percent that the activity expects to lose.
- Policies on internal thefts must be enforced. Frequently, a thief is a long-term employee whose honesty has always been above question. In these cases, the temptation is very strong to let the offender go with a "slap on the hand." However, this is a clear signal to other employees that diversion of Army resources for personal use is acceptable to management. If an employee chooses to steal and is caught, he should be prosecuted.
- Publicity campaigns using posters and other media should be used to disseminate command policies on internal theft. Crime hot lines are also useful in increasing reports of employee thefts.

PILFERAGE

B-226. The protection of property, including the prevention of pilferage of government supplies and equipment, is one of the primary functions of the MP and civil service security forces. This function may include protecting supplies and equipment while in storage areas, during the issue process, or while they are in transit.

B-227. Pilferage is probably the most common and annoying hazard with which security personnel are concerned. It can become such a financial menace and detriment to operations that a large portion of the security force's efforts may have to be devoted to its control. Pilferage, particularly petty pilferage, is frequently difficult to detect, hard to prove, and dangerous to ignore.

B-228. Military property loss throughout the world would increase millions of dollars each year if subjected to uncontrolled pilferage. However, the risks incurred cannot be measured in terms of dollars alone. Loss of critical supplies for tactical units could result in loss of life or a danger to national defense. In some areas, losses could assume such proportions as to jeopardize an installation's mission. All installations and facilities can anticipate loss from pilferage. Actual losses will depend on such variable factors as the type and amount of materials, equipment, and supplies produced, processed, and stored at the facility; the number of persons employed; social and economic conditions in surrounding communities; command attitude; and physical-security measures used. Because these factors differ greatly in various types of installations and in different geographical locations, each must be considered separately.

B-229. To determine the severity of this hazard at any given installation or facility, it is necessary to determine the amount of loss that may be occurring.

Unfortunately, this is not always an easy task. Accounting methods may not be designed to pinpoint thefts; consequently, such losses remain undisclosed or they are lumped together with other shrinkages, thus effectively camouflaging them.

B-230. Inventory losses may be inaccurately labeled as pilferage for the following reasons:

- Failure to detect shortages in incoming shipments.
- Improper stock usage.
- Poor stock accounting.
- Poor warehousing.
- Improper handling and recording of defective and damaged stock.
- Inaccurate inventories.

B-231. In some cases, inventory losses may be impossible to detect because of the nature and quantities of materials involved. Stock inventory records may not be locally maintained, or there may be no method for spot checks or running inventories to discover the shortages.

PROFILE OF PILFERERS

B-232. Physical-security personnel must be able to recognize and counteract two types of pilferers—casual and systematic. A casual pilferer is one who steals primarily because he is unable to resist the temptation of an unexpected opportunity and has little fear of detection. There is usually little or no planning or premeditation involved in casual pilferage, and the pilferer normally acts alone. He may take items for which he has no immediate need or foreseeable use, or he may take small quantities of supplies for the use of family or friends or for use around his home. The degree of risk involved in casual pilferage is normally slight unless a very large number of persons are involved.

B-233. Casual pilferage occurs when the individual feels the need or desire for a certain article, and the opportunity to take it is provided by poor security measures. Though it involves unsystematic theft of small articles, casual pilferage is nevertheless very serious. It may have a great cumulative effect if permitted to become widespread, especially if the stolen items have a high cash or potential value.

B-234. There is always the possibility that casual pilferers, encouraged by successful theft, may turn to systematic pilferage. Casual pilferers are normally employees of the installation and usually are the most difficult to detect and apprehend.

B-235. A systematic pilferer is one who steals according to preconceived plans. He steals any and all types of supplies to sell for cash or to barter for other valuable or desirable commodities. He may work with another person or with a well-organized group of people, some of whom may be members of a cleaning team or even be in an advantageous position to locate or administratively control desired items or remove them from storage areas or transit facilities.

B-236. The act of pilferage may be a one-time occurrence, or such acts may extend over a period of months or even years. A large quantity of supplies with great value may be lost to groups of persons engaged in elaborately planned and carefully executed systematic-pilferage activities. Systematic pilferers may or may not be employees of the installation; if they are not, they frequently conspire with employees.

TARGETS FOR PILFERAGE

B-237. Both casual and systematic pilferers have certain problems to overcome to accomplish pilferage objectives. These problems include the following:

- A pilferer's first requirement is to locate the item or items to be stolen. For the casual pilferer, this may be accomplished through individual search or even accidental discovery. In systematic pilferage, more extensive means are generally used. These may consist of surveilling by members of the group or checking shopping and storage areas or documents by those who have access to them.
- The second requirement is to determine the manner in which he can gain access to and possession of the desired item. This may involve something as simple as breaking open a box, or it may be as complex as surveying security factors (such as physical safeguards or security procedures) for weaknesses. It may also involve attempting to bribe security forces, altering or forging shipping documents or passes, or creating disturbances to divert the attention of security personnel while the actual theft is taking place.
- The third requirement is to remove the stolen items to a place where the thief may benefit from his act. Articles of clothing may be worn to accomplish this. Small items may be concealed in many possible places on the thief's body or in vehicles. Through falsification of documents, whole truckloads of supplies may be removed from their proper place without immediate discovery.
- Finally, to derive any benefit from his act, the pilferer must use the item himself or dispose of it in some way. The casual pilferage of supplies is intended primarily to satisfy the need or desires of the thief. The systematic pilferer usually attempts to sell the material through "fences," pawnbrokers, or black-market operations.

B-238. Detection of use or disposal can help prevent similar pilferage through investigation and discovery of the means used to accomplish the original theft. Similarly, each of the problems faced by would-be pilferers offers an opportunity for constructive preventive measures. Careful study of the possible opportunities for the pilferer to solve his problem is essential in security work.

B-239. The primary concern of a systematic pilferer in selecting a target is its monetary value. Since he steals for personal profit, the systematic pilferer looks for items from which he can realize the greatest financial gain. This means he must also have or be able to find a ready market for items he may be able to steal. He pilfers small items of relatively high value (such as drugs, valuable metals, or electronic items). However, we cannot discount the possibility that a systematic pilferer may, if the profit is substantial, select a target of great size and weight. As a rule, bulk-storage areas contain most of the material that may be selected by systematic pilferers.

B-240. The casual pilferer is likely to take any item easily accessible to him. Since he normally will remove the item from the installation by concealing it on his person or in his POV, size is also an important consideration. Monetary value and available markets are not of any great concern to the casual pilferer, because he usually does not have any idea of selling the property he steals. Storage areas containing loose items are more likely to tempt casual pilferers than bulk-storage areas.

METHODS OF PILFERAGE

B-241. There are many ways that pilfered items may be removed from installations. Because the motives and targets likely to be selected by systematic and casual pilferers are very different, the methods of operation for each are very different.

B-242. As stated above, the casual pilferer steals whatever is available to him and generally removes it from the installation by concealing it on his person or in his automobile. The methods of the systematic pilferer are much more varied and complex. The means he may use are limited only by his ingenuity.

B-243. Shipping and receiving operations are extremely vulnerable to systematic pilferage. It is here that installation personnel and truck drivers have direct contact with each other and readily available means of conveyance. This offers a tempting opportunity for collusion.

B-244. One individual must not have control of all shipping and receiving transactions. Obviously, this procedure invites manipulation of government bills of lading and inaccurate storage and movement procedures through failure of one activity to compare bills and invoices with another activity. The opportunities for monetary kickbacks increase without a sound system of checks and balances.

B-245. Railway employees assigned to switching duties on the installation can operate in a similar manner. However, this operation is more difficult because a railway car normally cannot be directed to a location where stolen property can be easily and safely removed.

B-246. Tanker trucks used for shipping petroleum products may be altered to permit pilferage of the product. Trash- and salvage-disposal activities offer excellent opportunities to the systematic pilferer to gain access to valuable material. Property may be hidden in waste materials to be recovered by a accomplice who removes trash from the installation.

B-247. Other methods that may be used by systematic pilfers to remove property from military installations include—

- Throwing items over fences to be retrieved later by themselves or by accomplices.
- Packaging property and sending it to outside addresses through mail channels.
- Conspiring with security personnel.

- Wearing loose-fitting clothing to conceal small items.
- Removing items by using vehicles belonging to outside contractors and vendors.

CONTROL MEASURES FOR CASUAL PILFERAGE

B-248. Specific measures for preventing pilferage must be based on careful analyses of the conditions at each installation. The most practical and effective method for controlling casual pilferage is to establish psychological deterrents. This may be accomplished in a number of ways, such as—

- Searching individuals and vehicles leaving the installation at unannounced times and places.
- Conducting spot searches occasionally to detect attempts of theft.
- Making employees aware that they may be apprehended if they attempt to illegally remove property.

B-249. Care must be taken to ensure that personnel are not demoralized nor their legal rights violated by oppressive physical control or unethical security practices. An aggressive security-education program is an effective means of convincing employees that they have much more to lose than to gain by engaging in acts of theft. Case histories may be cited where employees were discharged or prosecuted for pilferage. Care must be taken in discussing these cases to preclude the identification of individuals because of possible civil suits for defamation of character. Also, it is generally poor policy to publicize derogatory information pertaining to specific individuals. It is important for all employees to realize that pilferage is morally wrong no matter how insignificant the value of the item taken.

B-250. It is particularly important for supervisory personnel to set a proper example and maintain a desirable moral climate for all employees. All employees must understand that they have a responsibility to report any loss to the proper authorities. Adequate inventory and control measures should be instituted to account for all material, supplies, and equipment. Poor accountability, if it is commonly known, provides one of the greatest sources of temptations to the casual pilferer.

B-251. Identifying all tools and equipment by some mark or code (where feasible) is necessary so that government property can be identified. Installation tools and equipment have counterparts on the civilian economy and cannot otherwise be identified as government property. Another control method requires individuals to sign for tools and equipment. The use of the signature control method reduces the temptation to pocket the item.

B-252. In establishing any deterrent to casual pilferage, physical-security officers must not lose sight of the fact that most employees are honest and disapprove of theft. Mutual respect between security personnel and other installation employees must be maintained if the facility is to be protected from other more dangerous forms of human hazards. Any security measure that infringes on the human rights or dignity of others will jeopardize rather than enhance the overall protection of the installation.

CONTROL MEASURES FOR SYSTEMATIC PILFERAGE

B-253. Unlike the casual pilferer, the systematic thief is not discouraged by psychological controls. Nothing short of active physical-security measures are effective in eliminating loss from this source. Some of these measures include—

- Establishing security surveillance at all exits from the installation.
- Establishing an effective package- and material-control system.
- Locating parking areas for POVs outside of the activity's perimeter fencing.
- Eliminating potential thieves during the hiring procedure by careful screening and observation.
- Investigating all losses quickly and efficiently.
- Establishing an effective key-control system.
- Establishing adequate security patrols to check buildings, grounds, perimeters, and likely locations for clandestine storage of property removed from its proper location.
- Installing mechanical and electrical intrusion-detection devices where applicable and practical.
- Coordinating with supply personnel to establish customer ID to authenticate supply-release documents at warehouses and exit gates.
- Establishing appropriate perimeter fencing, lighting, and parking facilities and effective pedestrian, railway, and vehicle gate-security control.

HOW TO STOP EMPLOYEE THEFT

B-254. No matter what it is called—internal theft, peculation, embezzlement, pilferage, inventory shrinkage, stealing, or defalcation—thefts committed by employees are behind at least 60 percent of crime-related losses. So many employees are stealing so much that employee theft is the most critical crime problem facing businesses today.

B-255. Although employee theft results in part from factors beyond control, the extent of employee theft in any business is a reflection of its management—the more mismanagement, the more theft. An effective stop-employee-theft policy must include at least the following:

- Pre-employment screening.
- Analysis of opportunities for theft.
- Analysis of how employees steal.
- Management-employee communication.
- Prosecution of employees caught stealing.

B-256. Each employer must reduce losses as much as possible. A police state need not be created. Large monetary expenditures need not be made.

B-257. The best way to stop employee theft is to simply not hire those employees inclined to steal. The best way is also impossible. The employer must set up a screening process that will weed out obvious security risks. Many experts believe that personnel screening is the most vital safeguard against internal theft. The following are some basic guidelines for the employer:

- Have the applicant fill out a written application. Ensure that the written application does not discriminate and that it conforms to any applicable laws.
- Solicit references, but keep in mind that those contacted will give favorable opinions. Ask primary references for secondary references. In contacting the latter, make it clear that the applicant did not refer them.
- Interview. During an interview, assess the applicant's maturity and values. Observe his gestures.
- Use psychological deterrents. Inform the applicant that your business routinely runs a background security check or that fingerprints will be taken. The hope is that the dishonest applicant will not be back.
- Obtain credit-bureau reports, but only after following guidelines set forth in the Fair Credit Reporting Act.

Opportunities, Methods, and Control

B-258. Cases of employee theft have been documented in almost every conceivable phase of business operations, from theft of petty cash to theft of railroad cars. An infinite variety of methods have been used. Some of the areas that are most vulnerable are—

- Shipping and receiving.
- Inventory.
- Accounting and record keeping.
- Cash, check, and credit transactions.
- Accounts payable.
- Payroll.
- Facility storage units.

B-259. Some of the methods used include—

- Pilfering (one item at a time).
- Stealing from cash registers or altering cash-register records.
- Issuing false refunds.
- Using the back door or trash containers.
- Taking advantage as a supervisor.
- Avoiding package controls.
- Embezzling.
- Forging checks.
- Stealing credit cards.
- Manipulating computers and stealing computer time.
- Conspiring with night cleaning crews.
- Duplicating keys or using a master key that is not properly controlled.
- Conspiring with outsiders (such as inflating insurance claims).

B-260. Too many opportunities exist for employees to exploit. Reduce these opportunities and the losses will be reduced. Reduce opportunities by using the following controls:

- Perform random spot checks on all phases of business in addition to regular, comprehensive audits.
- Check the payroll. Make sure that you are not paying a fictitious or dead employee.
- Take physical inventory seriously.
- Know what you own. Be able to identify it.
- Do not allow one employee to perform all functions. Separate receiving, purchasing, and accounts payable. Separate accountants from cash.
- Control payment authorizations.
- Keep blank checks secured. Do not presign or use uncoded, unnumbered checks.
- Reconcile cancelled checks with original invoices or vouchers.
- Secure exits. Restrict employees to one exit, preventing exit from the rear of buildings.
- Establish strict package control.
- Inspect cash-register receipts daily. Inspect the tape and ensure that the employee is identified on the slip. Deposit money daily.
- Issue ID badges to decrease employee presence in unauthorized areas.
- Simplify red tape; make it harder for the employee to disguise theft.
- Locate employee parking away from the business area.
- Establish a usage schedule of supplies to isolate irregularities.

Management-Employee Communication

B-261. Leadership must be firm, yet reasonable. Most employees pattern their values after their leaders, so a good example must be set. If you expect employees to remain honest, do not take home office supplies or goods.

B-262. Train new employees, advising them of the company's values and the standards by which they will be expected to perform. Explain all security procedures, stressing their importance. Emphasize that any deviations will be thoroughly investigated.

B-263. Establish grievance procedures; give your employees an outlet for disagreement and be receptive to all grievances submitted. Ensure that employees are aware of grievance procedures and that no reprisals are taken.

B-264. Regularly evaluate employee performance and encourage employees to evaluate management. Unrealistic performance standards can lead either to desperation and anger (resulting in dishonesty) or to get-even attitudes. Regularly review salaries, wages, and benefits—do not force employees to steal from you.

B-265. Delegate responsibility. Unless decision making exists among lower and middle levels, there is a tendency for development of an "it's-us-against-them" attitude. Delegate accountability as well; no decision is valid if it is lost in a "pass-the-buck" routine.

SECTION IV — ARMY PROPERTY AT THE LOCAL LEVEL

B-266. Proper accountability by commanders and subordinate personnel cannot be overemphasized. To ensure accountability of property, commanders must establish, implement, and supervise an installation's, activity's, or organization's security program.

B-267. Weaknesses in security procedures at the installation, activity, or organizational levels involving military property create vulnerability supported by criminal activity. Criminal activity includes—

- Theft.
- Fraud.
- Property diversion.
- Property manipulation.

B-268. Commanders and subordinate personnel must conduct a risk analysis; identify military property and; in the interest of monetary value and mission accomplishment, design mandatory security measures for specific property.

B-269. Security doctrine (as outlined in this manual) should be used to the maximum extent in securing Army property vulnerable to theft, destruction, or manipulation. Certain categories of property (such as in Table B-2, page B-62) must be assessed for security vulnerability and protective treatment. Security protective measures addressing this military property should be documented in the installation's physical-security plan. If the security measures recommended in Table B-2 are implemented using established doctrine, they should eliminate or reduce the property's vulnerability. This will reduce the incidents of theft, pilferage, and manipulation at the installation.

MOTOR VEHICLES

B-270. Security of tactical vehicles should be based on a uniform and cost-effective approach. For example, to ensure that a tactical vehicle without a locking device is properly secured, install a clamp, chain, and locking device as illustrated in Figure B-7, page B-63. To install the security device properly while maintaining safety, refer to Technical Bulletin (TB) 9-2300-422-20. Army motor-vehicle security should also incorporate the use of the following:

- Key/lock security and accountability.
- Protective lighting.
- Fencing.
- Walking patrols, as appropriate.
- Frequent observation and visits by mobile patrols or unit personnel (such as the charge of quarters [CQ], the staff duty officer [SDO], and the staff duty noncommissioned officer [SDNCO]).

Table B-2. Recommended Security Measures

Property	Inventoried by—					Inspected by—							
	Security Plan	Hand Receipt/Property Book	Secured by IDS	PS Plan	Regulation	Unit Leaders	PS Officer	SDO	Maint Officer	Supply Officer	Dining-Facility Officer	CQ	SDNCO
Arms/ammunition	x	x	x	x	x	x	x	x	x	x			x
Small arms	x	x	x	x	x	x	x	x	x	x			x
Explosives	x	x	x	x	x	x	x	x	x	x			x
Communications/electronic equip	x	x	x	x	x	x	x	x	x	x			x
Hand tools, tool sets/kits, and shop equip	x	x	x	x	x	x	x		x	x			x
Subsistence items	x	x		x	x	x				x	x		x
Controlled substances, precious metals, and tax-free items	x	x	x	x	x	x	x	x		x			x
Accounts	x	x	x	x	x	x							
POL products	x	x		x	x	x	x	x	x	x		x	
Repair parts	x	x	x	x	x	x	x		x				
Aircraft	x	x	*	x	x	x	x	x	x				
Vehicles	x	x		x	x	x	x	x	x	x		x	x
Towed weapon systems/components	x	x	x	x	x	x	x	x	x				
Carriage-mounted weapon systems	x	x	x	x	x	x	x	x	x				x
Construction material	x		*	x		x	x			x			x
Special-issue clothing (CTA)	x	x	x	x	x	x	x	x			x		
Individual clothing and equipment	x	x	x	x	x	x		x		x			
Organizational equip/components	x	x	*	x	x	x	x	x	x	x		x	x
Compasses, binoculars, and flashlights	x	x		x	x	x	x			x			x
Medical-unique items	x	x	x	x	x	x	x	x		x			x
Housekeeping supplies and equipment	x	x	x	x	x	x				x		x	
Housing furniture	x	x		x	x	x				x		x	
Mess equipment	x	x	x	x	x	x	x			x	x		x
Office machines	x	x	x	x	x	x	x			x			x
Expendable/consumable supplies	x		x	x	**	x		x		x	x		x

*** Depends on facility availability and cost effectiveness**
**** Depends on local policy**

Figure B-7. Typical Clamp and Chain Installation

CONSUMER OUTLETS

B-271. The lack of initiative at the management level within operational consumer outlets does little to prevent or reduce pilferage. Such shortcomings are identified as—

- Failure to present a professional image.
  - Lack of continuing interest, motivation, and direction.
  - No alertness to internal control of pilferage.
- Failure to institute and implement methods of operational effectiveness and efficiency. These methods include—
  - A clearly defined delegation of responsibility.
  - Insistence on stringent accountability
  - Orientation and training programs for subordinate supervisors and current and new employees.
- Failure to emphasize and enforce established criteria for continual employment.
  - Rules of conduct.
  - Standards of job performance. (Officially request appropriate action for employees guilty of criminal acts or infractions conducive to criminal acts.)
  - Inattentive job attitudes of subordinate supervisors.
  - Inadequate personal checks of established accounting and inventory procedures. **NOTE: Checks on both a regular and unannounced basis tend to control access to official stock records and to ensure careful and organized storage or stocking of merchandise.**
  - Infrequent observation of an employee's job performance.

  - Failure to report misconduct, criminal or otherwise, to superiors or responsible law-enforcement personnel in the activity.
  - Failure to implement recommendations made during physical-security inspections or crime-prevention surveys.

PILFERAGE

B-272. Pilferage may be accomplished by individual employees, by a team of employees, or by employees and patrons in collusion. These actions can be greatly reduced by tightening supervision and security in the following areas:

- **Merchandise display or dispensing areas.** The following measures can reduce merchandise pilferage in display areas:
  - Detecting unauthorized price reductions.
  - Preventing or making it difficult to alter price tags.
  - Checking procedures for declaring merchandise old, shopworn, damaged, or salvage.
  - Providing more unpackaged items for personal consumption.
  - Discouraging the careless waste of foods and other perishable items.
- **Cash registers.** The following are methods of pilferage from cash registers:
  - Stealing directly from an unattended register.
  - Rerunning register tapes at lower figures (this is preventable if the reset key is maintained by the supervisor).
  - Clearing the register at a lower total figure than actual receipts for the operational period.
  - Reporting overrings and refunds falsely.
- **Theft of merchandise.** The following methods enable merchandise to be pilfered:
  - Underringing.
  - Reusing cash-register tapes. This occurs when employees fail to provide patrons with tapes or when patrons allow employees to retain tapes. The tapes allow employees to package merchandise and remove it from the premises.
  - Removing items from bags or containers carried out by employees.

SHOPLIFTING

B-273. Shoplifting is usually confined to sales areas and is committed by casual and systematic pilferers. Items that are most frequently pilfered—

- Are relatively small in size.
- Have a high degree of consumer desirability.
- Are easily carried in pocketbooks or secreted on the person.

B-274. Amateur, adult shoplifters share the following characteristics:

- Theft is from a sudden temptation (impulse theft). There is success in the initial theft, which leads to more temptation, stronger impulses, and more thefts.
- They rarely have a genuine need for the item.

- They usually have enough money to pay for the item.
- They display symptoms of nervousness and uneasiness.

B-275. Juvenile shoplifters have the following traits:

- They act on a dare or "to belong."
- They may be coached or directed by an adult.

B-276. Professional shoplifters share these characteristics. They—

- May be talkative and are usually polite and deliberate.
- Look for opportunities continually.
- Do not take many chances.
- Are very capable of spotting security personnel.
- Steal for resale.
- Have "fences" (usually).
- Steal "to order" (often). They may have a list describing the items to be pilfered.
- Use innovative techniques.

B-277. Genuine cases of kleptomania are rare. Kleptomaniacs share the following characteristics. They—

- Take items without regard to their value or use.
- Steal compulsively and often openly.
- Are nervous and shy.

B-278. Narcotic addicts as shoplifters have the following characteristics. Only MP/security personnel should attempt apprehension. These types of shoplifters—

- Are desperate for money and they fear imprisonment.
- Take big chances.
- Take the merchandise and exit the premises quickly.
- Steal when they are at their lowest physical and/or psychological ebb.
- Are dangerous if you try to apprehend them (sometimes violent).

B-279. Alcoholics and vagrants as shoplifters share the following traits. They—

- Usually steal because of need.
- Are often under the influence of liquor at the time of theft.
- Take the merchandise quickly, then exit the premises.
- Are less likely to steal regularly at a single location.

B-280. The most pilferage occurs when employee coverage is low or when employees are untrained, inexperienced, or indifferent to the issue. The ineffective use of floor space aids shoplifters by creating congestion in the patron's traffic flow. Allowing an emphasis on small rooms or partitioned areas causes congestion that clusters, isolates, or partially hides displays. Shoplifting involves the use of one or more of the following means to obtain items:

- Palming or placing an open hand on a small article, squeezing the muscles of the hand over the article to grasp it, and lifting the still open and apparently empty hand.
- Using fitting rooms to put on tight or close-fitting garments under clothing worn into the store.
- Trying on hats, gloves, sweaters, and jackets, then exiting the store.
- Stepping around counters and removing items from unlocked showcases.
- Handling several items at once and replacing all except the item pilfered.
- Using accomplices to create a diversion of employee attention when secreting items on the person. Such items include:
  - Clothing.
  - Pocketbooks or handbags.
  - Umbrellas.
  - Various items placed in packages or paper sacks containing merchandise paid for at other departments.

ARSON

B-281. On an installation there is little incentive for a professional arsonist to operate since the government owns the buildings and insurance fraud in collusion with the property owner is not possible. However, arson can still be a problem. In the civilian community, most deliberate fires are not set for profit. They are set to "get even" with the property owner or just for the excitement of watching something burn. Military installations are susceptible to this type of crime. On many Army posts, there is a large number of empty, wooden structures that are ideal targets for revenge seekers or vandals.

B-282. Arson is an easy crime to perpetrate, and it is relatively difficult to collect the information needed to convict an arsonist. However, this does not mean we are helpless in combating this crime. Several major cities, including Seattle, Denver, Houston, and Philadelphia, have developed successful programs to reduce the number of arsons. Some of the successful, proactive measures that have been developed are—

- Securing or disposing of materials that could be used to start fires. Ensuring that the regulations on the storage of gasoline, paint, and solvents are enforced and that paint lockers are locked.
- Enforcing command policies on the police of the post. Removing piles of trash, scrap lumber, and other material that burns easily.
- Securing empty buildings (especially empty wooden structures) and posting them as off-limits areas.
- Patrolling areas susceptible to arson.
- Encouraging participation in neighborhood watches, taxi patrols, and other community programs that increase surveillance.
- Encouraging the reporting of suspicious activities through the establishment of crime hot lines.

- Establishing a close working relationship between fire fighters and law-enforcement personnel to ensure that fires of suspicious origin are reported and thoroughly investigated.
- Offering rewards for information leading to the apprehension of the arsonist (if there is an outbreak of arson).

SECTION V — COMMUNITY CRIME-PREVENTION PROGRAMS

B-283. There are a number of crime-prevention programs set up in communities to help deter and detect crime. These programs are supported and often headed by members of the community in conjunction with local law-enforcement agencies. The programs are as diverse in nature as they are in number, yet they complement one another.

NEIGHBORHOOD WATCH PROGRAM

B-284. A neighborhood watch program is an organized network of citizens interacting with other neighbors and the police in preventing and detecting crime in their neighborhood. Law-enforcement efforts to reduce crime cannot be accomplished effectively without the support and cooperation of all citizens. A strong community involvement with neighbors helping themselves and other neighbors in becoming more alert to activities in the neighborhood, protecting their property, and reporting suspicious activities is essential to an effective crime-prevention program.

B-285. The Army Neighborhood Watch Program is designed to encourage Army service members and their families to actively participate in protecting their own property and the property of their neighbors, joining community crime-prevention programs, and reporting suspicious activities to MP officers. The program is designed to develop the following:

- The awareness of community crime trends and prevention efforts.
- The knowledge of quarters' security procedures.
- A cooperative system of surveillance over each neighbor's property.
- Accurate observation and reporting of suspicious activities.
- The establishment of a reliable, two-way information link between the community and MP forces.

B-286. Most neighbors know the routines of the other families that live near them. They know what cars are normally parked in the neighborhood and when families are on vacation or out of the area. Neighbors are in a very good position to recognize burglars and other intruders. Also, residents are in a good position to recognize safety hazards and crime-conducive conditions near their homes.

B-287. To capitalize on these advantages, neighborhood watch programs organize blocks in family-housing areas or floors in troop billets to improve police and community interaction. They also disseminate information on crime problems and countermeasures.

BLOCK CLUBS

B-288. Block clubs are the basic components of an installation-wide neighborhood watch program. The geographical size of a block club may vary widely depending on the population's density and the nature of the terrain. The key factor is that the terrain organized into a single block club should promote a feeling of unity and mutual assistance. In a troop billet area, block clubs could be organized along company lines, by individual barracks, or by floor. In a family-housing area, a block club could cover one high-rise apartment building, one block, or one or more streets that are so situated that the residents identify themselves as a subcommunity within the housing area.

Organization

B-289. Individual residents, community-service organizations, or MP units can initiate block clubs. Existing organizations (such as the PTO) have frequently established contacts within the community and have sponsored the organization of block clubs. Regardless of the approach used to organize the installation, every family or resident in the area should be contacted and encouraged to attend a block club (see Figure B-8).

B-290. A block captain and a deputy block captain should be elected at the initial block meeting. These individuals serve as "spark plugs" to sustain interest in their geographic area. They also represent their block club at district meetings. The effectiveness of the presentation at the initial block-club meeting is critical. A PMO representative should explain the crime problems on the installation and clearly outline the functions of a block club and how it can affect the crime problem. The main objective should be to generate enthusiasm and build a foundation upon which an effective neighborhood watch can be built. The use of camcorders or other audiovisual programs can generate interest. Association with programs in the civilian community may also help to generate interest.

B-291. The functions of a block club should be clearly identified. Block clubs—

- Serve as the "eyes and ears" of the police.
- Encourage the implementation of individual countermeasures, such as marking personal property.
- Disseminate crime-prevention information.
- Improve police and community relations

Block-Club Districts

B-292. A block-club district should be organized as an intermediate link between individual block clubs and the installation's crime-prevention council. These districts should cover at least one housing area and should be composed of the block captains and deputies from each block club in the district's area. District leaders should be elected from the block captains and should serve as members of the installation's crime-prevention council. The primary functions of districts are to transmit information from the crime-prevention council to the block clubs and to develop incentive awards to recognize effective participation by the clubs.

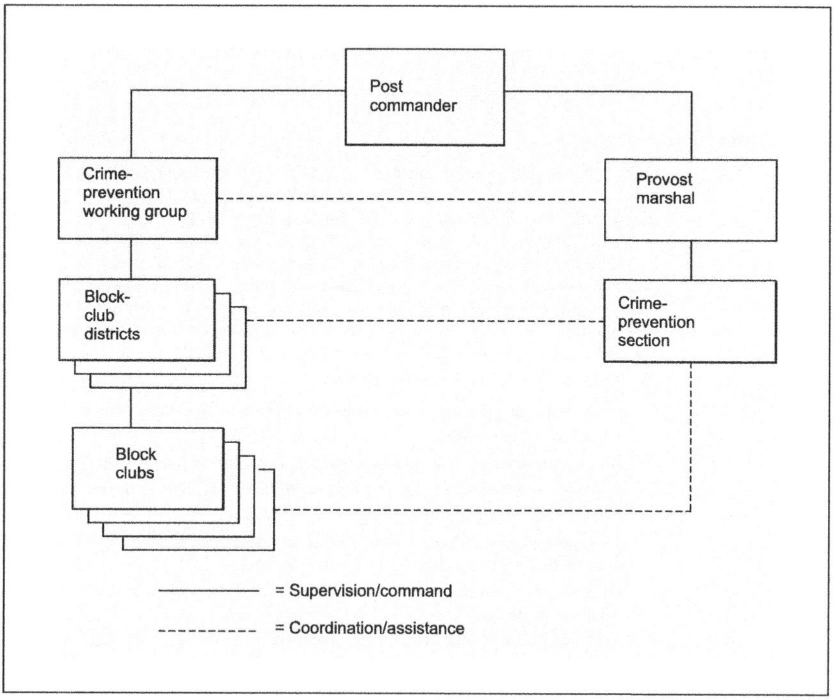

Figure B-8. Block-Club Organization

Maintaining Interest

B-293. Maintaining interest in neighborhood watch programs is a major problem. Typically, a particular incident generates the level of interest required to initially organize a system of block clubs and districts; but as the particular problem is overcome, interest wanes and the clubs gradually dissipate. A review of programs that have maintained their effectiveness for extended periods indicates that successful programs have the following characteristics:

- They are a formal organization with elected block captains and district leaders.
- They address a wide range of problems.
- They feature an incentive award program to recognize individuals and clubs that participate effectively.

- Their members attend periodic workshops to train leaders on various aspects of crime prevention.
- They have clear-cut, attainable objectives. The members need to see accomplishments.
- They are actively supported by the local police.

Block-Club Operations

B-294. Block clubs may become umbrella organizations that conduct the entire range of community crime-prevention programs, including foot and mobile patrols. However, each of these programs will be discussed in a separate section. In their most basic form, block clubs disseminate information on residential security and report suspicious activity in their neighborhood. Members of a neighborhood watch program should exchange names, addresses, and telephone numbers to enhance communication among neighbors. They must learn how to record and report suspicious activities. Some activities that block-club members should be alert for and, when observed, report to the MP officers are—

- A stranger entering a neighbor's house when it is unoccupied.
- Someone screaming.
- An offer of merchandise at a ridiculously low price (it could be stolen).
- Persons entering or leaving a place of business after duty hours.
- The sound of breaking glass or an explosion.
- A person going door to door and then going into a back or side yard, or a person trying a door to see if it is locked.
- A person loitering around a school, a park, a secluded area, or in the neighborhood.
- A person carrying property at an unusual hour or in an unusual place.
- A person exhibiting unusual mental or physical symptoms (he may be in need of medical help).
- A vehicle being loaded with valuables when parked by a closed business or untended residence.
- A business transaction conducted from a vehicle.

ARMY NEIGHBORHOOD WATCH MEETINGS

B-295. A PMO representative should attend the initial neighborhood block meeting to explain the relationships of crime prevention and the neighborhood watch programs. The meeting should be publicized with handout invitations announcing the time, location, and purpose of the meeting. If possible, the meeting should be held in the neighborhood (such as the training room in the troop billets, a local school, or a residence in family quarters). In addition to explaining the program concepts, the initial meeting could include the following:

- Installation and community criminal statistics concerning the nature and volume of housebreakings, larcenies, and other crimes.
- An exchange of names and telephone numbers (if applicable) of attendees. This information should also be placed on a neighborhood block sheet that is a geographical diagram of the block showing the

location of each room, apartment, and building address number in the block. The names and phone numbers of participants would be added to the address number of each residence drawn on the diagram. This block sheet should be distributed to block members at a subsequent meeting.

- Determination of the second meeting date and location.

B-296. Other agenda items and activities for subsequent meetings could include the following discussions:

- Residential security procedures and the conduct of quarters security inspections.
- Operation ID—marking and recording of personal property.
- The observation and report of suspicious activities.
- Other crime-prevention measures and procedures being implemented at the installation (such as rape prevention and citizen escort patrols).
- Fire prevention, personal safety, and other related activities.

OPERATION ID

B-297. Operation ID, which entails marking of property to make it identifiable and traceable to its owner if lost or stolen, was initiated by the Monterey Park, California, Police Department in 1963 and has been adopted by more than 80 percent of the police departments in the US. Operation ID is a low-cost, highly effective crime-prevention program. However, its success is contingent upon the willingness of individuals and communities to actively participate in marking and identifying their personal property.

B-298. Operation ID is designed to encourage Army service members and their families to mark their personal property with a standard Army-wide, owner-applied number. This numbering system permits the positive ID of the property and determines the location of the owner in case of theft or loss.

B-299. The principal advantages of Operation ID are theft deterrence and recovery of personal property. Marked stolen property is more difficult to dispose of, and illegal possession can result in the prosecution of a thief. Recovered lost or stolen property can only be returned if there is some means of identifying and locating the rightful owner.

METHODS OF IDENTIFYING PROPERTY

B-300. Various methods of establishing positive ID and ownership of property in case of loss are available for individuals. Each method has advantages and limitations, and a combination of these methods would be required to ensure the ID of all high-value personal property.

- Inscribing the owner's applied number with an etching or engraving tool would allow the recovering agency to visually identify the number inscribed on the property for notification and subsequent return of the property. Electrostatic markers are available for use at no cost to the individual. However, some personnel are reluctant to use this method since it can mar the property. The inscription should be made in a location that can be readily seen by the recovering agency but which would not deface the property's appearance or reduce its value. Some high-value personal items such as coins, jewelry, and silver cannot be inscribed with an owner's applied number. Another method of identifying these items would be required.

- Using invisible fluorescent ink, powder, or paste to mark the property can make it easier for the agency recovering the property to use an ultraviolet light and identify the owner. This marking will not mar the property. However, the fluorescent markings and ultraviolet light are an additional cost, and many agencies do not have ultraviolet lights to inspect recovered property.
- Using a laser photographic process to identify diamonds. Every diamond emits a unique reflection when penetrated by a low-level laser light. A laser photographic process has been developed to record a diamond's pattern of light reflection on film. Several jewelers throughout the country have the laser photographic equipment available. To register a diamond, two photographs are taken. One is provided to the owner and the other to a central registry. If a diamond is lost or stolen, the recovering agency can take the diamond to a jeweler that maintains the laser photographic equipment for print. This photograph would then be forwarded to the registry for owner ID.
- Photographing the personal item. Individuals can photograph personal high-value items that cannot be engraved. Although the agency recovering the lost or stolen items could not identify the owner, use of photographs could assist in verifying ownership if it is known that the item has been recovered. In addition, a photograph would assist in submitting claims against the government or private insurance companies, as appropriate.

RECORDING PERSONAL PROPERTY

B-301. Individuals should record identifying data (such as brand name, model, serial number, and value of the personal items), even if they use other methods of identifying property. This information would assist in determining what items may be lost, stolen, or damaged through fire, explosion, or other hazards. This information can also be used in claims against the government or private insurance companies, as appropriate.

IDENTIFICATION NUMBERING

B-302. There are various types of owner-applied numbering systems used to mark personal high-value items. Criteria that should be considered in determining which owner-applied numbering system to incorporate include—

- Uniqueness, where no two people have the same identifier.
- Permanence, so that the owner-applied number will not change.
- Ubiquity, so that an identifying number is available to any individual who desires one.
- Availability, where the identifying number can be easily obtained and remembered.
- Indispensable, so that there are incentives requiring an individual to have the number.

- Privacy, so that the number is not a means of infringing upon an individual's right to privacy.
- Uniformity, so that the owner-applied number would be readily recognized by law-enforcement agencies who handle or come into contact with the recovered property.
- Traceability, so that the property owner can be identified and located.

B-303. The most commonly used owner-applied numbering systems include—

- Driver's license number with the issuing state abbreviation prefix.
- Social security number.
- Personal numbers assigned to individuals by a local law-enforcement agency.
- Personal numbers with the marking agency's National Crime Information Center originating-agency ID number.
- A private numbering system maintained by a commercial organization.

STANDARD ARMY NUMBERING SYSTEM

B-304. To ensure that criteria for property ID numbering are met, the standard Army-wide, owner-applied numbering system is designated as the service member's social security number with a "USA" prefix. Upon recovery of lost or stolen property by other military or civilian law-enforcement agencies, the USA prefix, owner-applied number would alert the recovering agency that the property belongs to a member of the Army. The recovering agency can then contact the nearest Army installation's PMO/security office concerning the property and the owner-applied number inscribed. The service number can then be identified and located through the Army worldwide locator system. Since this locator system lists only social security numbers and locations of active Army service members, family members should also use the service member's social security number with the USA prefix when marking their personal items.

IDENTIFYING AND LOCATING OWNERS OF RECOVERED PROPERTY

B-305. The installation's PMO/security officer should be the initiator of tracer actions to identify and locate the owner of recovered property marked by the standard Army Operation ID numbering system. Continuous liaison should be maintained with local civilian law-enforcement agencies and other military installations to ensure that they are cognizant of the standard Army owner-applied numbering system and will contact the PMO/security office upon recovery of private property marked with this system. The PMO/security officer should readily accept custody of the private property if the recovering agency is willing to release the property and it is not required as evidence for criminal prosecution.

B-306. Upon notification of recovered property and the inscribed Army standard owner-applied number, the PMO/security officer should contact the servicing military personnel office (MILPO) for assistance in determining the name and location of the property owner by using the Army's worldwide locator microfiche.

B-307. If the owner-applied social security number is not listed on the Army's worldwide locator microfiche and the servicing MILPO is unable to provide the requested information, the PMO/security office should contact the following offices:

- For enlisted members: USA Enlisted Records and Evaluation Center, ATTN: PCRE-RF-L, Fort Benjamin Harrison, Indiana 46249.
- For officers: PERSCOM, ATTN: TAPC-MSR-S, 200 Stovall Street, Alexandria, Virginia 22332-0444.
- For warrant officers: PERSCOM, ATTN: TAPC-OPW, 200 Stovall Street, Alexandria, Virginia 22332-0444.

B-308. In case the owner-applied number cannot be identified by either the servicing MILPO or the USA Enlisted Records and Evaluation Center, the Army Reserve Personnel Command (AR-PERSCOM) should be contacted for assistance. The AR-PERSCOM retains files on separated, retired, and Reserve component Army members. Written requests should be forwarded to: Commander, AR-PERSCOM, ATTN: ARTC-PS, 1 Reserve Way, St. Louis, Missouri 63132.

B-309. When the name and location of the service member associated with the owner-applied number inscribed on the property is verified, the PMO/ security office should notify the service member in writing that the property has been recovered. The notification should ascertain if the recovered property belongs to the service member, if the service member ever reported the property as lost or stolen, and if a claim was submitted to the SJA claims service for loss or theft of the property. Notification should also state where the property is located and a point of contact that the owner can deal with for the return of the property. If the property is to be retained as evidence for legal proceedings, the owner should be informed that the property will be returned upon completion of the proceeding. A copy of this letter should be provided to the US Army Claims Service, Fort Meade, Maryland 20755. The Army Claims Service will advise the PMO/security office if a claim was or was not submitted.

B-310. If the owner cannot be located, recovered property in the custody of the PM/security offices should be disposed of according to AR 190-22 and DOD 4160.21-M.

USE OF THE NATIONAL CRIME INFORMATION CENTER

B-311. Stolen articles may be entered into the National Crime Information Center (NCIC) file if a theft report has been made, if the item is valued at $500 or more, and if it has a unique manufacturer's assigned serial number and/or owner-applied number. Entering stolen personal property items meeting the above criteria into the NCIC or other police information systems as outlined in AR 190-27 is encouraged.

NEIGHBORHOOD WALKS

B-312. While most people are unwilling to participate more actively than as observers in a neighborhood watch, there are some individuals who want to become more actively involved in securing their neighborhoods. For this segment of the population, the organization of "neighborhood walks" provides a welcome opportunity to make a more active contribution. The basic idea is simple; residents patrol on foot through their own neighborhoods to observe and report crime. In practice, it is a little more complicated; however, neighborhood walks can have a dramatic impact on the crime rate, so the effort expended is worthwhile. The points that must be considered are—

- **Patrol composition.** Both adults and teenage children can volunteer to participate in neighborhood walks. However, there should be at least one adult in each party. There should be at least two people in each patrol group. Groups of four to six individuals are desirable since, by their numbers alone, they discourage attacks on the walkers. Larger groups are also more fun, and this is important when volunteers are providing the manpower.
- **Times/patrol duration.** As with other crime-prevention programs, maintaining a high level of interest can be a problem. A successful walk program in Philadelphia schedules groups for one 2-hour patrol per month. More frequent tours caused high drop-out rates among participants. Neighborhood walks should be conducted only during those times when the crime rate is the highest. Normally, there are not enough volunteers to conduct walks at times other than peak crime periods.
- **Functions.** Members of neighborhood-walk groups must understand that they are to observe only and not actively intervene in criminal acts. Participants and the government are legally liable for their actions during walks. When a crime or a suspicious activity is spotted, the neighborhood walkers should report it to the police. In Philadelphia, the walkers are equipped with horns. When a crime is spotted, they activate their horn and go to the nearest house to call the police. When residents hear the walker's horn, those in the immediate area turn on all of their lights and sound their own horns. The noise and increased lighting invariably causes the criminal to flee.
- **Neighborhood escorts.** In addition to observing and reporting criminal activity, neighborhood patrols can escort children and older persons between community-service facilities and residences. They can also request that owners secure property when they find it unsecured; for example, when there are unsecured bicycles parked on a front lawn.

VIGILANTISM

B-313. While active community participation is essential, vigilantism must be discouraged at all costs. Both formal law codes and US common law offer few protections for private citizens who take the law into their own hands. Block captains and installation crime-prevention officers must be alert for indications that neighborhood patrols are doing more than observing and take swift remedial action when required. Of course, all nonpolice participants must be prohibited from carrying weapons of any type while engaged in crime-prevention programs. Experience in neighborhoods having much higher violent crime rates than found on Army installations has demonstrated that passive devices like horns or whistles were adequate to discourage attacks.

These devices, plus the assignment of four to six individuals to a neighborhood patrol, provide sufficient protection.

MOBILE PATROLS

B-314. Some communities with high street-crime rates have been successful in organizing private citizens into mobile-patrol programs. Like neighborhood foot patrols, these mobile patrols serve as the "eyes and ears" of the police but do not actively intervene when they spot a crime in progress.

B-315. In a typical program, block captains or police crime-prevention officers assign specific patrol areas to each private mobile patrol. In addition, each patrol receives training on the patrol's functions, communications procedures, and emergency actions. Normally, patrols are instructed to blow their auto horns steadily when they observe a crime in progress. This is usually sufficient to drive off the criminal.

B-316. Most private citizen patrols use cellular telephones or citizen's band (CB) radios as direct communications links with the supporting police department. Installation CB radio clubs are often willing to sponsor anticrime patrols under police supervision. Commercial taxi companies that operate on military installations are also excellent candidates to organize into patrols. The cab drivers normally cover most of the high-crime-rate areas on the installation. Because of the frequency with which they cover them, they are familiar with the routine conditions in each area and are quick to spot suspicious activity.

B-317. The government cannot provide gasoline for POVs used for anticrime patrols; however, it may often provide magnetic signs to affix to the vehicle for identifying it as part of the police-sponsored, neighborhood-patrol program. As in the case of foot patrols, the installation crime-prevention officer must be alert for signs of vigilantism and must take positive action to discourage it if it appears.

PROJECT LOCK

B-318. Nearly one million automobiles are stolen in the US every year. The total value of cars stolen is around the billion dollar mark, making auto theft the nation's costliest crime involving property. Of even greater importance is the social impact of auto theft. For a growing number of young people each year, stealing cars represents the first step toward a life of crime.

B-319. Police agencies have been diligent in the apprehension of auto thieves and the recovery of stolen vehicles, but the auto-theft problem seems to be more amenable to improvement through prevention rather than punishment. Barring strict security, auto theft is one of the easier crimes to commit. All vehicles left unattended are vulnerable, and widespread prevention by police surveillance is a physical impossibility.

B-320. The problem has grown to serious proportions despite determined law enforcement because motorists continue to be negligent or unaware of their responsibility. As long as people invite theft by leaving their cars unlocked or leaving the key in the ignition, auto thefts will continue to climb. Almost half of all stolen cars each year had been left with keys in the ignition; nine out of ten of the stolen vehicles had been left unlocked.

B-321. If a significant reduction is to be made, the motorists themselves must make it. Widespread adoption of accepted and effective prevention practices by motorists presents the most logical and immediate improvement to this growing problem.

B-322. In 1963, the Boston police department conducted a broad information campaign with the assistance of the National Automobile Theft Bureau and the Insurance Information Institute. Since then, more than 525 "Lock Your Car" campaigns have been held in about 400 communities in 49 states.

B-323. In the months following these campaigns in such cities as Denver, Chicago, Atlanta, and San Francisco, significant reductions in the number of auto thefts (ranging from 9 to 54 percent) have been recorded. Undoubtedly, the "Lock Your Car" campaigns contributed to reducing auto-theft statistics.

B-324. Project Lock is designed to permit sponsoring groups to conduct one-day or one-week "Lock Your Car" campaigns. Its purpose is twofold; it—

- Alerts the public on the importance of locking cars and removing keys as a deterrent to auto theft.
- Contributes to the welfare of youths by preventing the commission of a first crime.

B-325. The following materials can be ordered to support Project Lock:

- Windshield flyers.
- Identifying insignia to be worn by inspectors.
- Tally cards for noting cars left with keys in ignitions or unlocked doors.

B-326. Groups considering sponsorship of this campaign should consider installation areas known to have car-theft problems. The PMO should assign patrolmen to accompany the teams on inspection day. Usually, the teams' routes can be arranged to fit the regular patrols of the MP officers. If uniformed police will not be available, the campaign should not be held.

COMMUNITY ORIENTATION

B-327. If the commander desires, a community orientation meeting might be held a month or so in advance of the campaign. This would be more desirable for a week-long rather than a day-long campaign. If such a meeting is planned, the PMO should issue invitations to representatives of the installation crime-prevention council, service clubs, women's clubs, PTOs, high schools, and churches.

B-328. The meeting should be opened with an introduction of the installation commander. After appropriate comments, the commander would read a proclamation setting the date for Lock-Your-Car Week. If possible, a representative of the National Automobile Theft Bureau should be asked to address the meeting. An alternative would be to have the PM review national and local trends in auto thefts stressing the importance of the forthcoming campaign. In conclusion, a representative of the sponsoring group might review the schedule of activities for the campaign.

PROGRAM ACTIVITIES

B-329. If the campaign is conducted in one day, a kick-off breakfast might be substituted for the orientation meeting. The program and the attendance might be similar if a weeklong campaign is planned and the orientation meeting were held. The kick-off breakfast might have a simple program with attendance limited to the sponsoring group and police representatives. An alternative to the breakfast would be a luncheon at which results to the moment are reported. Whether the campaign will last a week or a day, the general activities will be similar.

B-330. With the assistance of the MP officers, the installation should be zoned according to the established MP patrol areas where possible. The size and number of the zones will depend on the number of police and volunteer personnel that will be available. However, the zones should cover most of the installation's service areas, troop billet areas, and family-housing areas.

B-331. The inspection teams are each composed of a uniformed patrolman and three to five sponsoring group members, cover their assigned zones and place flyers under the windshield wipers of all cars found to have keys in the ignition or to be unlocked. Under no circumstances should the flyers be placed inside the cars, even through open windows.

B-332. It is suggested that flyers be ordered early enough to allow a local printer to inscribe an overleaf statement such as "This public service is provided as a courtesy of the MP force." The tally cards should be used to record the number of cars inspected, the number that were unlocked, and the number with keys in the ignition.

SCHOOL PROGRAM

B-333. The crime-prevention office might sponsor a poster contest for art students. It should be announced at least a month before the campaign to allow the entries to be placed in public in advance of the date. The PMO should present awards in an office ceremony. The ceremony should take place at noon on the campaign day or midweek if the campaign is longer.

PUBLIC ADDRESSES

B-334. Close to or during the campaign date, addresses by crime-prevention professionals should be scheduled for programs of service clubs, women's clubs, PTOs, and other civic groups. Well in advance of the campaign, the PAO should be visited by the PM and the other sponsors to develop comprehensive internal-information programs in support of the installation's Project Lock. Plans for the public-affairs program should include advance publicity for the campaign, coverage of events during the campaign, and wrap-up coverage following its completion. During the inspection days, findings on how many cars were unlocked and how many had keys in the ignition should be reported regularly to a headquarters (preferably to the MP unit).

CONCLUSION

B-335. Project Lock has been outlined to provide basic suggestions for a "Lock Your Car" day or week. No procedure can be designed to fit all needs or circumstances, and variations often will be desirable. However, if the suggestions and the materials contained in this section are used, Project Lock will not be difficult to organize and conduct. It provides better publicity opportunities than most public-service projects. It presents an opportunity for active participation by a number of crime-prevention groups. It has been field-tested and found to be an outstanding success.

SECTION VI — EVALUATION

B-336. Evaluations of installation crime-prevention programs (see Figure B-9, page B-80) do not rely on control groups, a tight control of variables, or elaborate statistical analyses to produce worthwhile results. Most often, the resources to conduct a formal evaluation that will stand up to rigorous academic scrutiny are not available.

CRIME-PREVENTION PROGRAMS

B-337. Several inherent difficulties in data collection on crime-prevention programs make it difficult to determine with 100 percent accuracy that a particular reduction in the crime rate was a result of a particular crime-prevention measure (unless a very elaborate analysis and control system are used). These difficulties include the control of variables, the displacement effect, and unreported crime.

CONTROL OF VARIABLES

B-338. In its simplest form, the type of evaluation most commonly used in academic or scientific settings seeks to determine the relationship between two variables. By varying the independent variable (for example, the dosage of a drug), the effect on the dependent variable (for example, a pulse rate) is determined while all other variables (such as food intake) are held constant.

B-339. Although more complex in form, the same model can be used to evaluate complex programs. Experimental and control groups can be selected. The "treatment" can be administered by researchers or those taught by researchers. The results can be analyzed for their statistical significance. However, since crime-prevention programs deal with human subjects, certain complications arise. The degree of success may have nothing to do with the efficacy of the program, but only with the way it was introduced or with the personal predilections of the groups involved. There is no standard population; human beings are not standardized as mice are for laboratory purposes. A program found successful at one installation may be a failure in another.

B-340. These considerations also apply in the evaluation of crime-control programs. This evaluation is further complicated by another problem—the people whose behavior is to be modified (the offenders) cannot be treated directly or separated into experimental and control groups; they will not stand up and be counted. Although public-health programs often encounter this problem, they often deal with physical cause-and-effect links between treatment and improvement. The same is not true for crime-control programs.

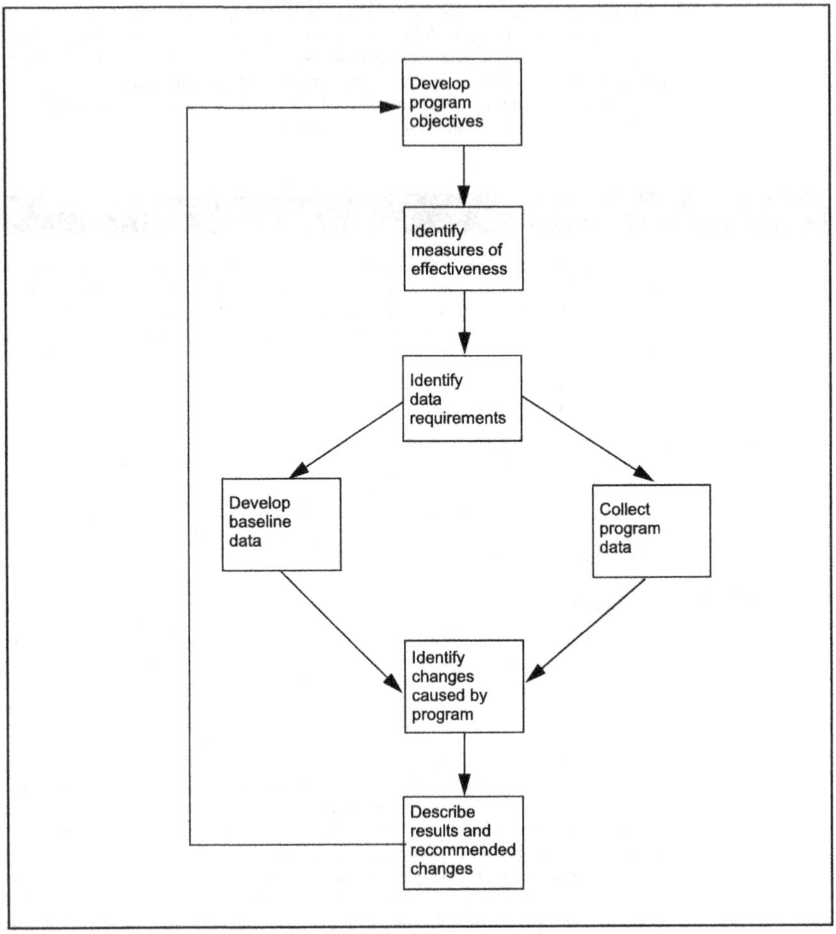

Figure B-9. Crime-Prevention Program Evaluation

The effectiveness of these programs is normally determined by looking at statistics of reported crime and arrests, which are more indirect indicators.

B-341. In a crime-control program, it may be impossible to classify variables as dependent and independent; they may all affect and be affected by each other. Furthermore, because of the difficulty in determining why people behave the way they do, a number of intervening and antecedent variables may go unnoticed. Police programs designed to reduce crime may have their most direct effect on the behavior of the general public toward the police, which in turn affects the crime rate.

B-342. Evaluations are not necessarily restricted to the analysis of objective crime data; they can also include subjective considerations and perceptions. These subjective evaluations can be of significant benefit in augmenting the statistical analyses of the program's results. They are especially helpful in assessing why and how a program worked and whether a statistical outcome is actually evidence that the program was successful. Interviews of participating agency personnel and residents of the program's area are usually used to supply this information. They can give the evaluator new insight into the actual program operation.

DISPLACEMENT EFFECT

B-343. In many cases where crime reductions have been measured and attributed to programs, it is unclear whether there has been an actual reduction in crime or whether the crime has been displaced. The amount of displacement depends to an extent on the offender's characteristics. An opportunistic offender can be pictured as having a relatively elastic demand—if the risk is too high, he will forgo the crime. An addict offender is typically pictured as having a relatively inelastic demand for the product because of his inelastic demand for drugs—despite the risks, he needs the product.

B-344. Deterrents may have little effect on perpetrators of expressive crimes. These are crimes in which the perpetrator is emotionally involved and is expressing these emotions. Most assaults and homicides fit this category. On the other hand, deterrents may have a strong effect on instrumental crimes, those that are seen by the offender only as a means to an end (usually money). If alternative avenues to the same end are made more attractive by comparison, the offender may well be deterred. Deterrence may produce a diversion to legal alternatives to crime; it also may cause displacement to illegal alternatives.

Displacement to Other Crimes

B-345. There is no immutable law that says a burglar cannot hold up a liquor store and a robber cannot burglarize a warehouse. If a specific crime or a set of crimes is the target of a crime-control program, offenders may decide to avoid the target crimes and ply their trade in other ways. Some offenders will be deterred from all crime if their crime specialty is the object of a crime-control program, but the extent of this deterrence should not be overestimated. The statutory categories of crime should not be confused with categories that serve to classify offenders.

B-346. In some cases, the result of displacing offenders to other crimes is beneficial. If the targeted crimes are more serious than the ones to which offenders are diverted, the net effect on the program may be the reduced danger to society. Of course, the converse may also be true; closing off the vulnerable and more easily protected targets of crime may cause an offender to commit more serious crimes with a net increase in the danger to society.

B-347. In some instances, the individual effect may be substantial but the overall effect may be negligible. Protecting a small fraction of premises against burglary will reduce the number of crimes committed against them, but the burglary rate against unprotected premises may go up.

Displacement to Other Tactics and Targets

B-348. Offenders can change their manner of committing a crime when a new program is established to counter their activity. One example of this took place in 1969 in a section of the Bronx that was having a rapid increase in outdoor crime. The crimes took place primarily in the evening hours when people were returning from work. The program instituted by the police consisted of intensive sweeps of randomly selected city blocks, coupled with plainclothes police officers patrolling the streets. It succeeded in reducing the number of offenses committed during the evening hours, but at the expense of increasing the number taking place in the late afternoon when patrolmen were taking their lunch hours or were occupied with school crossings or shift changes.

Displacement to Other Areas

B-349. The most frequently discussed type of crime displacement is from one area to another. For instance, it has been cynically suggested that the goal of the New York subway police is to chase crime into the streets where it belongs. More seriously, some recent police-helicopter-program evaluations have been questioned because they did not consider possible area displacements.

B-350. One type of boundary of interest to crime displacement is the jurisdictional boundary between the installation and surrounding cities. It has been conjectured that the crime reduction experienced in some central cities has been at the expense of the surrounding suburbs that have experienced increased crime rates.

B-351. An initial study of crime displacement was performed for the Washington, DC, area. It concluded that, although the decrease in Washington's crime rate was concurrent with an increase in the suburban crime rate, there is no evidence that the reduction in reported crime in Washington, DC, has resulted in a corresponding crime increase in the nearby suburbs.

B-352. The area-displacement effect can be measured with some degree of reliability. Three zones can be defined for the purposes of the measurement—the area containing the crime-control program (zone 1), a border around the area (zone 2), and the area chosen as the control area (zone 3). The width of the border may depend on the type of program implemented. If the program involves police helicopters, a quarter-mile-wide border may be necessary; for a patrol car, one or two blocks may suffice.

B-353. Crime rates before program initiation should be determined for all three zones. If zone 2 records a greater increase in crime than zone 3 while zone 1's crime rate decreases, then the increase in zone 2 can be attributed to two factors—

- The general increase in crime rate verified by any increase in zone 3.
- The increase caused by a displacement of crime from zone 1.

B-354. A displacement of this crime does not mean that the program is ineffective. It may suggest that the program should be expanded for all three zones.

UNREPORTED CRIME

B-355. Crime statistics are based on crimes reported to the police. It is well known that many crimes go unreported. Victimization studies can determine the extent of unreported crime and its change from year to year by area of the country or the reasons for failure to report them. These victimization studies are best suited to determining long-term effects. They are not well suited to most crime-control evaluations in which short-term changes must be assessed.

B-356. The amount of unreported crime is important but not for planning crime-control programs affecting police activities. The extent of unreported crimes is of little significance unless a program affects it. If a program encourages reported crime, the reported-crime rate may increase despite the program's effectiveness.

B-357. Ironically, a lowered reported-crime rate may be the direct result of an increase in the actual crime rate. Taking reports from victims of crimes occupies a substantial amount of a patrolman's time. Many of these crimes are minor and have no potential for solution. In an effort to increase the police department's time on patrol, some police chiefs have stopped the practice of sending a patrolman to get reports from the victim of a minor crime. This requires the victim to travel to the police station to report the crime. If the crime is minor or is seen by the victim to be unsolvable or the theft is not covered by insurance, the victim may decide not to inconvenience himself by going to the police station to report the crime. Therefore, the number of crimes reported to the police may drop. This may lead to a larger number of unreported crimes and prevent a complete picture to the crime-prevention counsel. Conversely, an actual decrease in crime due to the increased effectiveness of the police may produce an increase in the reported crime rate.

CRIME RATES

B-358. It has been pointed out that the crime rates, as presently calculated, do not reflect the true situation. For example, the rape rate should be calculated by dividing the annual number of rape cases by the number of women (since they are the population at risk). You would expect that the rate of commercial burglaries would be less in a residential area than a commercial area. When calculated on the basis of "per thousand people," this would be true; however, these rates should be obtained by dividing the number of cases by the number of commercial establishments (the population at risk) in each area.

B-359. The victim or the target is only one aspect of the crime. The offender can also be calculated into the rate. For example, the potential offenders in stranger-to-stranger crimes are usually considered to be males between 16 and 25 years of age. Therefore, one would expect fewer of these crimes in a city full of pensioners and retirees than in a city of the same population but with a higher proportion of young men. This fact is of minor importance in evaluating crime-control programs, since the age distribution of people in a city or a section of a city does not normally change greatly over the evaluation period. However, the former factor (the population at risk of becoming victimized) can be misleading if it is not taken into account. If possible, crime rates in experimental and control areas should be compared to the population that risks becoming victims of the target crimes.

MEASURES OF EFFECTIVENESS

B-360. The goals of the program determine the criteria that are used to measure its effectiveness. These goals and criteria should not be seen as confining; the evaluator should be amenable to broadening the criteria, especially if the program to be evaluated is a new one. For example, the program might be beneficial in some unforeseen way, wholly outside the original criteria. Conversely, the program may be an overall failure but a success according to the evaluation. It may be that the specified measures were the wrong ones to use for the program or should not have been used alone.

B-361. Programs aimed at controlling crime should not be evaluated solely for their effect on crime. Most programs cannot, by their very nature, focus on one specific objective alone. They normally are multifaceted in their effect and should be evaluated with respect to all of their facets. Similarly, the measures of effectiveness discussed in this section may not be adequate for every crime-control program, but they comprise some of the more useful measures that can be used.

B-362. This section concentrates on the two evaluation types—internal and external. Internal and external refer to whether the evaluation is conducted on the program's inner workings and logic or the external effect of the program (which depends on the program type). An internal evaluation of a crime-control program involving the use of new police patrol techniques would include the analysis of police response time and how it was effective in controlling crime or why it was successful in one area and not in another. The external evaluation would focus only on the effectiveness of the program in reducing crime rates or solving crimes, not on how or why or the conditions under which the results were achieved.

B-363. Evaluating how well a program achieved its goals is not the only purpose of an evaluation; how and why the results were achieved are of equal importance. External measures relate to the former evaluation; internal measures are concerned with the latter. The following examples will further serve to highlight the differences between these measures:

- Many crime-control programs depend on good community relations in order to achieve their goals. In these cases, a public-affairs campaign is often instituted concurrent with the crime-control program. The success of the public-affairs campaign should not be interpreted as program success. It may be a necessary part of the program, but it does not substitute for the results of the program in controlling crime. Testimonials from people involved in the program should also be considered only as a supplement to the evaluation based on external measures.

- A study undertaken for the President's Crime Commission showed that for certain types of incidents, the probability of arrest increased as the response time decreased. As a result of this finding, many police departments purchased new equipment or tried novel techniques to reduce response time without first determining whether their workloads included enough of the incidents for which quick response is useful. If this measure (response time) is to be used, it should be recognized as an internal measure and not substituted for the external evaluation.

INTERNAL MEASURES

B-364. Each program will have its own internal measures of effectiveness based on the logical elements of which it is constituted. This section covers only the internal measures of effectiveness that are common to most crime-control program evaluations. The measures covered include the crime rate, the clearance rate, the arrest rate, the crime-seriousness index, and the fear of crime.

CRIME RATE

B-365. The crime rate (the number of a specified type of crime committed per resident in a specified time period) is normally considered to be a measure of deterrence. If the crime rate decreases, it is presumed that potential offenders have modified their behavior to some extent and have committed fewer crimes. This is based on the assumption that the program has made the target crimes unattractive by increasing the actual or perceived risk of apprehension, by reducing the expected return from the crime, or by making alternative forms of behavior more attractive than the target group of offenses.

B-366. These deterrent effects use different means for their accomplishment. Most crime-control programs are police-oriented and concentrate on the risk-related aspects of deterrence. Victim-oriented programs focus on reducing the expected return. Many social and recreational programs deal with making alternatives more attractive. Regardless of the orientation of the programs, their deterrent effects are determined by measuring reported crime rates.

B-367. Reported crime rates can be changed by a number of factors, some of which are misleading. The public may feel that the police are becoming less effective in dealing with crimes and, therefore, report them less often. Conversely, if the public perceives that the police are becoming more effective, they may begin to report crimes that previously would have gone unreported. Another apparent crime-rate reduction may be due to the police not recording crimes that have been reported to them. Displacement effects that can produce misleading crime-rate reductions were discussed earlier.

B-368. There may also be an actual reduction in crime due to a program's deterrent effect. In some cases, the reduction in crime can be attributed to psychological deterrence. That is, the police department may have instituted some change (such as painting all police cars canary yellow) in preexisting patterns of operation that may cause a change in the behavior patterns of potential offenders. This type of deterrence is rarely long-lived.

B-369. On the other hand, there may have been a change instituted by the police that had the desired effect of increasing the actual risk of apprehension and, therefore, reducing the number of target offenses. An example of this is the police-operated burglar alarm of commercial establishments. In the experimental program, the number of alarms were increased almost tenfold compared to the (nonalarmed) control establishments. There was 1 capture in 36 control-group burglaries (2.8 percent), while there were 12 captures in 46 experimental-group burglaries (26 percent). Crime displacements to other crimes, tactics, targets, and areas reduced the actual effectiveness of the program, but this example shows that a significant change can be made in the actual risk of apprehension. Preliminary results indicate that the rate of increase of commercial burglaries has been decreased from about 15 percent per year to about 0 percent (at the expense of a greater increase in residential burglaries).

B-370. It is difficult but useful to distinguish between actual deterrence (due to an actual increase in risk) and deterrence that is purely psychological in nature (due to a perceived increase in risk). If it is suspected that part of the deterrent effect may be transient, a long-term study would be of benefit. In this way, the half life of the psychological deterrence can be gauged, which can give some indication of the extent to which resources should be committed to the program.

B-371. Some forms of psychological deterrence are almost entirely counterproductive. They may appear effective to those who would not commit a crime and ineffective to those who are "in the business" and study the presumed deterrent more closely. For example, a tear-gas pen may give a person a sense of security that is entirely without foundation. It may be dangerous to him if he actually attempts to use it when faced with an assailant.

B-372. One investigator has pointed out that for given criminal situations, nondelinquents perceive a higher risk of apprehension than do delinquents; in all probability, the delinquents have a more realistic assessment of the situation. A purely psychological deterrent may have the unfortunate effect of making only a cosmetic improvement. This gives the general population the impression that there has been a change for the better, while in reality the situation may not have changed or may have changed for the worse because of the division of resources to a nonexistent solution.

B-373. The crime rate can be used as a measure of effectiveness. However, the evaluator should delve into the determination of the crime rate to see if any change in the rate reflects a change in reporting procedures or the deterrent effect (with tangible evidence).

CLEARANCE RATE

B-374. The clearance rate is normally considered to be a measure of the ability of police to solve crimes. A cleared crime is one in which the police have identified the offender and have sufficient evidence to arrest him. The clearance rate is the percentage of total crimes that were cleared.

B-375. This measure of effectiveness should be used with care. A decreasing clearance rate may not mean that a police department is becoming less effective, and an increasing clearance rate may not mean that it is becoming more effective. This is due to a number of factors, primarily the public's conception of the role of the police with respect to crime and the present method of collecting crime data.

B-376. Often overlooked in discussion about crime is the role of the public in assisting the police. Police rely on community support to legitimize their authority as well as to help them carry out their work. If a segment of the community becomes alienated from the police (for whatever reason) and offers them little assistance in pursuing offenders, crime rates in these areas may rise. However, it is not only alienation of community groups that reduces the ability of the police to deal with crime; the profit motive is also to blame. Many store owners that have been robbed refuse to give their clerks time off (with pay) to help the police in their investigation. They absorb the loss of a robbery easily (it rarely comes close to the amount lost from shoplifting, employee theft, and damaged goods) and are unwilling to increase it by helping the police. They may feel that the chances of apprehending the offender are too slim, or they may be afraid of retribution if the offender discovers their assistance. They may also be afraid that their insurance will be cancelled.

B-377. If a police department begins a drive to increase its clearance rate, the increase may be forthcoming without any real change in police effectiveness. A survey of three police departments found that arrests for felonies were not made by the police in about 43 percent of the cases, in which there was probable cause, while the police were accompanied by witnesses. Making arrests in such instances would inflate the clearance rate quite easily. However, it should be noted that the police officer has a great deal of discretion in the exercise of his power of arrest. He may feel that the arrest charges will not hold up. One measure of the arrest quality is the percentage of arrests that leads to prosecutions.

B-378. In summary, clearance rate can be a measure for determining the effectiveness of crime-control programs. Its use can be increased by careful selection and specification of the crime categories that are studied, by determining the manner in which the crimes were cleared, and by determining if there has been a change in where the police draw the line in the exercise of their discretion.

ARREST RATE

B-379. Another measure of effectiveness that is often used as a determinant of crime-control effectiveness is the arrest rate, calculated either per police officer or per resident for a specified time period. Most of the considerations concerning the clearance rate (discussed above) also apply to the arrest rate. However, the arrest rate is distinguished from the clearance rate by an additional factor—it is not related to the total number of offenses. For example, the number of arrests for drug violations has risen considerably over the past few years. However, this increase is indicative of the extent of the problem, not of the effectiveness of the solution. It has been described how drug arrests may be traded off against arrests for other offenses and vice versa, especially when informal arrest quotas are established. Therefore, the use of the arrest rate by itself does not appear to be appropriate as a measure of the effectiveness for most crime-control programs.

CRIME-SERIOUSNESS INDEX

B-380. Among the many criticisms of crime statistics is the contention that, even if the data were reliable and complete, we would still have only a count of the number of incidents without an indication of their relative seriousness. The crime-seriousness index was proposed to include some of the major disutilities of crime typically committed by juveniles. Crimes are weighted according to the degree and nature of injury to the victims—whether they were intimidated and the nature of the intimidation or whether premises were forcibly entered, and the kind and value of property stolen. The weights were determined by requesting a sample of people to estimate the relative seriousness of various crimes.

B-381. All of the factors used to determine the weight are (or should be) included in offense reports. It would not be difficult to calculate an incident-seriousness score based on these reports, either for a specific evaluation or as a matter of course. Use of the seriousness index has also been proposed to measure the relative performance of law-enforcement agencies.

B-382. The crime-seriousness index is not the ultimate weighting scheme. The seriousness appears to be calculated more from the offender's viewpoint and the event than from the victim or society's viewpoint. For example, most people would consider the murder of a robbery victim by his assailant to be more serious than the murder of one spouse by the other. With regard to property loss, there is a difference between loss suffered by an individual who is insured and one who is not covered.

B-383. The loss relative to the individual's income is also an important factor; the theft of a $100 television from a low-income family has a much greater impact than the loss of $10,000 of jewels from a wealthy family. Perhaps a better index of the relative value of property loss to the victim would be the value of the loss in relation to the amount of the individual's discretionary income (that is, income not used for the basic necessities of life). Of course, such information is not available on police crime reports.

FEAR OF CRIME

B-384. It has been pointed out that the perceived risk of crime is greater than the actual risk of crime, and that this perceived risk does not seem to be correlated with the actual crime rate. Unless the public feels safer in proportion to its increased actual safety, the full potential of improvements will not have been reached. Therefore, the goal of a crime-control program can be broadened to include not only improved public safety (deterrence), effectiveness (clearance rate), and reduced crime impact (seriousness); but also improved, more accurate, public perceptions of safety as well.

B-385. Measurements of perceived safety can be both direct and indirect. Public-opinion surveys with regard to perceptions about crime and safety have been made frequently. It is also possible to gauge the effect of this fear using indirect measures by observing what people do rather than what they say. The number of patrons of movie theaters and restaurants at night (or other observations of this type of activity) could be used to gauge the fear of crime.

B-386. A reliable measure of the public's perception of public safety has not been developed. Additional research is being done and needs to be done before this type of measure of effectiveness can be used with confidence.

Appendix C

Intelligence, Counterintelligence, and Threat Analysis

Intelligence and counterintelligence make up the first line of defense in an antiterrorism program. A well-planned, systematic, all-source intelligence and counterintelligence program is essential. The role of intelligence and counterintelligence in antiterrorism is to identify the threat. Additionally, counterintelligence provides a warning of potential terrorist attacks and provides information for counterterrorism operations. This appendix provides the elements of the intelligence cycle that have particular importance in a viable antiterrorism program. Effective intelligence and counterintelligence support requires effort, planning and direction, collection and analysis, production, investigation, and dissemination. The entire process provides decision makers with information and timely warnings upon which to recommend antiterrorism actions.

INFORMATION SOURCES

C-1. The primary sources of intelligence and counterintelligence for the antiterrorism program are open-source information, criminal information, government intelligence and counterintelligence, and local information.

- **Open-source information.** This information is publicly available and can be collected, retained, and stored without special authorization. The news media is an excellent open source of information on terrorism. The media reports many major terrorist incidents and often includes in-depth reports on individuals, groups, or various government counterstrategies. Government sources include congressional hearings; publications by the Defense Intelligence Agency (DIA), the FBI, the Central Intelligence Agency (CIA), the Department of State (DOS), and the national criminal justice reference services. Additionally, there are private data services that offer timely information on terrorist activities worldwide. Terrorist groups and their affiliates may also have manuals, pamphlets, and newsletters that reveal their objectives, tactics, and possible targets.
- **Criminal information.** Both military and civil law-enforcement agencies collect criminal information. Because terrorist acts are criminal acts, criminal information is a major source for terrorist intelligence. Commanders must work through established law-enforcement liaison channels because the collection, retention, and dissemination of criminal information are regulated. Local military criminal investigative offices of the CID; the Naval Investigative Service Command (NISCOM); the Air Force Office of Special Investigations (AFOSI); and Headquarters, US Marine Corps,

Criminal Investigations Division, maintain current information that will assist in determining the local terrorist threat.

- **Government intelligence and counterintelligence.** The Community Counterterrorism Board (CCB) is responsible for coordinating with the national intelligence agencies concerning combating international terrorism. These agencies include the CIA (the lead agency), the DIA, the National Security Agency (NSA), the DOS, the Department of Justice (DOJ), the FBI, the Department of Energy (DOE), the Federal Aviation Administration (FAA), the Department of Transportation (DOT) (including the USCG), and the DOD. Service intelligence and counterintelligence production organizations include the US Army Counterintelligence Analysis Center; the Navy Antiterrorism Analysis Center (NAVATAC); Headquarters, US Marine Corps, Counterintelligence; and the US AFOSI Operations Center. These organizations compile comprehensive intelligence and counterintelligence for distribution on a need-to-know basis throughout the services. In combatant commands, the J2 is responsible for the intelligence fusion center. The Counterintelligence Support Officer (CISO) provides interface between the combatant command, the component commands, and the joint staff.
- **Local information.** Other valuable sources of information are the soldiers, civil servants, family members, and individuals with regional knowledge (such as college faculty or members of cultural organizations). Local crime or neighborhood watch programs can be valuable sources of information and can serve as a means to keep individuals informed in dispersed and remote areas. Intelligence exchanges with local government agencies through cooperative arrangements can also augment regional information.

RESPONSIBILITIES OF US GOVERNMENT LEAD AGENCIES

C-2. The FBI is responsible for collecting and processing domestic terrorist information. Overseas, terrorist intelligence is principally a CIA responsibility; but the DOS, the DIA, and the HN are also active players. The MI activities are conducted according to Presidential executive orders, federal law, status of forces agreements (SOFAs), memorandums of understanding (MOUs), and applicable service regulations. Responsibilities of intelligence activities include the following:

- The combatant commander (through the commander's J2 and the CISO in coordination with the DIA, the CIA, the embassy staff, the country team, and applicable HN authorities) obtains intelligence and counterintelligence specific to the AO. The commander issues intelligence and counterintelligence reports, advisories, and assessments to the units within his command or those operating within his command's AO. This network is the backbone for communicating intelligence and counterintelligence information and advisories and for warning of terrorist threats throughout the region.
- The Secretaries of the military departments were asked (in DOD Directive 2000.12) to ensure that a capability exists to receive and evaluate data from a service perspective and that the capability exists

to disseminate all relevant data on terrorist activities, trends, and indicators of imminent attack. To accomplish this task, each Secretary appoints an MI agency (INSCOM, NISCOM, or AFOSI) to conduct intelligence and counterintelligence activities directed against terrorists and to detect, neutralize, or deter terrorist acts. To accomplish this mission, the military department intelligence agency establishes counterintelligence offices on an area basis to collect and disseminate information to combatant commanders. Each military department's intelligence agency—

- Coordinates with appropriate US and HN agencies.
- Provides overall direction and coordination of the service counterintelligence effort.
- Operates a 24-hour operations center to receive and disseminate worldwide terrorist threat information to and from the combatant command's J2, applicable service staff elements, subordinate commands, and national agencies.
- Provides service commanders with information on terrorist threats concerning their personnel, facilities, and operations.
- Investigates terrorist incidents for intelligence, counterintelligence, and force-protection aspects (with the FBI or HN authorities).
- Provides terrorist threat information in threat briefings.
- Conducts liaison with representatives from federal, state, and local agencies (as well as HN agencies) to exchange information on terrorists.
- Provides international terrorism summaries and other threat information to supported commanders. On request, provides current intelligence and counterintelligence data on terrorist groups and disseminates time-sensitive and specific threat warnings to appropriate commands.

- Service criminal investigative services (such as the CID, the NISCOM, and the AFOSI) collect and evaluate criminal information and disseminate terrorist-related information to supported installation and activity commanders as well as to the service lead agency. As appropriate, criminal investigative elements also conduct liaison with local military or security police and civilian law-enforcement agencies.
- Intelligence staff elements of commanders at all echelons—
  - Report promptly all actual or suspected terrorist incidents, activities, and early warnings of terrorist attacks to supported and supporting activities, to the local counterintelligence office, and through the chain of command to the service lead agency.
  - Initiate and maintain liaison with the security police or PMO; local military criminal investigative offices; local counterintelligence offices; security offices; HN agencies and; as required, other organizations, elements, and individuals.
  - Develop and present terrorism threat-awareness briefings to all personnel within their commands (in cooperation with the local counterintelligence offices).
- Law-enforcement staff elements will—

  - Report all actual or suspected terrorist incidents or activities to their immediate commander, supported activities, and service lead agency through established reporting channels.
  - Initiate and maintain liaison with local counterintelligence offices and military criminal investigative offices.
  - Maintain liaison with federal, HN, and local law-enforcement agencies or other civil and military antiterrorism agencies as appropriate.
- Installation, base, unit, and port security officers—
  - Report all actual or suspected terrorist incidents or activities to their immediate commander, supporting military law-enforcement office, other supported activities, local counterintelligence office, and local military criminal investigation office.
  - Conduct regular liaison visits with the supporting military law-enforcement office, counterintelligence office, and local criminal investigation office.
  - Coordinate with the supporting military law-enforcement office and counterintelligence office on their preparation and continual updating of the threat assessments.
  - Assist in providing terrorism threat-awareness training and briefings to all personnel and family members as required by local situations.

INFORMATION REQUIREMENTS

C-3. To focus the threat analysis, intelligence and counterintelligence officers develop information requirements to identify targets using the following terrorist considerations—

- Organization, size, and composition of group.
- Motivation.
- Long- and short-range goals.
- Religious, political, and ethnic affiliations.
- International and national support (moral, physical, and financial).
- Recruiting methods, locations, and targets (students).
- Identity of group leaders, opportunists, and idealists.
- Group intelligence capabilities and connections with other terrorist groups.
- Sources of supply and support.
- Important dates, such as religious holidays.
- Planning ability.
- Internal discipline.
- Preferred tactics and operations.
- Willingness to kill.
- Willingness for self-sacrifice.
- Group skills, demonstrated or perceived (for example, sniping, demolitions, masquerade, industrial sabotage, airplane or boat

operations, tunneling, underwater, electronic surveillance, poisons, or contaminants).

- Equipment and weapons (on hand and required).
- Transportation (on hand and required).
- Medical-support availability.
- Means and methods of C^2.
- Means and method of communication.

THREAT ANALYSIS AND ASSESSMENT

C-4. The preparation of the terrorist threat analysis is a continual process of compiling and examining all available information to identify terrorist targeting of US interests. A vulnerability analysis is a continual process of compiling and examining information on a facility's security posture. The threat analysis is then paired with the facility's vulnerability analysis to create the threat and vulnerability assessment. Threat analysis is an essential step in identifying the probability of a terrorist attack. To enhance the capability to collect and analyze information from many sources, the DIA maintains a terrorism database. The combatant command's J2 and CISO (in coordination with the DIA) focus this database information and regional information toward the intelligence and counterintelligence needs specific to the security of the command. However, this terrorism database is limited to foreign terrorist groups because of limitations on US intelligence-collection operations. A country's threat assessments, information and biographies about terrorist organizations, and incidents in the database can be disseminated to commands. Commands at all echelons then augment or refine the DIA's threat analysis to focus on their area of interest. This process, operative across the full range of military operations, promotes coordination between all levels of the intelligence, counterintelligence, and law-enforcement communities; broadens acquisition channels; and enhances timely distribution of information to the supported commander.

C-5. Several factors complicate intelligence and counterintelligence collection and operations. The small size of terrorist groups, coupled with their mobility and cellular organization, make it difficult to identify the members. Unlike other criminals, terrorist cadres often receive training in counterintelligence and security measures from foreign intelligence agencies or other terrorists. Additionally, the traditional orientation of police organizations is toward individual criminals, while MI organizations focus on conventional forces. Terrorist activity, therefore, requires some degree of reorientation for police and MI and counterintelligence collection and operations.

C-6. An intelligence system's ability to provide critical and timely information to the user depends not only on efficient collection and processing, but also on the ability to organize, store, and retrieve this information rapidly. This capability, coupled with early warning, careful observation, and assessment of threat activity, enhances the probability of accurately predicting the types and timing of terrorist attacks.

C-7. Commanders must carefully exercise judgment in estimating both the existing terrorist threat and the need for changes in antiterrorism measures. Key questions are—

- What has changed (mission, political climate, installation and unit personnel or equipment, terrorist capabilities)?
- What affect will the changes have on the security posture?

C-8. Extraordinary security measures, unless part of a deliberate deception during critical or high-threat situations, draw attention and detract from mission accomplishment. Sound physical security, personnel who are aware, accurate threat and vulnerability assessments, and well-rehearsed response plans reduce the probability of a successful terrorist venture. The goal is to make an attack too difficult or the level of risk unacceptable to the terrorist.

DETERMINATION OF THE THREAT LEVEL

C-9. This threat-analysis methodology is used by the DIA, the joint staff, and the unified and specified commands for selecting the level of threat for an installation. It is applicable in an overseas setting, but fails to address issues unique to the sustaining base within CONUS. In CONUS there is a lack of intelligence and counterintelligence information from the CIA, the DIA, or the military services for the CONUS-based threat. That information must be extracted from law-enforcement channels at the local, regional, and national levels. In that context, the factors mentioned in this appendix are not as clear as they are within the intelligence process in place overseas. A modified version of this appendix should be considered for assessing threat levels in CONUS.

C-10. Threat levels within CONUS historically have been either low or negligible. This trend will most likely continue at least through the next decade. Domestic groups not covered by DOD intelligence reporting pose the greatest threat to the CONUS-based military. Due to the lack of reporting and information on these groups, the domestic terrorist groups are not currently factored into the current terrorist threat program. Therefore, the law-enforcement community must become a key player in establishing the threat levels in the context of the recommended model. A possible methodology would be the establishment of an additional threat level between low and medium—one that allows local commanders more flexibility in implementing additional security measures. Table C-1 shows the procedure for determining the threat level.

Table C-1. Threat Levels

Threat Level	
Critical	Factors 1, 2, and 5 are present. Factors 3 or 4 may be present.
High	Factors 1, 2, 3, and 4 are present.
Medium	Factors 1, 2, and 4 are present.
Low	Factors 1 and 2 are present. Factor 4 may be present.
Negligible	Factors 1 and/or 2 may be present.
Explanation of Factors	
Factor 1: Existence. A terrorist group is present, assessed to be present, or able to gain access to a given locale. Factor 2: Capability. The acquired, assessed, or demonstrated level of capability to conduct terrorist attacks. Factor 3: Intentions. Recent demonstrated anti-US terrorist activity or stated and/or assessed intent to conduct such activity. Factor 4: History. Demonstrated terrorist activity over time. Factor 5: Targeting. Current credible information on activity indicative of preparations for specific terrorist operations and/or specific intelligence that shows an attack is imminent.	

Appendix D

Crisis-Management Plan

The following pages highlight areas of concern in crisis-management planning. This plan is not meant to be all-inclusive or rigidly followed. Figure D-1 is a sample format only. It does not reflect a format developed and approved for use with OPLANs or contingency plans (CONPLANs) prepared by the CINCs to fulfill tasks assigned in the Joint Strategic Capabilities Plan (JSCP) or as otherwise directed by the Chairman of the Joint Chiefs of Staff. Figure D-2, page D-4, is a sample of the Crisis-Management-Plan Checklist, which is Annex A or Appendix H to the crisis-management plan. This checklist will help ensure that the plan is sound.

Copy No. ________
Issuing Headquarters
Place of Issue
Date of Issue

Crisis-Management Plan

Ref: Maps, charts, and other relevant documents.

Time Zone: X

Task Organization: List units organized to conduct antiterrorism operations. Include attachments, supporting roles, and the delegation of operational control as necessary.

1. Situation. Identify essential information to understand ongoing events.

a. Terrorist force. Identify the terrorist's composition, disposition, methods of operation, and estimated strengths and capabilities that could influence the crisis-management operation. Refer to an appropriate annex.

b. Response force. Explain the response force's abilities and responsibilities. The response force's abilities can influence the crisis-management mission.

c. Attachments and detachments. Address here or refer to an annex.

d. Assumptions. Provide assumptions used as a basis for this plan (for example, the strength of the response force to be supported and the support available from other agencies).

Figure D-1. Sample Crisis-Management Plan

(1) Tactical-situation possibilities. Obtained from the commander's planning guidance.

(2) Personnel situation. Provided by the personnel officer.

(3) Logistics situation. Provided by the logistics officer.

(4) Legal-situation possibilities. Provided by the SJA.

2. Mission. Identify the antiterrorism mission (for example, detect, deter, contain, and neutralize terrorist threats and actions aimed at the disruption of the installation).

3. Execution.

a. Concept of operations. State the commander's tactical plan. The purpose is to inform. It may address how the commander will conduct combat-terrorism operations. It provides enough detail to ensure proper action by subordinates in the absence of specific instructions. If the required details are extensive, address them in an annex. If an operation involves two or more distinct phases, designate each phase and use subparagraphs (for example, Phase I and Phase II).

b. Tasks. Identify specific tasks for each command element charged with executing a crisis-management mission. When giving multiple instructions, itemize and indicate the priority or sequence.

c. Coordinating instructions. Include coordination and control measures applicable to two or more command elements.

4. Service Support. Provide a statement of service-support instructions and arrangements supporting the crisis-management operation. Use the following subparagraphs as required:

a. General. Outline the general plan for service support.

b. Materiel and services. Address supply, transportation, labor, and services required.

c. Medical evacuation and hospitalization. Provide the plan for evacuating and hospitalizing sick, wounded, or injured personnel. Address evacuation responsibilities and the air-evacuation policy.

d. Personnel. Provide required information and instructions to supporting unit personnel.

(1) Maintenance of unit strength.

(a) Strength reports. Provide instructions for submitting status reports. Include requirements for routine and special reports.

(b) Replacements. Address validating existing personnel requisitions, instructions for submitting requisitions, and instructions for processing and removing replacements.

(2) Personnel management. Address military and civilian personnel and civilian detainee management procedures.

(3) Development and maintenance of morale.

(a) Morale and personnel services. Provide postal and financial services, religious activities, personal hygiene, and special services activity information.

(b) Graves registration. Include evacuation procedures and handling personal effects.

Figure D-1. Sample Crisis-Management Plan (continued)

(4) Maintenance of discipline, law, and order. Obtain this guidance from the PMO/security officer.

(5) Miscellaneous. Include personnel administrative matters not specifically assigned to another coordinating staff section or included in preceding subparagraphs.

e. Miscellaneous. Provide special instructions or special reports not covered in preceding paragraphs.

5. Command and Signal. Provide instructions for the command and operation of communications-electronics equipment. Communications-electronics instructions may refer to an annex but should list the index and issue number of the command, control, and communications (C^3) operation instructions in effect. If not already issued, give instructions for the control, coordination, and establishment of priorities in the use of electromagnetic emissions. Command instructions include subordinate and higher unit CP locations and designated alternate CPs.

6. Acknowledgement Instructions.

/s/

Commander

Annexes as applicable

Distribution:

Figure D-1. Sample Crisis-Management Plan (continued)

Annex A

Crisis -Management Plan Checklist

Yes	No	
		1. Intelligence and/or Counterintelligence.
____	____	Does the plan allow for the threat-analysis process (collection, analysis, production, and dissemination) to help identify the local threat?
____	____	Does the plan consider restrictions placed on the collection and storage of information?
____	____	Does the plan indicate an awareness of sources of information for the threat-analysis process (MI, counterintelligence, federal agencies, and state and local authorities)?
____	____	Does the plan allow for liaison and coordination of information (such as establishing a committee)?
		2. Threat Assessment.
____	____	Does the plan identify the local threat (immediate or long term)?
____	____	Does the plan identify other threats (such as national and international groups that have targeted or might target US installations)?
____	____	Does the installation incorporate factors for assessing the threat? Does it address —
____	____	Geography of the area concerned?
____	____	Law-enforcement resources?
____	____	Population cultural resources?
____	____	Communication capabilities?
____	____	Does the plan establish a priority of identified weaknesses and vulnerabilities?
____	____	Is the threat assessment periodically updated?
		3. Security Countermeasures.
____	____	Does the plan have specified THREATCONs and recommended actions?
____	____	Do security countermeasures include a combination of physical operations and sound-blanketing security measures?
____	____	Do the THREATCONs correspond to DOD 0-2000.12-H, Appendix BB?

Figure D-2. Sample Crisis-Management-Plan Checklist

Yes	No	
		4. OPSEC.
____	____	Have procedures been established that prevent terrorists from readily obtaining information about plans and operations (for example, not publishing the commanding general's itinerary and safeguarding classified material)?
____	____	Does the plan allow for in-depth coordination with the installation's OPSEC program?
____	____	Has an OPSEC annex been included in the CONPLAN?
		5. Personnel Security.
____	____	Has the threat analysis identified individuals vulnerable to terrorist attacks?
____	____	Has a training program been established to educate both military and civilian personnel in the proper techniques of personnel protection and security commensurate with the local threat and the type of position held?
		6. Physical Security.
____	____	Are special-threat plans and physical-security plans mutually supportive?
____	____	Do security measures establish obstacles to terrorist activity (such as guards, HN forces, lighting, and fencing)?
____	____	Does the special-threat plan include the threats identified in the threat statements of higher headquarters?
____	____	Does the physical-security officer assist in the threat analysis and corrective action?
____	____	Does the installation have and maintain detection systems and an appropriate assessment capability?
		7. Security Structure.
____	____	Does the plan indicate that the FBI has primary domestic investigative and operational responsibility in the US and US territories?
____	____	Has coordination with the SJA been established?
____	____	Does the plan allow for close cooperation between principal agents of the military, civilian, and HN communities and federal agencies?
____	____	Does the plan clearly indicate parameters for the use of force, including briefing any elements augmenting MP assets?
____	____	Is there a mutual understanding between all local agencies (military, local, FBI resident or senior agent-in-charge, HN forces, and local law enforcement) that might be involved in a terrorist incident on the installation regarding authority, jurisdiction, and possible interaction?

Figure D-2. Sample Crisis-Management-Plan Checklist (continued)

Yes	No	
___	___	Has the SJA considered the ramifications of closing the post (such as possible civilian union problems)?
___	___	Does the plan identify the DOS as having primary investigative and operational responsibilities overseas?
		8. Operations-Center Training.
___	___	Has the operational command and coordination center been established and exercised?
___	___	Is the operations center based on the needs of the installation while recognizing manpower limitations, resource availability, equipment, and command?
___	___	Does the plan include a location for the operations center?
___	___	Does the plan designate alternate locations for the operations center?
___	___	Does the plan allow for the use of visual aids (chalkboards, maps with overlays, and bulletin boards) to provide status reports and countermeasures?
___	___	Does the plan create and designate a location for a media center?
___	___	Have the operations and media centers been activated together within the last quarter?
___	___	Does the operations center have SOPs covering communications and reports to higher headquarters?
___	___	Does the operations center offer protection from a terrorist attack?
		9. Reaction-Force Training.
___	___	Has the force been trained and exercised under realistic conditions?
___	___	Has corrective action been applied to shortcomings and deficiencies?
___	___	Has the reaction force been formed and mission-specified trained (for example, building entry and search techniques, vehicle assault operations, countersniper techniques, and equipment)?
___	___	Has the reaction force been tested quarterly (alert procedures, response time, and overall preparedness)?
___	___	Has responsibility been fixed for the negotiation team? Has the negotiation team been trained and exercised under realistic conditions?
___	___	Does the negotiation team have the proper equipment?
		10. General Observations.
___	___	Was the plan developed as a coordinated staff effort?
___	___	Does the plan outline reporting requirements (logs, journals, and after-action reports)?

Figure D-2. Sample Crisis-Management-Plan Checklist (continued)

Yes	No	
____	____	Does the plan address the media's presence?
____	____	Does the plan include communication procedures and communication nets?
____	____	Does the plan consider the possible need for interpreters?
____	____	Does the plan consider the need for a list of personnel with various backgrounds to provide cultural profiles on foreign subjects and victims as well as to assist with any negotiation efforts?
____	____	Does the plan provide for and identify units that will augment MP assets?
____	____	Does the plan delineate specific taskings for each member of the operations center?
____	____	Does the plan provide for a response force for each phase of antiterrorism activity (initial response, negotiation, and assault)?
____	____	Does the plan designate service-support communications?
____	____	Does the plan make provisions for the notification of an accident-and-incident control officer?
____	____	Does the plan provide for EOD support?
____	____	Does the plan take into consideration the movement from various locations, including commercial airports, of civilian and military advisory personnel with military transportation assets?
____	____	Does the plan allow for the purchase or use of civilian vehicles, supplies, and food (if needed)? (This includes items used to satisfy a hostage demand.) Does the plan make provisions for paying civilian employees overtime if they are involved in a special-threat situation?
____	____	Does the plan take into consideration the messing, billeting, and transportation of civilian personnel?
____	____	Do appropriate personnel have the necessary language training?
____	____	Is WMD support available?

Figure D-2. Sample Crisis-Management-Plan Checklist (continued)

Appendix E

Office Security Measures

The office environment should afford executives the greatest degree of physical security. Executives usually work in facilities where attackers must pass by guards, security checkpoints, office workers, aides, or secretaries before reaching them. Unfortunately, the high media value of attacking executives in security strongholds where they are clearly associated with government activity increases the value of such attacks to terrorists. Hence, there may be a need to add security measures to offset the escalating capability of attack on more secure office areas by terrorist groups.

PHYSICAL-SECURITY SURVEY

E-1. A thorough physical-security survey of an office facility should be conducted. Offices of defense components attached to US embassies abroad should have these surveys performed by the DOS. Other DOD facilities should have surveys performed by the cognizant physical-security and facilities-engineering staffs. The best way to approach a physical-security site survey is to think like an intruder. Consider how approaches to the installation or facility could be made, how access to the building that houses executive offices could be gained, and how attacks on offices or other frequently used facilities could be mounted.

SECURITY-ENGINEERING ASSESSMENT

E-2. The next step in evaluating the need for supplemental physical-security measures is a thorough and detailed assessment of the weapons and tactics that terrorists might use to attack the structure in which DOD executives work. Security engineers and architects need technical threat data or assessments containing the following information—

- The mode of attack, such as—
  - Standoff weapons (man-portable AT/antiaircraft weapons, sniper rifles, rock grenades, and mortars).
  - Close combat weapons (submachine guns, pistols, knives, and garrotes).
  - Contact weapons (bombs, incendiary devices, and mines).
- Perimeter penetration aids (such as power tools, hand tools, or explosives), if used.
- The time of attack.
- The attacking force's size.
- The anticipated degree of outside support or autonomy.

E-3. Engineering design requirements are developed from the security engineering assessments. The data is used to—

- Assess the ability of building components to resist the effects of the threat.
- Identify appropriate security window-glazing materials and window treatments to determine what is required to achieve the desired penetration resistance times for anticipated threats.
- Calculate the total amount of delay time. This time is achieved by using camouflage, deception, barriers, and security devices to permit response forces to reach the scene of a terrorist attack in time to thwart the attack, capture or eliminate the terrorists, and rescue executives and their staffs or dependents.

TECHNICAL ASSESSMENT OF RESPONSES

E-4. After establishing a basic-design threat, engineers need data on the anticipated performance of response forces to be arrayed against the design threat and the expected or desired behavior of the protected executive. Some specific information needed includes—

- The response force's size, capability, supporting weapons, response time, and estimated effectiveness against the range of attacks.
- The desired options for the executive's protection—evacuate on warning, on detection, only if attacked, or only if forced to capitulate or do not evacuate.

E-5. Security planners need to know how long the structure that houses executives can withstand an attack before help arrives. Matching threat capabilities and anticipated operations by response forces establishes significant physical-security-system performance parameters. These can be quantified and used to develop detailed plans, drawings, and physical-security equipment-acquisition plans.

PHYSICAL-SECURITY ENHANCEMENT MEASURES

E-6. Several physical-security measures intended to provide additional protection for executives can be considered based on the requirements defined through the detailed analyses outlined above. The primary purpose of such measures should be to increase the time required by persons outside an installation to reach the executives housed at an installation. A secondary purpose of such measures should be to reduce or eliminate hazards to executives that might result from violence in the vicinity. Examples of physical-security measures to consider are—

- Increase the threat-detection time by installing sensors on perimeters and barriers. This includes—
  - Combining surveillance systems including seismic, acoustic, and IR sensors at or beyond the outer perimeter.
  - Supplementing surveillance systems with CCTV/imaging IR systems tied into the alert response-force staging area.

  - Extending restricted areas or exclusion zones and relocating access-control points from the executive office area to a point closer to the installation's boundary.
  - Enlarging and extending intrusion-detection sensors from within the installation to its perimeter, allowing the IDS to collect additional data necessary and sufficiently classify and identify an intrusion before the response force arrives.
  - Enhancing both the number and the phenomenology of surveillance and detection systems within the executive office area as well as approaches leading to and from it in conjunction with measures listed below.
- Increase the threat's delay time between the perimeter and the executive office building. This includes—
  - Installing vehicle barriers and realigning roadways to eliminate straight, level stretches of road in excess of 50 meters in length.
  - Increasing concentric rings of fences, Jersey barricades, planters, bollards, and vehicle/personnel barriers.
  - Enhancing access-control areas supplemented by fire doors/security doors kept in a closed condition between the entrance to the building that houses executive offices and the executive office area.
- Confuse, camouflage, and deceive observers by hiding an executive's location. Accomplish this by—
  - Relocating executives to buildings not usually associated with office activities (barracks, motor pools, research and development [R&D] facilities, and so forth).
  - Constructing office areas in the barracks, motor pool, R&D facilities, and so forth.
  - Adding executive styles, decorative lighting, and window treatments to several different areas of office buildings to minimize the differences in external appearances between executive and nonexecutive offices.
- Increase the delay time between the entrance to the building that houses executives and the executive office area. Execute this by—
  - Adding fire doors, access-control points, dead-end corridors, and midcorridor physical barriers to complicate access to the executive office areas.
  - Adding security devices that, when activated, disrupt the intruder's ability to retain his thought processes (for example, flashing strobe lights, fog generators, noise generators, sirens, and fire-extinguishing systems).
- Increase the delay time by making access more difficult within the executive office structure. This may be accomplished by—
  - Substituting high-security doors and door frames for standard doors in areas leading to or from executive offices.
  - Installing high-security grating, wire mesh, or other materials to bar access to the executive office area through utility tunnels or conduits.

  - Strengthening walls, floors, and ceilings by substituting steel-plate, concrete-filled, steel-reinforced cinder blocks or other ballistic-resistant materials for plaster/lath or wallboard room dividers, thereby protecting against explosive devices that are used as tools to breach a barrier.
- Increase the protection for building occupants against weapons and explosives effects. This includes—
  - Substituting blast- or bullet-resistant panels for glass windows or adding a fragment-retention film at least 4 millimeters thick to the interior of glass windows.
  - Adding exterior screens/plates to cover window areas and protect against gunfire and grenade/bomb fragments.
  - Installing blast curtains, metal blinds, metal shutters, or other window treatments in executive offices to protect interior space from glass shards and other small projectiles.
  - Strengthening walls to resist weapons and explosives effects by adding steel plates, reinforced concrete, or other retrofitting measures.
  - Adding steel plates or other ballistic materials in crawl spaces above dropped ceilings or extending walls separating the executive office area from other portions of an office building from floor to floor, thereby preventing unobserved and undetected access to the space between dropped ceilings.
- Increase the hold time to contain penetrators by—
  - Adding positive-action controls to a facility's doors and gates so that gates default to a closed and locked condition unless manually released.
  - Adding positive-action controls to access-control areas so that persons inside an access-control area can neither advance nor withdraw without affirmative action by a security officer posted outside the access-control area.

E-7. These measures are used to facilitate the apprehension of terrorists. There may be some instances when defeating terrorist attempts to gain access to the executive enhances the security of the executive and the response force. This is accomplished by channeling the terrorists out of the facility and installation along one route, leaving alternative routes available to evacuate executives and other key personnel.

E-8. Install emergency executive-support facilities (including a safe haven and an emergency evacuation facility) by—

- Installing helicopter landing aids on a structure's roof or on an adjacent field far removed from parking areas.
- Installing a safe haven or other reinforced security structure adjacent to a helicopter landing facility to provide a secure waiting place for executives until a rescue helicopter with additional supporting air and ground units can extract the executives.

Appendix F

Physical-Security Plan

It is essential and in the best interest of security that each installation, unit, or activity maintains and uses a detailed physical-security plan. The plan should include at least special and general guard orders, access and material control, protective barriers/lighting systems, locks, and IDSs. All physical-security plans have the potential of being classified documents and must be treated accordingly. Figure F-1 depicts a sample physical-security plan.

Map Reference

Copy No. ________
Issuing Headquarters
Place of Issue
Date of Issue

Physical-Security Plan

1. **Purpose.** State the plan's purpose.

2. **Area Security.** Define the areas, buildings, and other structures considered critical and establish priorities for their protection.

3. **Control Measures.** Define and establish restrictions on access and movement into critical areas.

a. Categorize restrictions as to personnel, materials, and vehicles:

(1) Personnel access:

(a) Establishment of controls pertinent to each area or structure.

- Authority for access.
- Criteria for access.
  - Unit personnel.
  - Visitors.
  - Maintenance personnel.
  - Contractor personnel.
  - National guard.
  - Emergency response teams (police, fire, ambulance, and so forth).

Figure F-1. Sample Physical-Security Plan

(b) Identification and control.

- Description of the system to be used in each area. If a badge system is used, a complete description covering all aspects should be used in disseminating requirements for ID and control of personnel conducting business on the installation.

- Application of the system:

  - Unit personnel.
  - Visitors to restricted areas.
  - Visitors to administrative areas.
  - Vendors, tradesmen, and so forth.
  - Contractor personnel.
  - Maintenance or support personnel.
  - Fail-safe procedures during power outages.

(2) Material control.

(a) Incoming.

- Requirements for admission of material and supplies.
- Search and inspection of material for possible sabotage hazards.
- Special controls on delivery of supplies or personal shipments in restricted areas.

(b) Outgoing.

- Documentation required.
- Controls, as outlined in paragraph 3a(2a).
- Classified shipment not involving nuclear/chemical material.

(c) Nuclear/chemical material.

- Controls on movement of warheads/chemicals on the installation.
- Controls on shipments or movement of training warheads/chemicals.
- Controls on pickup or delivery of warheads/chemicals outside the installation.

(3) Vehicle control.

(a) Policy on search of military and privately owned vehicles.

(b) Parking regulations.

Figure F-1. Sample Physical-Security Plan (continued)

(c) Controls for entrance into restricted and administrative areas:

- Military vehicles.
- POVs.
- Emergency vehicles.
- Vehicle registration.

b. Indicate the manner in which the following security aids will be implemented on the installation:

(1) Protective barriers:

(a) Definition.

(b) Clear zones.

- Criteria.
- Maintenance.

(c) Signs.

- Types.
- Posting.

(d) Gates.

- Hours of operation.
- Security requirements.
- Lock security.
- Barrier plan.

(2) Protective lighting system:

(a) Use and control.

(b) Inspection.

(c) Action taken in case of commercial power failure.

(d) Action taken in case of failure of alternate power source.

(3) Emergency lighting system:

(a) Stationary.

(b) Portable.

Figure F-1. Sample Physical-Security Plan (continued)

(4) IDSs:

(a) Security classification.

(b) Inspection.

(c) Use and monitoring.

(d) Action taken in case of alarm conditions.

(e) Maintenance.

(f) Alarm logs or registers.

(g) Tamper-proof provisions.

(h) Monitor-panel locations.

(5) Communications:

(a) Locations.

(b) Use.

(c) Tests.

(d) Authentication.

(6) Security forces: General instructions that would apply to all security-force personnel (fixed and mobile). Detailed instructions such as special orders and SOP information should be attached as annexes. Security-force facets include—

(a) Composition and organization.

(b) Tour of duty.

(c) Essential posts and routes.

(d) Weapons and equipment.

(e) Training.

(f) Use of MWD teams.

(g) Method of challenging with signs and countersigns.

(h) Alert forces:

- Composition.
- Mission.
- Weapons and equipment.
- Location.
- Deployment concept.

Figure F-1. Sample Physical-Security Plan (continued)

(7) Contingency plans: Required actions in response to various emergency situations. Detailed plans for situations (counterterrorism, bomb threats, hostage negotiations, disaster, fire, and so forth) should be attached as annexes.

(a) Individual actions.

(b) Alert-force actions.

(c) Security-force actions.

(8) Use of air surveillance.

(9) Coordinating instructions. Matters that require coordination with other military and civil agencies such as—

(a) Adjacent installations or units.

(b) State and local agencies.

(c) Similar host-country agencies.

(d) Federal agencies.

The coordination/interaction allows for an exchange of intelligence information on security measures being used, contingency plans, and any other information to enhance local security.

On an installation, the host activity shall assume responsibility for coordinating physical-security efforts of all tenants, regardless of the components represented, as outlined in the support agreements and the host-activity security plan. Applicable provisions shall be included in, or be an appendix to, the support agreement. A formal agreement will contain definite assignment of physical-security responsibility for the items stored. The agreement should address—

- Maximum qualities to be stored.
- Physical safeguards to be used.
- Frequency of, and responsibility for, physical inventories or reconciliations.
- Reporting of losses for investigation.
- Lock and key control.
- The unit that has overall responsibility.

Procedures for authorization and ID of individuals to receipt for and physically take custody of Army property. The purpose of such coordination is protection in depth. Authority, jurisdiction, and responsibility must be set forth in a manner that ensures protection and avoids duplication of effort.

/s/ Commander

Figure F-1. Sample Physical-Security Plan (continued)

ANNEXES

F-1. Annexes to the plan should include, but are not limited to, the following. More information can be found in AR 190-13.

- **Annex A.** The installation threat statement (intelligence). This annex should contain the Terrorism Counteraction Plan (refer to AR 190-13).
- **Annex B.** A bomb-threat plan. As a minimum, the bomb-threat plan should provide guidance for—
  - Control of the operation.
  - Evacuation.
  - Search.
  - Finding the bomb or suspected bomb.
  - Disposal.
  - Detonation and damage control.
  - Control of publicity.
  - After-action report.
- **Annex C.** An installation closure plan.
- **Annex D.** A natural-disaster plan. This plan will be coordinated with natural-disaster plans of local jurisdictions. At a minimum, the natural-disaster plan should provide guidance for—
  - Control of the operation.
  - Evacuation.
  - Communication.
  - Control of publicity.
  - After-action report.
- **Annex E.** A civil-disturbance plan. It is the commander's responsibility to formulate a civil-disturbance plan based on local threats. (For example, commanders of chemical facilities should anticipate the need to develop crowd-control procedures to handle antichemical demonstrations.)
- **Annex F.** A resource plan to meet the minimum-essential physical-security needs for the installation or activity.
- **Annex G.** A communication plan. This plan is required to establish communications with other federal agencies and local law-enforcement agencies to share information about possible threats. The communications plan should address all communication needs for annexes B through F above.
- **Annex H.** A list of designated restricted areas.
- **Annex I.** A list of installation MEVAs.
- **Annex J.** A contingency plan. In most instances, it will be necessary to increase security for AA&E and other sensitive property, assets, and facilities during periods of natural disasters, natural emergencies, or increased threat from terrorists or criminal elements. Therefore, CONPLANs should include provisions for increasing the physical-security measures and procedures based on the local commander's assessment of the situation. Such contingencies may include hostage negotiations, protective services, and special-reaction teams. These

provisions should be designed for early detection of an attempted intrusion, theft, or interruption of normal security conditions.

- **Annex K.** Work-stoppage plan. This is a requirement for conducting a physical-security survey.

TACTICAL-ENVIRONMENT CONSIDERATIONS

F-2. In a tactical environment, the development of a physical-security plan is based on METT-TC (using the OPORD format and the higher headquarters' order). The order may be specific about the tasks the unit will perform. Time available may be limited and the scheme of maneuver may be dictated, but the leader must still evaluate the mission in terms of METT-TC to determine how MP elements can best carry out the commander's order.

F-3. Consider each of the following factors and compare courses of action to form a base for the physical-security plan. When the plan is firm, issue it as an order.

- Concepts for reconnaissance, coordination with adjacent and/or supporting units, and troop movement.
- Physical-security installation configurations and facilities. Areas to consider may include drop zones, landing zones, ranges, and training areas.

MISSION

F-4. The mission is usually the emplacement of defensive security rings to protect the populace against insurgents. The number of defensive security rings depends on the particular site and situation. The following questions must be evaluated:

- What is the mission?
- What specific and implied tasks are there to accomplish the mission?
- What is the commander's intent?

ENEMY

F-5. The commander identifies insurgent units operating in the area and tries to determine the type and size of the unit; the enemy's tactics, weapons, equipment, and probable collaborators; and the inhabitants' attitudes toward the insurgents. The following questions must be evaluated:

- What is known about the enemy?
- Where is the enemy and how strong is he?
- What weapons does the enemy have?
- What is the enemy doing?
- What can the enemy do in response to MP actions?
- How can we exploit the enemy's weaknesses?

TERRAIN AND WEATHER

F-6. The commander can use observation and fields of fire, cover and concealment, obstacles, key terrain, and avenues of approach (OCOKA) to plan for the physical-security defensive sites. The following questions must be evaluated:

- How will the terrain and weather affect the operation?
- How fast can movement be accomplished, and how much space does the terrain and unit formations take up?
- Will the weather affect the terrain or personnel?
- Has the weather already affected the terrain?

TROOPS

F-7. The commander must consider available equipment, the reaction time, reaction forces, communication assets, organization of troops, and medical support (if available). The following questions must be evaluated:

- What are the present conditions of vehicles and personnel?
- What is the status of ammunition and supplies?
- Who is best able to do a specific task?
- How much sleep have the soldiers had in the past 24 hours?
- What other assets are available to support the mission?
- How many teams/squads are available?
- What supplies and equipment are needed?
- What fire support is available and how can it be obtained?

TIME AVAILABLE

F-8. This factor is critical since the inhabitants must be ready to respond to an insurgent attack with little or no warning. The following questions must be evaluated:

- How much time is available to conduct planning?
- How long will it take to reach the objective?
- How long will it take to prepare the position?
- How much time do subordinates need?
- How long will it take the enemy to reposition forces?

CIVILIAN CONSIDERATIONS

F-9. The commander also must consider nonbelligerent third parties (such as dislocated civilians, personnel of international businesses and relief organizations, and the media). Every commander must prepare a site overlay that shows, as a minimum, the following:

- The attitude of the HN toward US forces.
- The population density near the objective.
- The condition of the local civilians.
- The possible effect of refugees and dislocated civilians on the mission.

Appendix G

Personal-Protection Measures

Terrorists frequently emulate military organizations as they develop, plan, train, and carry out terrorist attacks against DOD assets. Terrorists have a critical need for information regarding the whereabouts, habits, working environments, home environments, and other potential points of leverage against their targets. The three intelligence-collection methods used by terrorists against potential targets are human intelligence (HUMINT), photographic intelligence (PHOTINT), and signal intelligence (SIGINT).

PERSONAL PROTECTION

G-1. The measures that follow are useful in providing personal protection for US government employees and DOD civilian contractors in CONUS or OCONUS facilities.

OVERCOME ROUTINES

G-2. The reduced probability of success in kidnapping or killing a target makes the target far less desirable. Perform the following measures to prevent daily routines from being observed:

- Vary your route to and from work and your arrival and departure times.
- Vary your exercise schedule, using different routes and distances. It is best not to exercise alone.
- Do not divulge family or personal information to strangers.
- Enter and exit buildings through different doors, if possible.
- Avoid other routines.

MAINTAIN A LOW PROFILE

G-3. Americans are easy to identify in an overseas area. Perform the following measures to reduce easy ID:

- Dress and behave in public in a manner consistent with local customs. Items that are distinctively American should not be worn or displayed outside American compounds.
- Reduce visibility in the local community.
- Avoid flashing large sums of money, expensive jewelry, or luxury items.
- Avoid public disputes or confrontations, and report any trouble to the proper authorities.
- Ensure that personal information (home address, phone number, or family information) is not divulged.

PREPARE FOR UNEXPECTED EVENTS

G-4. All DOD personnel, contractors, and their family members should implement the following general measures:

- Get into the habit of checking in with friends and family.
- Know how to use the local phone system.
- Know the locations of civilian police, military police, government agencies, and the US embassy.
- Know certain key phrases in the local language.
- Set up simple signal systems that can alert family members or associates that danger is present.
- Carry ID showing your blood type and any special medical conditions.
- Keep personal affairs in good order.
- Avoid carrying sensitive or potentially embarrassing items.

WORKING ENVIRONMENT

G-5. The working environment is not immune from attempted acts by criminals or terrorists. DOD installations in CONUS and OCONUS usually provide a level of basic security comparable or superior to the basic level of security provided in the surrounding community. The following are general practices that can help reduce the likelihood of a terrorist attack:

- Establish and support an effective security program.
- Discourage the use of office facilities to store objects of significant intrinsic value unless it is mission essential.
- Train personnel to be alert for suspicious activities, persons, or objects.
- Arrange office interiors so that strange or foreign objects left in the room will be recognized immediately.
- Provide for security systems on exterior doors and windows.
- Ensure that access-control procedures are rigorously observed at all times for access to—
  - The installation.
  - Buildings within an installation.
  - Restricted or exclusion areas within buildings.
- Use an ID badge system containing a photograph.
- Identify offices by room number, color, or object name and not by rank, title, or the name of the incumbent.
- Avoid using nameplates on offices and parking places.

OFFICE PROCEDURES

G-6. In an office, the following steps can be taken to make intelligence collection and targeting more difficult for terrorists:

- Telephone and mail procedures:
  - When answering the telephone, avoid using ranks or titles.
  - When taking telephone messages, do not reveal the whereabouts or activities of the person being sought.

  - When leaving telephone messages, place them in unmarked folders; do not leave them exposed for observers to identify caller names and phone numbers, persons called, and messages left.
  - When opening mail, use a checklist to help identify letter bombs or packaged IEDs.
- Visitor-control procedures:
  - Place strict limitations on access to the executive office area.
  - Lock doors (from the inside) from the visitor-access area to executive offices or other restricted areas of a facility.
  - Ensure that receptionists clear all visitors before they enter inner offices.
  - Permit workmen or visitors access to restricted areas or exclusion areas under escort and only with proper ID. Confirm the work to be done before admitting workmen to restricted areas of the facility.
  - Limit publicity in public waiting areas to information that does not identify personnel by name, position, or office location.
  - Avoid posting unit rosters, manning boards, or photo boards where visitors or local contractors can view them.
  - Restrict the use of message boards, sign-in/-out boards, and other visual communications to general statements of availability.
- General working procedures:
  - Avoid carrying attaché cases, briefcases, or other courier bags unless necessary.
  - Avoid carrying items with markings that identify the owner by rank or title, even within the office environment.
  - Avoid working alone late at night and on days when the remainder of the staff is absent.
  - Ensure that office doors are locked when the office is vacant for any lengthy period, at night, and on weekends. If late-night work is necessary, work in conference rooms or internal offices where outside observation is not possible.
  - Ensure that the security office retains the office keys.
  - Ensure that papers, correspondence, communications materials, and other documents are not left unattended overnight.
  - Ensure that maintenance activity and janitorial services in key offices, production offices, or maintenance facilities are performed under the supervision of security personnel.
  - Prohibit the removal of property, material, or information stored on any media from the facility without proper written authorization.
  - Consider prohibiting the importation of property, material, or information stored on any media into the facility unless such items have been properly inspected.
  - Lock offices not in use to prohibit unauthorized access of stored materialthatcouldbeusedtohideIEDsorintelligence-collectiondevices.
  - Minimize the use of vehicles or vehicle markings that make it possible to readily identify the vehicle and its occupants as US-government or DOD-contractor personnel.

- Ensure that all personnel have access to some sort of duress alarm to annunciate and warn of a terrorist attack.
- Ensure that secretaries and guard posts are equipped with covert duress alarms that can be used to alert backup forces.
- Avoid placing office furnishings directly in front of exterior windows.

SPECIAL PROCEDURES FOR EXECUTIVE ASSISTANTS

G-7. The following suggestions are intended to be a guide for secretaries and executive assistants who may find themselves performing personnel-security duties as collateral duty. Executive assistants and security personnel should regularly train and exercise procedures used in case they must evacuate mission-critical personnel to safe havens.

- Request the installation of physical barriers (such as electromagnetically operated doors) to separate offices of senior executives from other offices.
- Request the installation of a silent trouble-alarm button with a signal terminating in the security department.
- Admit visitors into the executive area when they are positively screened in advance or are personally recognized.
- Do not inform unknown callers of an executive's whereabouts, home address, or telephone number.
- Store a fire extinguisher, a first aid kit, and an oxygen bottle in the office area.
- Remain calm and listen carefully when receiving a threatening call.
- Do not accept packages from strangers until satisfied with the individual's identity and the nature of the parcel.
- Keep travel itineraries for all personnel confidential.
- Distribute daily schedules for senior officers and civilian officials on a limited basis.

HOME ENVIRONMENT

G-8. The following discussion is intended to assist personnel in formulating plans to obtain housing outside US government compounds or DOD facilities. Personnel assigned to government housing may also find the antiterrorism and security tips presented below helpful in reducing the threat of violence and loss of property.

G-9. For general residential-security routines, discuss with family members the importance of—

- Varying routines in their daily activities.
- Blending in with the local environment.
- Avoiding unnecessary publicity and photographs that identify individual family members.
- Being alert to individuals, parked or abandoned vehicles, unusual utility work, or gatherings of people inconsistent with the residential environment.

SECURITY PRACTICES AT HOME

G-10. The following measures are specifically recommended for residential implementation. These measures are an extension of office antiterrorism-security practices.

- Do not use nameplates or uniquely American symbols on the exterior of residences occupied by DOD personnel overseas.
- Do not use nameplates on parking places, and avoid parking private or government vehicles in the same location day after day.
- Ensure that all family members answer the telephone politely but that they provide no information as to the name of the occupants until the caller's identity has been established.
- Treat all telephone conversations as though anyone who wanted to listen in was doing so.
- Examine carefully all mail delivered to the residence.

SOCIAL AND RECREATIONAL ACTIVITIES

G-11. DOD personnel are encouraged to participate in many social and recreational activities. The following precautions are recommended:

- Respond to formal social invitations in person (where possible) or by direct telephone contact.
- Be attentive to the security environment of social gatherings.
- Avoid the development of patterns with respect to time of arrival or departure at social events.
- Avoid prolonged presence at social functions where there is a high concentration of persons thought to be terrorist targets.
- Refrain from excessive use of alcohol at social functions; remain clearheaded and unimpaired.
- Vary routes to and from social events held at a central facility.
- Minimize appearances in uniform or formal attire.
- Decline invitations to appear in publicity photos.
- Participate in recreational activities within the American compound or at a DOD installation whenever possible.

NOTE: Refer to DOD 0-2000.12-H, Graphic Training Aid (GTA) 19-4-3, and Joint Services (JS) Guide 5260 for further guidance and explanation regarding protective measures.

Appendix H

Bombs

Terrorists have frequently used homemade devices or IEDs to carry out their attacks against DOD personnel, facilities, and assets. The IEDs are ideal terrorist weapons. They are relatively inexpensive to make, and the components of many IEDs are common items that can be obtained from many sources and are difficult to trace. The IEDs can be large or small and be designed so that they are transported to the attack site in components for last-minute assembly. Such design concepts make detection more difficult and provide an additional increment of personal safety to the terrorists.

GENERAL

H-1. The use of IEDs can enhance the violence that gives terrorist groups their ability to intimidate or coerce a target population. The detonation itself creates a highly visual, newsworthy scene, even hours after the detonation occurs. Bombs can detonate anywhere, without apparent reason and without warning. The use of bombs in a terror campaign emphasizes the authorities' inability to safeguard the public and maintain law and order. Bombs are ideal weapons because they can be designed to give terrorists opportunities to escape from the scene of their crimes.

CONCEALING BOMBS

H-2. Given the question, "Where have terrorists placed bombs in the past, and where should we look for them?" results in no easy answer. Table H-1, page H-2, lists a few obvious locations that should be examined. Terrorists who use bombs as their weapons of choice can be very creative in designing and placing their weapons.

H-3. Bombs can be found anywhere people can place them. Without becoming paranoid and seeing a bomb under every rock and behind every tree, the practical answer to the above questions is: "Where they can be easily placed without the bomber being caught."

DAMAGE AND CASUALTY MECHANISMS

H-4. The IEDs and other explosive devices inflict casualties in a variety of ways, including the following:

- Blast over pressure (a crushing action on vital components of the body; eardrums are the most vulnerable).
- Falling structural material.
- Flying debris (especially glass).
- Asphyxiation (lack of oxygen).

Table H-1. Potential IED Hiding Places

Outside Areas	
• Trash cans • Dumpsters • Mailboxes • Bushes	• Street drainage systems • Storage areas • Parked cars
Inside Buildings	
• Mail parcels or letters • Inside desks/storage containers • Ceilings with removable panels • Areas hidden by drapes or curtains • Recent repaired/patched segments of walls, floors, or ceilings	• Restrooms • Trash receptacles • Utility closets • Boiler rooms • Under stairwells
In Plain Sight	

- Sudden body translation against rigid barriers or objects (being picked up and thrown by a pressure wave).
- Bomb fragments.
- Burns from incendiary devices or fires resulting from blast damage.
- Inhalation of toxic fumes resulting from fires.

H-5. It is impossible to calculate a single minimum safe distance from an IED or other explosive device. The safe distance varies with each device and its placement. As a rule, the farther away from a bomb, the safer the intended or collateral targets are. Blast effects, fragmentation injuries, and injuries resulting from flying debris diminish greatly as the distance between a bomb and possible targets increase. The amount of material in the device, the type of explosive material, the manner in which the device is constructed, and the location or the container in which it is placed all have a bearing on the specific destructive potential for each IED.

H-6. The following are four general rules to follow to avoid injury from an IED:

- Move as far from a suspicious object as possible without being in further danger from other hazards such as traffic or secondary sources of explosion (such as POL storage).
- Stay out of the object's LOS, thereby reducing the hazard of injury because of direct fragmentation.
- Keep away from glass windows or other materials that could become flying debris.
- Remain alert for additional or secondary explosive devices in the immediate area, especially if the existence of a bomb-threat evacuation assembly area has been highly publicized.

H-7. Some terrorists have used two especially devious tactics in the past to intensify the magnitude of casualties inflicted by bombing attacks. In some instances, they have detonated a small device to lure media attention and curiosity seekers to the site; a larger, more deadly device has detonated some time after the first device, thereby inflicting a large number of casualties.

H-8. Other terrorists have used a real or simulated device to force the evacuation of a facility only to detonate a much more substantial device in identified bomb-threat evacuation assembly areas. These attacks are especially harmful because the evacuation assembly areas often concentrate government or commercial office workers more densely than they are when dispersed throughout their usual workplaces.

TELEPHONIC THREATS

H-9. When receiving a telephonic threat, treat the call seriously. Often, an anonymous telephone call is made regarding a bomb or an IED. See Figure H-1, page H-4, for information to record/obtain when receiving these calls

H-10. When an anonymous warning or threat is received, initiate the bomb-threat data card and notify the PMO, security police, security forces, or other law-enforcement/security offices immediately. Local SOPs will determine subsequent actions. Immediate action may include a search without evacuation, the movement of personnel within the establishment, a partial evacuation, or a total evacuation. The following criteria helps determine what immediate action to take:

- Factors favoring a search before the movement of personnel:—
  - There is a high incidence of hoax telephone threats.
  - Effective security arrangements have been established.
  - Information in the warning is imprecise or incorrect.
  - The caller sounded intoxicated, amused, or very young.
  - The prevailing threat of terrorist activity is low.
- Factors favoring movement of personnel before searching:
  - The area (post or base) is comparatively open.
  - Information in the warning is precise as to the matters of location, a description of the device, the timing, and the motive for the attack.
  - A prevailing threat of terrorist activity is high.

EVACUATION DRILLS

H-11. Evacuation and search drills should be performed periodically under the supervision of the installation's or unit's senior officer. The drills should be held in cooperation with local police if possible. Personnel in adjacent buildings should be informed of drills to avoid causing unnecessary alarm.

H-12. Evacuation procedures depend on the circumstances. Prepare, publicize, and rehearse evacuation plans in advance. Address alarm systems, assembly areas, routes to assembly areas, personnel-evacuation responses, building and area clearances, and evacuation drills.

PERSONNEL-EVACUATION RESPONSE

H-13. The bomb-threat alarm system should be easily distinguished from the fire alarm. When the alarm sounds, personnel should—

- Lock up or secure all classified materials.
- Conduct a quick visual search of their immediate working area.

Instructions: Be calm. Be courteous. Listen, do not interrupt the caller. Notify supervisor/security officer by prearranged signal while caller is on line.

Name of Operator____________________ Time________ Date________

Caller's Identity

Sex: ❑ Male ❑ Female ❑ Adult ❑ Juvenile   Approximate age: Years________

Origin of Call

❑ Local   ❑ Booth   ❑ Internal (From within bldg)
❑ Long Distance   If internal, leave line open for tracing the call.

Voice Characteristics		Speech		Language	
❑ Loud	❑ Soft	❑ Fast	❑ Slow	❑ Excellent	❑ Good
❑ High Pitch	❑ Deep	❑ Distinct	❑ Distorted	❑ Fair	❑ Poor
❑ Raspy	❑ Pleasant	❑ Stutter	❑ Nasal	❑ Foul	❑ Other ____
❑ Intoxicated	❑ Other____	❑ Slurred	❑ Lisp		
			❑ Other ____		

Accent	Manner		Background Noises	
❑ Local	❑ Calm	❑ Angry	❑ Factory Machines	❑ Trains
❑ Not Local	❑ Rational	❑ Irrational	❑ Bedlam	❑ Animals
Region____	❑ Coherent	❑ Incoherent	❑ Music	❑ Quiet
❑ Foreign	❑ Deliberate	❑ Emotional	❑ Office Machines	❑ Voices
Race____	❑ Righteous	❑ Laughing	❑ Mixed	❑ Airplanes
			❑ Street Traffic	❑ Party Atmosphere

Bomb Facts

Pretend difficulty with your hearing.   Keep caller talking.

If caller seems agreeable to further conversation, ask questions like —
When will it go off? Certain Hour - Time Remaining - What kind of bomb? - Where are you now?
How do you know so much about the bomb? - What is your name and address?

If building is occupied, inform caller that detonation could cause injury or death.
Did caller appear familiar with plant or building by his description of the bomb location?

Write out the message in its entirety and any other comments on a separate sheet of paper and attach to this checklist.

Action To Take Immediately After Call

Notify your supervisor/security officer as instructed. Talk to no one other than as instructed by your supervisor/ security officer.

Figure H-1. Sample Bomb-Threat Data Card

- Open windows (wherever possible).
- Leave the building, taking only valuable personal belongings.
- Leave doors open and immediately proceed to the assembly area.

H-14. Opening the building will reduce internal damage due to blast effects. It will also somewhat mitigate the extent of debris flying out of or falling from the building should a detonation occur.

ASSEMBLY AREAS

H-15. Choose the routes to the assembly area so that personnel do not approach the IED at any time. Preselect the routes to the assembly area, but devise a system to inform personnel of the location of the suspected IED and alternate routes. Routes prevent confusion and bunching and avoid potential hazards (such as plate glass, windows, and likely locations of additional IEDs).

H-16. Assembly areas should be preselected and well known to personnel. Establish a clearly defined procedure for controlling, marshaling, and checking personnel within the assembly area. If buildings or establishments are in a public area, coordinate the assembly areas with local police. Assembly areas are selected using the following criteria:

- Locate assembly areas at least 100 meters from the likely target or building (if possible).
- Locate assembly areas in areas where there is little chance of an IED being hidden. Open spaces are best. Avoid parking areas because IEDs can be easily hidden in vehicles.
- Select alternate assembly areas to reduce the likelihood of ambush with a second device or small-arms fire. If possible, search the assembly area before personnel occupy the space.
- Avoid locating assembly areas near expanses of plate glass or windows. Blast effects can cause windows to be sucked outward rather than blown inward.
- Select multiple assembly areas (if possible) to reduce the concentration of key personnel. Drill and exercise personnel to go to different assembly areas to avoid developing an evacuation and emergency pattern that can be used by terrorists to attack identifiable key personnel.

BUILDING AND AREA CLEARANCE

H-17. Establish procedures to ensure that threatened buildings and areas are cleared. Prevent personnel from reentering the building. Establish a cordon to prevent personnel from entering the danger area. Establish an initial control point (ICP) as the focal point for the PMO and for MP control.

H-18. Cordon suspicious objects to a distance of at least 100 meters, and cordon suspicious vehicles to a distance of at least 200 meters. Ensure that nobody enters the cordoned area. Establish an ICP on the cordon to control access; relinquish ICP responsibility to the PMO or local police upon their arrival. Maintain the cordon until the PMO, security police, security forces, or local police have completed their examination or stated that the cordon may stand down.

SEARCHING FOR A SUSPECTED IED

H-19. Searches are conducted in response to a telephonic threat or a report of an unidentified object on or near premises occupied by DOD personnel. The following types of searches may be used when searching for a suspected bomb or IED:

- An occupant search is used when the threat's credibility is low. Occupants search their own areas. The search is completed quickly because occupants know their area and are most likely to notice anything unusual.
- A team search is used when the threat's credibility is high. The search is very thorough and places the minimum number of personnel at risk. Evacuate the area completely, and ensure that it remains evacuated until the search is complete. Search teams will make a slow, thorough, systematic search of the area.

H-20. The following procedures should be followed if a search for explosive devices must be conducted before qualified EOD teams arrive:

- Make an audio check, listening for unusual sounds.
- Sweep the area visually up to the waist, then sweep up to the ceiling. Do not forget the tops of cabinets and cupboards.
- Perform a thorough and systematic search in and around containers and fixtures.
- Pass search results as quickly as possible to the leader responsible for controlling the search area. Do not use a radio; it may detonate the explosive.

H-21. Circumstances might arise in the case of a very short warning period. In other instances, a threat of a bomb against some facilities (if true) might necessitate the evacuation of a very large area. In these circumstances, searching for the presence of an explosive device to identify its location, appearance, and possible operating characteristics may be warranted.

H-22. Personnel who have not been trained in IED search and ID techniques should not search for explosive devices. Two types of errors are very common—the false ID of objects as IEDs and the incorrect ID of IEDs as benign objects. Depending on the devices used to arm and trigger an IED, the search process could actually result in an explosion.

SEARCH ORGANIZATION

H-23. The person controlling the search should have a method of tracking and recording the search results (such as a diagram of the area). Delegate areas of responsibility to the search-team leader, who should report to the person controlling the search when each area has been cleared. Pay particular attention to entrances, toilets, corridors, stairs, unlocked closets, storage spaces, rooms and areas not checked by usual occupants, external building areas, window ledges, ventilators, courtyards, and spaces shielded from normal view.

DISCOVERY OF A SUSPECTED IED

H-24. When a suspicious object has been found, report its location and general description immediately to the nearest law-enforcement or supervisory person. Do not touch or move a suspicious object. Instead, perform the following steps:

- If an object appears in an area associated with a specific individual or a clearly identified area—
  - Ask the individual/occupant to describe objects they have brought to work in the past few days.
  - Ask for an accounting of objects.
  - Ask for a verbal description/ID of objects.
- If an object's presence remains inexplicable—
  - Evacuate buildings and surrounding areas, including the search team.
  - Ensure that evacuated areas are at least 100 meters from the suspicious object.
  - Establish a cordon and an ICP.
  - Inform personnel at the ICP that an object has been found.
  - Keep the person who located the object at the ICP until questioned.
  - Avoid reentering the facility to identify an object that may or may not be an IED.

REACTING TO AN EXPLODED IED

H-25. The following procedures should be taken when an explosive/IED detonates at a DOD facility:

- For explosions without casualties—
  - Maintain the cordon. Allow only authorized personnel into the explosion area.
  - Fight any fires threatening undamaged buildings without risking personnel.
  - Report the explosion to the PMO, security police, security forces, or local police if they are not on the scene.
  - Report the explosion to the installation operations center even if an EOD team is on its way. Provide as much detail as possible, such as the time of the explosion, the number of explosions, the color of smoke, and the speed and spread of fire.
  - Ensure that a clear passage for emergency vehicles (fire trucks, ambulances, and so forth) and corresponding personnel is maintained.
  - Refer media inquiries to the PAO.
  - Establish a separate information center to handle inquiries from concerned friends and relatives.
- For explosions with casualties—
  - Select a small number of personnel to help search for casualties.
  - Assign additional personnel the responsibility for maintaining the cordon to keep additional volunteers searching for casualties.

Maintain the cordon until the EOD team verifies no further presence of bombs/IEDs at the site and the fire marshal determines that risk of additional injury to searchers from falling debris is acceptable.

- Prepare a casualty list for notification of next of kin; delay publication of the list until its accuracy is determined.
- Arrange for unaffected personnel to contact their next of kin immediately.

H-26. Civilian management officials and subordinate military commanders continue to have important personal roles to fulfill during a bomb/IED attack on DOD personnel, facilities, and assets. Perform the following procedures when reporting an attack:

- Pass available information to the operations center.
- Avoid delaying reports due to lack of information; report what is known. Do not take risks to obtain information.
- Include the following information in the report:
  - Any warning received and if so, how it was received.
  - The identity of the person who discovered the device.
  - How the device was discovered (casual discovery or organized search).
  - The location of the device (give as much detail as possible).
  - The time of discovery.
  - The estimated length of time the device has been in its location.
  - A description of the device (give as much detail as possible).
  - Safety measures taken.
  - Suggested routes to the scene.
  - Any other pertinent information.

H-27. Perform the following procedures when providing emergency assistance to authorities:

- Ensure that the PMO, security police, security forces, and other emergency-response units from local police, fire and rescue, and EOD teams are not impeded from reaching the ICP. Help maintain crowd control and emergency services' access to the site.
- Evacuate through the doors and windows of buildings.
- Assist the on-scene commander by obtaining a building diagram showing detailed plans of the public-service conduits (gas, electricity, central heating, and so forth), if possible. If unavailable, a sketch can be drawn by someone with detailed knowledge of the building.
- Locate, identify, and make witnesses available to investigative agency representatives when they arrive on the scene. Witnesses include the person who discovered the device, witnessed the explosion, or possesses detailed knowledge of the building or area.

H-28. Performing the above steps will provide substantial assistance to the crisis-management team and give other personnel constructive, supportive actions to take in resolving the crisis. Care must be exercised, however, that additional explosive devices are not concealed for detonation during the midst of rescue operations. These attacks add to the physical damage and emotional devastation of bomb/IED attacks.

H-29. The use of bombs and IEDs during terrorist attacks against DOD personnel, facilities, and assets is a common occurrence. The procedures outlined in this appendix are intended to help a DOD facility respond to an attack before an explosive device detonates. The procedures are also intended to help mitigate the consequences of an attack in case efforts to find an explosive device and render it inoperable are not successful. Incurring the costs to DOD facilities and installations of detecting an explosive device and terminating a terrorist incident before the device can detonate are almost always preferable rather than exercising plans and options to respond to a detonation. Several of the security measures discussed will help reduce the likelihood of a successful bomb/IED attack against DOD assets.

Appendix I

Executive Protection

DOD Directive 2000.12 recognizes a need to provide protection to military officers and DOD civilians who are assigned to high-risk billets, who are (by the nature of their work) high-risk personnel, or who are assigned to facilities identified as high-risk targets. The directive defines these terms as follows:

- High-risk billet. Authorized personnel billet (identified and recommended by appropriate authority) that because of grade, assignment, travel itinerary, or symbolic value may make personnel filling them an especially attractive or accessible terrorist target.
- High-risk personnel. US personnel and their family members whose grade, assignment, travel itinerary, or symbolic value may make them an especially attractive or accessible terrorist target.
- High-risk target. US facilities and material resources that, because of mission sensitivity, ease of access, isolation, or symbolic value may be an especially attractive or accessible terrorist target.

NOTE: For purposes of this appendix, the term executive will be applied to all persons requiring additional security protection who are assigned to high-risk billets, designated as high-risk personnel, or identified as high-risk targets.

SUPPLEMENTAL SECURITY MEASURES

I-1. The specific supplemental security measures that may be furnished to executives are subject to a wide range of legal and policy constraints. US law establishes stringent requirements that must be met before certain security measures may be implemented. DOD regulations, instructions, and legal opinions may further constrain the implementation of some protective measures described in this chapter. The SOFAs and MOUs between the US and a foreign government will also limit the use of some supplemental security measures. Leases and other conditions imposed by contract for purchase of land or buildings by the US for DOD use may also limit the application of certain security techniques. All of these constraints should be carefully considered when conducting security surveys, developing plans, and implementing additional security measures to protect high-risk personnel.

EXECUTIVE PROTECTION GOALS

I-2. In the discussion that follows, several measures are outlined that can afford senior military officers and DOD personnel additional protection against terrorist acts. The purpose underlying these measures is to—

- Increase the interval of time between detecting a threat and the onset of hostile action against executives and their dependents.
- Increase the amount of time required by terrorists to gain physical access to executives from the onset of hostile actions, whether the executives are at home, at the office, or in transit.

I-3. The implementation of supplemental security measures should strive to achieve the following prioritized goals:

- Enhancements should hold the terrorist threat at bay until a response force arrives (delay at a distance).
- Enhancements in physical security should enable executives to flee to safety (delay to permit flight).
- Enhancements should permit the executive to retreat into a safe haven of sufficient strength and survivability. This should enable a response force to wage an effective counterattack to liberate executives and others accompanying them to a safe haven, including family members at home and colleagues and visitors at work (delay, hold, and counterattack).

I-4. The following supplemental measures should be applied with care. There is a clear trade-off between increasing the level of physical security at the office and at home and preserving the anonymity of executives, thereby avoiding telltale signs of activity that point to prominence or criticality. These measures can be expensive. Expense can be measured not just in terms of dollars, but also in terms of changes to organizational routine. Therefore, three questions must be resolved before implementing bold, disruptive, and expensive supplemental security enhancements:

- What are the most cost-effective means of enhancing the security of executives at risk?
- How many changes in organizational routines and personal behaviors will have to be made for security measures to be effective in reducing the risk of terrorist attacks and the vulnerability of executives to such attacks?
- What are the anticipated costs of additional security measures in terms of dollars, organizational functionality, and mission capability?

I-5. Security enhancements can be made to improve the security of executives and can be even more effective if executives and their families take full advantage of and reinforce those measures. If executives do not change their behavior to accommodate additional security and protective measures, then the behaviors can effectively defeat the purpose of additional protection. Additional increments of security can be obtained to defeat virtually any threat. However, there is a point at which it is no longer cost-effective to add layer upon layer of protective measures to defeat a threat.

RESIDENTIAL SECURITY MEASURES

I-6. While terrorist groups conduct intelligence operations to identify targets, mistakes have been made in the past. DOD personnel should avoid leasing residences previously used by representatives of governments or organizations known to be targets of various terrorist groups. DOD personnel leasing residences formerly used by representatives of such governments may be placing themselves unnecessarily at risk of being attacked as a result of mistaken identity.

I-7. An executive's entire lifestyle should be included in security surveys used to assess the need for supplemental physical-security measures at the office. The executive's home and transportation from home to office and back should also be examined for risk and vulnerability. The same principles used to identify supplemental security improvements in an office environment apply to an executive's home environment as well. The purposes of physical-security enhancements are to—

- Increase the amount of time terrorists need to initiate and complete an attack on executives while at home, thereby giving response forces more time to rescue executives and their dependents.
- Reduce potential harm to executives and their families because of a terrorist assault mounted against the residence.

I-8. The goals of enhanced residential physical-security measures are to—

- Increase the amount of time between detection of a threat and the onset of hostile actions.
- Delay the terrorists as long as possible. Prevent terrorist access to executives and their family members and make it difficult to leave the scene to escape prosecution. These measures should not further jeopardize the lives of executives and their family members.
- Provide a safe haven where executives and their family members may flee for security pending the arrival of a response force on the scene.

I-9. The following measures can be implemented selectively to help security personnel achieve these objectives:

- Increase the time interval between threat detection and the onset of hostile terrorist acts by—
  - Ensuring that all door locks and window clasps are working.
  - Ensuring that all doors and windows are properly secured to their frames and that the frames are properly anchored to the residential structure.
  - Locking driveway gates with a security lock to prevent entry.
  - Installing a through-door viewing device or visitor intercom.
  - Installing security lights to aid in viewing entrances.
- Increase the number of physical barriers between the outer perimeter of the residence and the interior of the residence by—
  - Adding heavy, remotely operated gates to all fences, walls, and perimeter barriers consistent with the penetration resistance of the barrier between the residence, the street, and adjacent neighbors.
  - Creating a vestibule or air lock between living quarters and the exterior of a residence, ensuring that no one can enter the residence directly from the outside.
  - Adding fire doors or security doors or gates between the residence's bedrooms and living areas.

- Increase the time required to penetrate exterior structural walls by explosives, hand-held power tools, and hand tools by—
  - Adding additional armor covered by aesthetically pleasing materials to exterior walls.
  - Adding a separate reinforced masonry wall around the residence.
- Increase the surveillance of the residence and decrease response time by—
  - Installing CCTV systems to permit remote viewing of all doors and windows accessible from the ground, nearby structures, trees, or easily acquired platforms (such as a van parked next to a wall).
  - Installing area IDSs between the residence's perimeter and the residence itself, varying the number and types of sensors, and adding backup communication channels between the IDS and a surveillance assessment/response dispatch center.
- Increase the residence's durability and survivability to a terrorist attack by—
  - Fitting windows with either venetian blinds or thick curtains to reduce the observability of activities within the residence and to reduce hazards of flying glass in case of nearby explosions or gunfire.
  - Installing backup power systems for security devices (surveillance systems, communication systems, and access-control systems).
  - Ensuring that backup communication is available with the installation or embassy's security department via a secure landline or two-way radio.
  - Fitting a panic-alarm bell to the outside of the house with switches on all floor levels. Such an alarm should also annunciate at the local police and cognizant DOD or DOS security office.
  - Installing a safe haven in the home.

TRANSPORTATION MEASURES

I-10. High-risk personnel are most accessible to terrorists while in transit in official or privately owned vehicles. Specific steps can be taken to reduce the vulnerability of executives in transit.

SPECIAL TRANSPORTATION IN TRANSIT FROM DOMICILE TO DUTY

I-11. As a rule, Congress has strongly opposed the provision of domicile-to-duty transportation by the federal government to its officers and employees. Only 16 officials are entitled by statute to such assistance. Congress did, however, grant authority to the President and the heads of executive agencies and departments to provide domicile-to-duty transportation under certain circumstances. According to the statute, "a passenger carrier may be used to transport between residence and place of employment an officer or employee with regard to whom the head of a Federal agency makes a determination, [provided] that highly unusual circumstances present a clear and present danger, that an emergency exists, or that compelling operational considerations make such transportation essential to the conduct of official business."

I-12. The phrase "highly unusual circumstances which present a clear and present danger" is understood to mean that—

- The perceived danger is real, not imaginary.
- The perceived danger is immediate or imminent, not merely potential.
- Proof is provided that the use of a government vehicle would provide protection not otherwise available.

I-13. Such a danger would exist where there is an explicit threat of terrorist attacks or riot conditions and such transportation would be the only means of providing safe passage to and from work.

I-14. The phrase "emergency exists" means that there is an immediate, unforeseeable, temporary need to provide home-to-work transportation for an agency's essential employees. The phrase "similarly compelling operational considerations" means that there is an element of gravity or importance for the need of government-furnished transportation comparable to the gravity or importance associated with a clear and present danger or an emergency. Congress suggested further, "in such instances, [it is expected] that home-to-work transportation would be provided only for those employees who are essential to the operation of the government."

I-15. The Secretary of Defense has the statutory authority to allow a CINC to use government-owned or -leased vehicles to provide transportation in an area outside of the US for members of the uniformed services and other DOD personnel under certain circumstances. These circumstances include and are limited to a determination by the CINC that public or private transportation in the area is unsafe or is not available. Under these circumstances, DOD may provide transportation (usually in government buses or passenger vans) to personnel and their family members if it will help the CINC and his subordinate commanders maintain the capability to perform or undertake assigned missions. This transportation is not intended for transporting personnel from their residences to their places of work. The Secretary of Defense and the Service Secretaries also have the statutory authority to provide transportation from home to duty stations and back on a limited basis. This authority is usually implemented by providing a nontactical armored vehicle (NTAV) to protect personnel.

I-16. It is a DOD policy to make NTAVs available where necessary to enhance the security of DOD personnel consistent with the requirements and limitations found in the statute. DOD issuances, service regulations, and CINC guidance stipulate detailed procedures by which DOD manages NTAV programs. The statute also establishes a procedure for Presidential waiver of the "buy American" requirement; DOD and service regulations provide for the delegation of Presidential authority from the President to the Secretary of Defense; to the Director, Defense Security Assistance Agency; and to the Director, DIA. DOD Instruction 5210.84 authorizes DOS acquisition and installation of light vehicle armoring to DOS specifications in local defense-component vehicles on a reimbursable basis. The level of protection provided to the Defense Component Office will comply with approved overseas security policy group armored-vehicle standards.

I-17. The DOD recognizes two classes of NTAVs—heavy and light. Heavy NTAVs are fully armored vehicles intended to protect occupants from attack by bombs; IEDs; grenades; and high-velocity, small-arms projectiles. Light NTAVs are less than fully armored vehicles and are intended to protect occupants from attack by medium-velocity, small-arms projectiles and at least some types of IEDs.

I-18. The dividing lines between heavy and light NTAVs have become less distinct over time as armoring techniques and materials have given greater capability to NTAVs that are not classified as heavy. As a practical matter, add-on vehicle-armoring kits are now in production which (when properly installed in an appropriately powered and suspended vehicle) will provide a level of protection approaching that of the heavy NTAVs.

Heavy NTAVs

I-19. Heavy NTAVs may be assigned to US personnel upon certification by a Service Secretary only under the following conditions:

- Highly unusual circumstances present a clear and present danger to the health and safety of a nominated protectee.
- Compelling operational considerations make such transportation essential to conducting official business.

I-20. If the physical-security survey concludes that a heavy NTAV is warranted, the nominated protectee's Service Secretary shall, on the advice and recommendation of a combatant commander, determine whether the use of a heavy NTAV is warranted. If so, the Service Secretary shall authorize the use of a vehicle for a renewable 90- to 360-day period. At the end of the period, the requirement will be reexamined and a recertification for the protection shall be issued by the Service Secretary.

I-21. Each of the services manages a portion of the DOD's NTAV program. Each service has issued supplementary mandatory guidance for processing requests for, as well as allocation and use of, these scarce assets.

I-22. Heavy NTAVs are complex systems requiring specialized maintenance and operation. Normally, they will be assigned to DOD personnel with a driver who has been properly trained in the operation and maintenance of the vehicle. The operator is not a chauffeur; he is an integral part of a supplemental security package provided by DOD to meet its obligations in protecting key assets.

Light NTAVs

I-23. Light NTAVs may also be provided to US employees and officers where highly unusual circumstances present a clear and present danger to the health and safety of a nominated protectee or compelling operational considerations warrant their use. This category of NTAV features add-on armoring. While they are a less-complex armoring system than those used in heavy NTAVs, light NTAVs afford substantial protection to occupants against a variety of threats. New developments in after-manufacture armoring kits for vehicles are occurring at a rapid pace, increasing the number of vehicle manufacturers and models for which other NTAV modifications are suitable. Each service and the DIA have instructions for implementing DOD policy that authorizes the use of other NTAVs to enhance personnel protection of high-risk persons.

PRIVATELY OWNED VEHICLES

I-24. High-risk personnel may wish to forego the use of POVs during periods of extreme risk. Considerations include selecting measures that—

- Deter secret entry, making undetected placement of IEDs in or under the vehicle difficult for terrorists to accomplish.
- Enhance the vehicle's ability to increase distance between it and pursuers.
- Assist response forces in case of an incident.
- Make the vehicle appear little different than its standard models.

INDIVIDUAL PROTECTIVE MEASURES

I-25. Executives can enhance their personal security in the office environment by—

- Discouraging staff members who are taking telephone messages from disclosing their whereabouts.
- Ensuring that caution is used when opening mail and being especially careful with letters or packages that might contain IEDs.
- Ensuring that access is strictly limited to their office area.
- Limiting publicity and keeping official biographies short. This includes using outdated photographs if a publicity photograph is essential.
- Ensuring that they are not working alone late at night and on days when the remainder of the staff is absent.
- Working in conference rooms or internal offices where outside observation is not possible if late-night work is necessary. Security officers should be notified of the work so that they can periodically look in.
- Ensuring that office furnishings are not placed directly in front of exterior windows.

OFFICIAL BUSINESS AWAY FROM THE OFFICE

I-26. The following suggestions reinforce efforts by executives to maintain the high level of security provided in the home or office environment while on official business outside of these locations:

- Discuss security requirements with the person planning the function.
- Travel to and from the function with escorts.
- Choose the route carefully.
- Avoid publicizing planned attendance at official functions (unless required).
- Attempt to sit away from both public areas and windows.
- Encourage the function's sponsor to close the curtains to minimize the likelihood that anyone outside will be able to see inside and determine who is attending the function. This is extremely important for an evening function, when a well-lit interior can be easily viewed from a darkened exterior.
- Request that external floodlights be used to illuminate the area around the building where an evening function will occur.

LOCAL OFFICIAL AND UNOFFICIAL TRAVEL

I-27. Executives can greatly enhance their personal security when conducting official and unofficial travel by following these general practices:

- Vary daily patterns, such as leaving and returning at different times.
- Consider escorts to and from work or travel with a neighbor.
- Establish a simple oral or visual duress procedure between executives and drivers (for example, a phrase or movement used by the executive or driver only if something is amiss).
- Vary taxi companies. Ensure that the ID photo on the license matches the driver. If uneasy for any reason, take another taxi.
- Attend social functions with others, if possible.
- Examine the car before entering to see if there has been any interference. A small mirror on a rod is a cheap and effective method to inspect underneath cars. Do not touch the vehicle until it has been thoroughly checked (inside, around, and under).
- Avoid leaving personal items exposed in the car (uniform items, service-issued maps, official briefcases, and so forth).

SECURITY PRACTICES WHILE DRIVING

I-28. Executives can take the following measures to enhance security while driving:

- Keep car doors locked. Do not open windows more than a few inches.
- Avoid overloading a vehicle, and wear seat belts.
- Park vehicles in parking areas that are either locked or monitored. Never park overnight on the street. Before entering vehicles, check for signs of tampering.
- Keep the trunk locked.
- Drive in the inner lanes to keep from being forced to the curb.
- Use defensive and evasive driving techniques. Drill with your driver by watching for suspicious cars and taking evasive action.
- Avoid driving close behind other vehicles (especially service trucks), and be aware of activities and road conditions two to three blocks ahead.
- Beware of minor accidents that could block traffic in suspect areas such as crossroads. Crossroads are preferred areas for terrorist or criminal activities because they offer escape advantages.

I-29. If a terrorist roadblock is encountered, use the shoulder or curb (hit at a 30- to 45-degree angle) of the road to go around it or ram the terrorist's blocking vehicle. Blocking vehicles should be rammed in a nonengine area, at a 45-degree angle, in low gear, and at a constant moderate speed. The goal is to knock the blocking vehicle out of the way. In all cases, do not stop and never allow the executive's vehicle to be boxed in with a loss of maneuverability. Whenever a target vehicle veers away from the terrorist vehicle, it gives adverse maneuvering room and presents a better target to gunfire.

INTERURBAN, NATIONAL, AND INTERNATIONAL TRAVEL SECURITY PRACTICES AND PROCEDURES

I-30. To enhance security in interurban, national, and international circumstances, executives should—

- Book airline seats at the last moment. Consider using an alias.
- Restrict the use of ranks or titles.
- Avoid allowing unknown visitors in the hotel room or suite.
- Keep staff and family members advised of the itinerary and subsequent changes. Clearly and emphatically restrict this information to those having a need to know.

HOME SECURITY PRACTICES AND PROCEDURES

I-31. To enhance security at home, executives should—

- Check the ID of persons entering the premises (electricians, plumbers, telephone-maintenance personnel, and so forth). When in doubt, call their office to verify their identity before allowing them in your home.
- Avoid opening the door to a caller at night until he is visually identified through a window or a door viewer.
- Close curtains in a room before turning on lights.
- Consider placing the telephone where you will not be seen from doors or windows when answering.
- Investigate the household staff (especially temporary staff members).
- Stay alert and be on the lookout for the unusual. Ensure that the home is locked and secure whenever the residence is unattended. Be cautious upon return and look for the movement of furniture or the placement of unusual wires.
- Note and report suspicious persons.
- Control house keys strictly.
- Park the car in a locked garage.
- Consider installing a panic-alarm bell to the outside of the house with switches located on all floor levels.
- Clear the area around the house of dense foliage or shrubbery.
- Test duress alarms (if available). Make certain that family members understand how they work as well as the importance of their use.
- Cooperate with law-enforcement personnel, and abide by their security recommendations.

SECURITY AT SOCIAL AND RECREATIONAL ACTIVITIES

I-32. The risk of terrorist incidents is always present for high-risk personnel or personnel assigned to high-risk billets. The following measures are intended to permit executives to live a close-to-normal life while still remaining mindful of the risks to their security.

- Ensure that the host is aware of and takes appropriate measures for your security.
- Have your personal staff assist a civilian host, if required.
- Arrange for visitors to be subject to adequate security control.

- Screen the invitation list, if possible.
- Vary the times of sporting activities (golfing, jogging, and so forth).

COMBATING-TERRORISM TRAINING FOR EXECUTIVES

I-33. Combatant commanders annually compile a list of high-risk billets in their AO. These lists are forwarded through the appropriate service personnel channels, enabling each service to identify, plan, and provide resources to meet training requirements. All personnel and adult family members en route to high-risk billets must attend the Individual Terrorism Awareness Course (INTAC) conducted at the US Army John F. Kennedy Special Warfare Center at Fort Bragg, North Carolina. During this one-week course, personnel will receive instruction in defensive-driving techniques and survival shooting as well as individual protective measures and hostage survival. These individuals should also attend the appropriate regional orientation course (Middle East, Asia/Pacific, Latin America, or Africa) offered at the US Air Force Special Operations School at Hurlburt Air Force Base (AFB), Florida. The service member whose duties will require frequent vehicle operation should attend an appropriate evasive-driving course. Information on current offerings may be obtained by contacting the service representative to the DOD Antiterrorism Coordinating Committee or the Combating Terrorism Branch in the Office of the Assistant Secretary of Defense (OASD) Special Operations/ Low-Intensity Conflict (SO/LIC).

TRAVEL TO POTENTIAL PHYSICAL-THREAT RISK AREAS

I-34. Personnel en route to potential physical-threat risk areas (as identified by the OASDSO/LIC) should attend one of the following courses:

- The Dynamics of International Terrorism Course conducted at the US Air Force Special Operations School at Hurlburt AFB, Florida. During this one-week course, personnel will receive lectures on threats by region (Europe, Middle East, Latin America, Asia/Pacific, and Africa), the history and psychology of terrorism, personnel combating-terrorism measures (vehicle, personal, airline, and physical security), and hostage survival.
- A Regional Orientation Course (Middle East, Latin America, Africa, Asia/Pacific) at the US Air Force Special Operations School at Hurlburt AFB, Florida. This one-week course offers personnel instruction in cultural, political/military, and individual security factors associated with the region.

I-35. Installation security personnel may also receive the above training if they have completed the Antiterrorism Instructor Qualification Course (AIQC) at Fort Bragg, North Carolina.

PROTECTIVE SECURITY DETAILS

I-36. Each service can provide bodyguards for key senior military officers, DOD civilians, other US officials, or foreign dignitaries requiring personal protection. Each Service Secretary is responsible for assigning protective security details (PSDs) to service members based on the recommendation of their counterintelligence and/or law-enforcement investigation staffs. The PSDs are assigned to DOD personnel who meet requirements established by service regulations. In general, PSDs may be assigned only to those personnel whose position or assignment places them at risk and whose continued availability to the National Command Authorities and the CINCs is vital to DOD's mission execution.

I-37. A PSD provides high levels of security to an executive by establishing a series of protective cordons around him. The establishment of defense in depth often means that the innermost protective layer is in close contact with the executive at all hours of the day and night.

I-38. A PSD is trained to maintain a low profile. It is concerned about the executive's visibility and its ability to blend into his surroundings. There is nothing more damaging to the security of an executive than the obvious, detectable presence of a PSD when all other measures to have him blend into the local environment have been successful. A PSD will strive to keep travel routes and means of transportation from being publicized. If this cannot be accomplished, the PSD may suggest editorial changes to the itinerary scheduled for release to limit details of planned travel from public disclosure. For example, routes to and from announced appointments usually do not need to be revealed.

I-39. During the course of a PSD's mission, its members may be asked to perform several different security functions. They may, for example, perform direct or indirect protection or escort duty. Direct protection is open and obvious; indirect is generally a surveillance measure. The security-guard unit may operate as an interior guard and may consist of one or more men stationed at fixed posts. A PSD's members should know the identity of each individual in the party of a protected official; executives can help by introducing them to each member of the official party.

I-40. The protected person's attitude is critical to the success of the PSD's mission. Executives have a right and a responsibility to make their wishes known with respect to their personal security. They also have an obligation to listen carefully to the head of the PSD who is trained and highly qualified to help make reasonable judgments about manageable risks. A PSD's members understand that their function is inherently intrusive and that executives can easily resent the loss of privacy that accompanies the protection offered. On the other hand, PSDs have jobs to do, not merely to protect executives, but to help safeguard mission-critical assets—senior military and civilian leaders.

I-41. One of the most demanding functions placed on a PSD is to limit the ability of individuals to circulate and approach the executive. This is often very frustrating to executives who wish to shake hands, engage in close conversations with visitors, and move freely and without impediment in a social situation. The PSDs are trained to strictly enforce limitations on the circulation of individuals, carefully checking each person for ID and ascertaining that he is authorized to be present at the occasion.

I-42. DOD personnel who are provided with PSDs and must conduct official business or hold social engagements in large rooms can take steps to minimize the disruptions to such functions. These steps include—

- Providing advance attendee lists to the head of the PSD.
- Having one or more members of the staff who know the attendees stand with PSD members and identify the attendees as they arrive.
- Informing attendees that they will be admitted only at specified entrances.

I-43. The PSD's members are highly trained security specialists. While in the company of executives, they will be accommodating and helpful. Executives should remember, however, that the primary function of the PSD's members is to protect them, not perform errands or carry out personal services. A PSD's members who are performing valet or other chores cannot effectively protect the senior officers or civilian officials to whom they have been assigned.

EXECUTIVE-PROTECTION SYSTEM INTEGRATION

I-44. This appendix has focused on supplemental security measures used to address terrorist threats to senior high-risk personnel within the DOD. Various methods and measures have been discussed that provide increments of security over and above the base level of security provided to all DOD personnel assigned to an installation, facility, activity, or unit. In making decisions to allocate protective resources to enhance the security of senior officers and senior DOD officials, it is essential to remember that measures must be applied systematically. Additional security measures implemented to protect high-risk personnel in the office environment must be carried over to official functions conducted outside the office. The security measures must also be extended to protected persons' private lives and, depending on the nature of the threat, the lives of their family members.

I-45. The converse is equally true. It makes no sense to provide domicile-to-duty transportation for a high-risk person and make no provision for additional protection at home, at the office, and at official business and social functions. In view of the total costs of security measured in dollars, time, and inconvenience to protected persons, their staffs, colleagues, and families, it may be more prudent to radically alter living and working arrangements than to try to augment security in a piecemeal manner. For example, it might be prudent to house high-risk personnel within a DOD installation rather than to try to secure a detached, private residence at a substantial distance from the operations base of a response force. The key to successful executive protection is to ensure that the level of protection afforded by physical-security measures, operational procedures in the office and at home, and PSDs is constant. The level of protection must be matched to the threat and must be sustainable. Executives have a special responsibility to set a personal example of combating-terrorism awareness; of attention to personal, family, office, information, and OPSEC concerns; and of combating-terrorism security measures implementation. By doing so, they make their colleagues and subordinates more aware and more conscious of their security environment and less likely to be victimized by terrorist attacks.

Appendix J

Resource Management

Programs need annual funding to operate. This funding is obtained by devising and documenting a resource program that looks seven years out. Looking ahead one or two years is generally not a problem. Commanders must be able to describe future requirements in the out years so that money will be available when the program arrives in future years.

FUNDING PROGRAMS

J-1. Physical security is dependent on integrated systems with budgetary constraints. This appendix serves to inform commanders of three basic funding programs contributing to physical security—

- The RJC6, which resources physical-security equipment purchase and maintenance.
- The QLPR, which resources law-enforcement operations (to include security guards and special-reaction teams).
- The VTER, which provides resources for projects and temporary programs that enhance any type of security due to an increase in terrorist threat.

PROJECTED REQUIREMENTS

J-2. Recurring VTER requirements are usually shifted to QLPR as standard requirements after a period of a few years. Installations send seven-year projected requirements to the MACOMs for submission into the Army's Program Objective Memorandum (POM), which is an annual significant event for resource managers at all levels. Installation requirements should be included in the POM build; otherwise, the installation's program loses visibility from the start of the funding process. Input format is determined locally, but a generic example of the format and the type of information requested is provided at Figure J-1, page J-2.

OBLIGATION PLAN

J-3. Budget execution deals with the current fiscal year (October through September), with the exception of some types of dollars (appropriations) that are multiyear (such as procurement dollars that are executable for three years). Once the resource manager notifies an installation of its available annual funding, an obligation plan by month or quarter is developed to display how and when the funds will be spent. This obligation plan is also used to forecast when the program will run out of money, which in turn will justify the submission of an unfinanced requirement (UFR) to obtain the proper level of resources. Examples of an annual obligation plan and UFR are located at Figures J-2 and J-3, page J-3.

Program Objective Memorandum (POM)

Installation: Fort McClellan, AL
Activity: Military Police School
Appropriation: Operation & Maintenance, Army (OMA)
Army Management Structure (AMS): 321731
Management Decision Package (MDEP): VTER
Element of Resource (EOR): 2500
Issue: Force Protection/Antiterrorism Courses

Resources	FY00	FY01	FY02	FY03	FY04	FY05	FY06
($000)	500	510	520	530	540	550	560

Justification/Impact:
Funding is required to continue the eight functional contract training courses conducted by the US Army Military Police School in Force Protection/Antiterrorism for the Army and selected DOD services and agencies. Courses focus on the five pillars of force protection and provide students with a variety of force-protection expertise to prevent, react to, and
contain a terrorist event. If funds are not provided, the projected annual student load of 2,500 cannot be trained.

Figure J-1. Sample POM

TYPES OF APPROPRIATIONS

J-4. The most common types of appropriations that managers will be involved with are day-to-day operations or mission (OMA) funds and procurement (OPA) dollars that must be used for major projects or equipment buys over the $100,000 threshold. Army Materiel Command (AMC) uses research, development, and engineering (RD&E) dollars for operations at US Army Test and Evaluation Command (TECOM) facilities and for testing new equipment. Construction projects over $500,000 use military construction (MILCON) funds.

J-5. Resources needed for physical security are the result of planning. Commanders include physical-security programs and improvements as a part of all annual budgets. When the situation changes based on METT-TC, physical-security programs are reviewed, updated and, when necessary, approved. Commanders allocate resources consistent with the threat. Force-protection requirements provide the fundamental reasons for resourcing physical security. Security managers, force-protection officers, and PMs help identify security requirements and prioritize expenditures. When physical-security improvements are not properly planned, integrated, and budgeted for, commanders accept risks for physical-security failures.

Annual Obligation Plan
FY 99

Installation: Fort McClellan, AL
Activity: Military Police School
Appropriation: Operation & Maintenance, Army (OMA)
Army Management Structure (AMS): 321731
Management Decision Package (MDEP): VTER
Element of Resource (EOR): 2500
Issue: Force Protection/Antiterrorism

Phased Monthly Obligation Plan (Cumulative):

Month	$000
Oct	40
Nov	80
Dec	120
Jan	160
Feb	200
Mar	240
Apr	280
May	320
Jun	360
Jul	360
Aug	360
Sep	360

Figure J-2. Sample Annual Obligation Plan

Unfinanced Requirement (UFR)
FY 99

Installation: Fort McClellan, AL
Activity: Military Police School
Appropriation: Operation & Maintenance, Army (OMA)
Army Management Structure (AMS): 321731
Management Decision Package (MDEP): VTER
Element of Resource (EOR): 2500
Issue: Force Protection/Antiterrorism

Justification/Impact:
Funds in the amount of $360K have been provided to conduct the eight Force-Protection/Antiterrorism contract courses in the US Army Military Police School. This amount will only support training through June 99. An additional $120K is required to conduct training during 4th quarter FY99. Unless additional funds are provided, 625 students will not be trained.

Figure J-3. Sample UFR

Appendix K

Vulnerability Assessment

After a commander has obtained a threat analysis, he proceeds to complete the analysis by conducting the vulnerability and criticality assessments. (This appendix will discuss only the VA.) This process considers a mission review and analysis of the installation, base, unit, or port in relation to the terrorist threat. The review should assess the cost of antiterrorism measures in terms of lost or reduced mission effectiveness. It should then assess the level of acceptable risk to facilities and personnel given the estimated erosion of mission effectiveness. Often the best operational method and routine may be the worst to counter potential terrorist activities. This review and analysis is performed routinely and particularly for deployment.

ASSESSMENT CONSIDERATIONS

K-1. The installation, base, unit, or port assessment is derived from the results of the vulnerability and criticality assessments. The assessment provides the staff with the overall vulnerability to terrorist attack. The staff then develops the crisis-management plan, which addresses all terrorist threat levels regardless of the present level. The THREATCONs are then applied according to the local threat. The considerations are—

- **Vulnerability.** The VA is a self-assessment tool used to evaluate its vulnerability to terrorist attack. The more vulnerable an installation, base, unit, or port is, the more attractive it becomes to terrorist attack.
- **Criticality.** The criticality assessment identifies key assets and infrastructures located on and adjacent to the installation, base, unit, or port. These assets are normally symbolic targets that traditionally appeal to a specific terrorist group (such as headquarters buildings and monuments). It addresses the impact of the temporary or permanent loss of key assets or infrastructures to the ability of the installation, base, unit, or port to perform its mission. The staff determines and prioritizes critical assets. The commander approves the prioritized list. The assessment—
  - Selects key assets.
  - Determines whether critical functions can be duplicated under various attack scenarios.
  - Determines the time required to duplicate key assets or infrastructure efforts if temporarily or permanently lost.
  - Determines the vulnerability of key assets or infrastructures to bombs, vehicle crashes, armed assault, and sabotage.
  - Determines the priority of response to key assets and infrastructures in case of fire, multiple bombings, or other terrorist acts.

- **Damage.** The damage assessment determines the ability of the installation, base, unit, or port to plan for and respond to a terrorist attack against key assets and infrastructures.
- **Recovery procedures.** The recovery-procedures assessment determines the capability to recover from the temporary or permanent loss of key assets and infrastructures. Based on this assessment, the staff establishes recovery procedures to ensure the continued ability to perform the mission.

THREATCON LEVELS

K-2. Specific security measures should be directly linked with THREATCON levels. These considerations are—

- **THREATCON Normal.** This THREATCON level exists when a general threat of possible terrorist activity exists but warrants only a routine security posture.
- **THREATCON Alpha.** This THREATCON applies when there is a general threat of possible terrorist activity against personnel and facilities (the nature and extent of which are unpredictable) and when circumstances do not justify full implementation of THREATCON Bravo measures. It may be necessary to implement measures from higher THREATCONs either resulting from intelligence or as a deterrent. The measures in this THREATCON must be capable of being maintained indefinitely.
- **THREATCON Bravo.** This THREATCON applies when an increased and more predictable threat of terrorist activity exists. The measures in this THREATCON must be capable of being maintained for weeks without causing undue hardship, affecting operational capability, or aggravating relations with local authorities. While in Bravo, the installation should bring manning levels and physical-protection levels to the point where the installation can instantly transition to THREATCON Charlie or Delta.
- **THREATCON Charlie.** The transition to THREATCON Charlie must be done on short notice. It is a result of an incident occurring or the receipt of intelligence indicating that some form of terrorist action against personnel and facilities is imminent. Charlie measures should primarily focus on manning adjustments and procedural changes. Security forces will usually enhance their security presence by acquiring additional manning or by adjusting the work-rest ratio (such as moving from a 3:1 to a 6:1 ratio). At Charlie, off-installation travel should be minimized.
- **THREATCON Delta.** Since the transition to THREATCON Delta is immediate, Delta measures should primarily focus on manning adjustments and procedural changes. THREATCON Delta applies in the immediate area where a terrorist attack has occurred or when intelligence has been received that terrorist action against a specific location or person is likely. The security force's manning level is usually peaked in Charlie; therefore, Delta's additional manning will usually come from an augmentation force. Once in Delta, nonessential operations will cease in order to enhance the security and response

posture. Normally, this THREATCON is declared as a localized condition.

K-3. With exception of THREATCON Normal, all THREATCON levels have certain measures associated with them. These measures are listed in JP 3-07.2, Appendix J, and AR 525-13.

ASSESSING VULNERABILITY

K-4. A VA addresses the consequences of terrorist attacks in terms of the ability of units, installations, commands, or activities to accomplish their assignments successfully, even if terrorists have inflicted casualties or destroyed or damaged DOD assets. The VA focuses on two broad areas—

- Preventing and, failing that, substantially mitigating the effects of a terrorist act.
- Maintaining emergency preparedness and crisis response.

K-5. The VA provides the commander with a tool to assess the potential vulnerability of an installation, base, unit, or port activity; but it is not a substitute for sound judgment. The VA must stand on its own and be supported by valid considerations. Typically, a small group of knowledgeable individuals develop the VA. The VA team consists of personnel with required areas of expertise. Some of these team members are the—

- Assessment-team chief.
- Physical-security specialist.
- Structural engineer.
- Infrastructure engineer.
- Operations-readiness specialist.
- Intelligence and/or counterintelligence specialist.

K-6. The functions and responsibilities of each team member are outlined in JP 3-07.2. The following paragraphs provide basic information regarding these areas:

- The assessment-team chief's key responsibilities include overseeing the management, training, and performance of the vulnerability-team members; finalizing the assessment-team out briefing; and preparing the population dynamics and risk assessment.
- The physical-security specialist is responsible for the security and safety of the installation, facility, and personnel.
- The structural engineer examines a variety of potential terrorist weapon effects and structural responses. This function serves to better protect personnel from shocks and blasts by reducing damage through the technically appropriate use of standoff measures, hardening, blast shielding, and shatter-resistant window film (such as Mylar®) as described in Chapter 3. The structural engineer's main responsibility is threat and damage assessment from terrorist weapons estimates and suggestions for threat protection or damage-mitigation measures.
- The infrastructure engineer examines protection against the effects of WMD, protection against terrorist-incident induced fires, and utility systems that can be used to minimize terrorist-incident casualties

(including elements of power, environmental control, and life-support systems). The primary responsibilities include infrastructure security and fire, safety, and damage control.

- The operations-readiness specialist examines plans, procedures, and capabilities for crisis response, consequence management, and recovery operations should a terrorist incident occur. The main responsibilities of this position include emergency-medical and individual-readiness assessments.
- The intelligence and/or counterintelligence specialist has the primary responsibility of performing logical analyses and preparing possible conclusions regarding terrorist targets and target vulnerabilities. These are based on processed intelligence information and knowledge of terrorist capabilities and methods in view of US installation, facility, and personnel safety and security practices.

CONDUCTING THE ASSESSMENT

K-7. Upon its arrival, the assessment team provides an in briefing for the commander, staff, and designated technical point of contact. Site personnel should conduct a site-familiarization briefing and tour. Administrative activities may include establishing the team support area, setting up equipment, scheduling team and technical points of contact meetings and discussions, ensuring classified-material control, establishing a personnel locator, and organizing materials for the out briefing and site folder. Each assessment-team member conducts the assessment based on the specific responsibilities for each function within his area.

POST-ASSESSMENT ACTIVITIES

K-8. Within 30 days of the visit, a summary narrative report and an annotated briefing should be delivered to the installation commander. Follow-on assistance for the commander may be applicable in areas of technical characteristics of improvement options, cost estimates, and generic sources of materials and equipment.

DRILLS AND EXERCISES

K-9. Multiechelon war gaming of possible terrorist attacks is the best test, short of an actual incident, to analyze the response of an installation, base, unit, or port. Drills and exercises test suspected vulnerabilities and antiterrorist measures. These exercises and drills also train the staff as well as reaction-force leadership and help maintain a valid threat assessment by identifying and adjusting to changing threat capabilities.

Glossary

AA&E arms, ammunition, and explosives

ABCS Army Battle Command System

AC alternating current

admin administration

ADP automated data processing

AF Air Force

AFB Air Force base

AFM Air Force manual

AFMAN Air Force manual

AFOSI Air Force Office of Special Investigations

AFR Air Force regulation

AIQC Antiterrorism Instructor Qualification Course

AIS automated information system

AL Alabama

AM amplitude-modulated

AMC Army Materiel Command

AMS Army management structure

AO area of operations

AP armor piercing

Apr April

AR Army regulation

AR-PERSCOM Army Reserve Personnel Command

ASP ammunition supply point

AT antitank

AT/FP antiterrorism/force protection

attn attention

Aug August

AWG American wire gauge

bldg building

BMS balanced magnetic switch

BTO barbed-tape obstacle

BUPERS Bureau of Naval Personnel

C^2 command and control

C^3 command, control, and communications

cav cavalry

CB citizen's band

CCB Community Counterterrorism Board

CCIR commander's critical information requirements

CCTV closed-circuit television

CD-ROM compact-disk, read-only memory

cdr commander

CG command guidance

chap chapter

CIA Central Intelligence Agency

CID Criminal Investigation Division

CINC commander in chief

CISO counterintelligence support officer

CMU concrete-masonry unit

CONEX container express

CONPLAN contingency plan

CONUS continental United States

CP command post

CPWG crime-prevention working group

CQ charge of quarters

CRIMP Crime Reduction Involving Many People

CTA common table of allowance

DA Department of the Army

DARE Drug Abuse Resistance and Education

DC direct current

DC District of Columbia

Dec December

DIA	Defense Intelligence Agency
DOD	Department of Defense
DOE	Department of Energy
DOJ	Department of Justice
DOS	Department of State
DOT	Department of Transportation
DS	direct support
DTM	data-transmission media
DTOC	division tactical operations center
EDM	emergency-destruct measures
EECS	electronic entry-control system
EOD	explosive-ordnance disposal
EOR	element of resource
EPW	enemy prisoner of war
equip	equipment
ESS	electronic security system
FAA	Federal Aviation Administration
FBI	Federal Bureau of Investigation
FCC	Federal Communications Commission
Feb	February
FIS	foreign-intelligence services

FM field manual

FM frequency-modulated

ft foot, feet

FY fiscal year

G2 Assistant Chief of Staff, G2 (Intelligence)

GHz gigahertz

GTA graphic training aid

HN host nation

HQ headquarters

HUD Housing and Urban Development Administration

HUMINT human intelligence

Hz hertz

IAW in accordance with

ICP initial control point

ID identification

IDS intrusion-detection system

IED improvised explosive device

IG inspector general

IID improvised incendiary device

in inch(es)

INSCOM US Army Intelligence and Security Command

INTAC	Individual Terrorism Awareness Course
IPB	intelligence preparation of the battlefield
IR	infrared
ISS	information systems security
J2	Intelligence Directorate (Joint Command)
Jan	January
JS	Joint Service
JSAT	Joint Security Assistance Training
JSCP	Joint Strategic Capabilities Plan
JSIIDS	Joint-Service Interior Intrusion-Detection System
Jul	July
Jun	June
K	one thousand
kHz	kilohertz
LED	light-emitting diode
liq	liquid
LOS	line of sight
LOTS	logistics over the shore
LP	listening post
LRA	local reproduction authorized
MACOM	major Army command

maint maintenance

Mar March

MCO Marine Corps order

MDEP management decision package

MDMP military decision-making process

METT-TC mission, enemy, terrain, troops, time available, and civilian considerations

MEVA mission-essential or vulnerable area

MI military intelligence

MILCON military construction

MILPO military personnel office

MILVAN military van

min minimum

mm millimeter(s)

MO modus operandi

MOU memorandum of understanding

MP military police

MPACS Military Police Automated Control System

MPMIS Military Police Management Information System

MPR military-police report

MS-DOS Microsoft®-disk operating system

MWD military working dog

N/A not applicable

naut nautical

NAVATAC Navy Antiterrorism Analysis Center

NBC nuclear, biological, and chemical

NCIC National Crime Information Center

NCO noncommissioned officer

NISCOM Naval Investigative Service Command

No. number

Nov November

NSA National Security Agency

NTAV nontactical armored vehicle

NVD night-vision device

O official

OASD Office of the Assistant Secretary of Defense

OCOKA observation and fields of fire, cover and concealment, obstacles, key terrain, and avenues of approach

OCONUS outside the continental United States

Oct October

OMA operations and maintenance, Army

OPA operations and procurement, Army

OPLAN operation plan

OPORD operations order

OPSEC operations security

Pam pamphlet

PAO public affairs office(r)

PD probability of detection

PERSCOM Personnel Command

PHOTINT photographic intelligence

PI police intelligence

PIR passive infrared

PM provost marshal

PMO provost marshal office

POL petroleum, oil, and lubricants

POM Program Objective Memorandum

POV privately owned vehicle

PS physical security

PSD protective security detail

PSI physical-security inspector

PTO Parent-Teacher Organization

PX post exchange

R&D research and development

RD&E research, development, and engineering

ref reference(s)

RF	radio frequency
RII	relevant information and intelligence
ROI	report of investigation
RORO	roll on/roll off
RPG	rocket-propelled grenade
/s/	signed
S2	Intelligence Officer (US Army)
SAW	squad automatic weapon
SDNCO	staff duty noncommissioned officer
SDO	staff duty officer
Sep	September
SIGINT	signal intelligence
SJA	staff judge advocate
SO/LIC	Special Operations/Low-Intensity Conflict
SOFA	status of forces agreement
SOP	standing operating procedure
St.	Saint
STANO	surveillance, target acquisition, and night observation
stat	statute
STC	sound-transmission coefficient
STD	standard
STU	secure telephone unit

TAACOM Theater Army Area Command

TB technical bulletin

TDY temporary duty

TECOM US Army Test and Evaluation Command

TEMPEST Terminal Electromagnetic-Pulse Emanation Standard

THREATCON threat conditions

TM technical manual

TMDE test, measurement, and diagnostic equipment

TNT trinitrotoluene

TRADOC US Army Training and Doctrine Command

TSC triple-standard concertina

UCMJ Uniform Code of Military Justice

UFR unfinanced requirement

US United States

USACE US Army Corps of Engineers

USACIDC US Army Criminal Investigation Command

USAMPS US Army Military Police School

USCG US Coast Guard

VA vulnerability assessment

VIP very important person

vol volume

WMD weapons of mass destruction

Bibliography

AR 12-15. *Joint Security Assistance Training (JSAT) Regulation (*Navy Instructions 4950.4; AFR 50-29). 28 February 1990.

AR 50-5. *Nuclear and Chemical Weapons and Material—Nuclear Surety.* 3 October 1986.

AR 50-6. *Nuclear and Chemical Weapons and Materiel, Chemical Surety.* 1 February 1995.

AR 190-11. *Physical Security of Arms, Ammunitions, and Explosives.* 30 September 1993.

AR 190-12. *Military Police Working Dogs.* 30 September 1993.

AR 190-13. *The Army Physical Security Program.* 30 September 1993.

AR 190-14. *Carrying of Fire Arms and Use of Force for Law Enforcement and Security Duties.* 12 March 1993.

AR 190-22. *Searches, Seizures and Disposition of Property.* 1 January 1983.

AR 190-27. *Army Participation in National Crime Information Center.* 28 May 1993.

AR 190-51. *Security of Unclassified Army Property (Sensitive and Nonsensitive).* 30 September 1993.

AR 190-56. *The Army Civilian Police and Security Guard Program.* 21 June 1995.

AR 190-59. *Chemical Agent Security Program.* 27 June 1994.

AR 380-5. *Department of the Army Information Security Program.* 25 February 1988.

AR 380-10. *Technology Transfer, Disclosure of Information and Contacts with Foreign Representatives.* 30 December 1994.

AR 380-19. *Information Systems Security.* 27 February 1998.

AR 380-67. *The Department of the Army Personnel Security Program.* 9 September 1988.

AR 381-20. *The Army Counterintelligence Program.* 15 November 1993.

AR 405-20. *Federal Legislative Jurisdiction.* 1 August 1973.

AR 525-13. *Antiterrorism Force Protection (AT/FP): Security of Personnel, Information, and Critical Resources.* 10 September 1998.

AR 600-8-14. *Identification Cards for Members of the Uniformed Services, Their Family Members, and Other Eligible Personnel* (AF 36-3026[I]; BUPERS I 1750.10A; MCO P5512.1B; CG M5512.1; Manual 29.2, Instructions 1 and 2). 1 March 1998.

AR 604-5. *Personnel Security Clearance, Department of the Army Personnel Security Program Regulation* 1 February 1984.

DA Form 2028. *Recommended Changes to Publications and Blank Forms.* 1 February 1974.

DA Form 2806-R. *Physical Security Survey Report.* April 1985.

DA Form 2806-1-R. *Physical Security Inspection Report (LRA).* April 1985.

DA Form 2819. *Law Enforcement and Discipline Report.* May 1988.

DA Form 3975. *Military Police Report.* May 1988.

DA Pam 190-12. *Military Working Dog Program.* 30 September 1993.

DA Pam 190-51. *Risk Analysis for Army Property.* 30 September 1993.

DOD 0-2000.12-H. *Protection of DOD Personnel and Activities Against Acts of Terrorism and Political Turbulence.* February 1993.

DOD 4160.21-M. *Defense Reutilization and Marketing Manual.* August 1997.

DOD Directive 2000.12. *DOD Antiterrorism/Force Protection (AT/FP) Program.* 13 April 1999.

DOD Instruction 2000.16. *DOD Combating Terrorism Program Standards.* 10 May 1999.

DOD Instruction 5210.84. *Security of DOD Personnel at US Missions Abroad.* 22 January 1992.

Federal Specification FF-L-2740. *Locks, Combination.* 12 October 1989.

FM 5-34. *Engineer Field Data.* 30 August 1999.

FM 19-4. *Military Police Battlefield Circulation Control, Area Security, and Enemy Prisoner of War Operations.* 7 May 1993.

FM 55-20. *Army Rail Transport Units and Operations.* 1 June 00.

GTA 19-4-3. *Individual Protective Measures.* July 1997.

JP 3-07.2. *Joint Tactics, Techniques, and Procedures for Antiterrorism.* 17 March 1998.

JS Guide 5260. *Service Member's Personal Protection Guide: A Self-Help Handbook to Combating Terrorism.* July 1996.

TB 9-2300-422-20. *Security of Tactical Wheeled Vehicles.* 27 August 1988.

TM 5-805-8. *Builders' Hardware.* 20 January 1992.

TM 5-820-4. *Drainage for Areas Other Than Airfields* (AFM 88-5, Chap 4). 14 October 1983.

TM 5-853-1. *(O) Security Engineering Project Development* (AFMAN 32-1071, Vol 1). 12 May 1994.

TM 5-853-2. *(O) Security Engineering Concept Design* (AFMAN 32-1071, Vol 2). 12 May 1994.

TM 5-853-3. *(O) Security Engineering Final Design* (AFMAN 32-1071, Vol 3). 12 May 1994.

TM 5-853-4. *Security Engineering Electronic Security Systems.* 12 May 1994.

USACE STD 872-50-01. *Entry Points for US Army Installations.* To be published within the next six months.

USACE STD 872-90-02. *FE5 Chain-Link Security Fence Details.* May 1992.

USACE STD 872-90-03. *FE6 Chain-Link Security Fence Details.* May 1992.

USACE STD 872-90-04. *FE7 Chain-Link Security Fence Details for Non-Sensored Fence.* May 1992.

USACE STD 872-90-05. *FE7 Chain-Link Security Fence Details for Sensored Fence.* May 1992.

USACE STD 872-90-06. *FE8 Chain-Link Security Fence Details for Sensored Fence.* May 1992.

USACE STD 872-90-07. *FE5 Chain-Link Fence Gate Details.* May 1992.

USACE STD 872-90-08. *FE6 Chain-Link Fence Gate Details.* May 1992.

USACE STD 872-90-09. *FE8 Chain-Link Fence Gate Details.* May 1992.

Index

Field Manual
No. 3-19.17

Headquarters
Department of the Army
Washington, DC, 6 July 2005

Military Working Dogs

Contents

Distribution Restriction: Approved for public release; distribution is unlimited.

Figures

Tables

Preface

The last Army dog field manual (FM) was published in 1977. It reflected military working dog (MWD) doctrine developed during the Vietnam era. Although useful at that time, much of the information has since become obsolete. Today, MWD teams are employed in dynamic ways never before imagined.

Today's MWD team is a highly deployable capability that commanders have used around the world from Afghanistan to Africa and from the Balkans to Iraq. These specialized teams aid commanders in stability and support operations as well as in warfighting. Being modular and mobile makes these teams very agile. As situations dictate, MWD teams are quick to arrive and able to conduct various operations. Their versatility allows for effective transformation at all echelons among readiness for deployment and operations on the ground, through redeployment and back to readiness.

The highly aggressive dog tactics of the 1960s and 1970s are long gone. Today's MWD program effectively employs expertly trained and motivated handlers coupled with highly intelligent breeds of dogs. These teams are continuously rotating between their assigned duties and deployments worldwide to perform joint operations, multiechelon tasks, and interagency missions.

This FM addresses the current capabilities of the Military Police Working Dog Program as well as the potential for future applications. As technology and world situations change, the MWD team will continue the transformation process and give commanders the full-spectrum capabilities needed to be combat multipliers on the battlefield as well as persuasive force protection and antiterrorism assets.

Appendix A complies with current Army directives, which state that the metric system will be incorporated into all new publications.

This publication applies to the Active Army, the Army National Guard (ARNG)/the Army National Guard of the United States (ARNGUS), and the United States Army Reserve (USAR).

The proponent of this publication is United States Army Training and Doctrine Command (TRADOC). Send comments and recommendations on *Department of the Army (DA) Form 2028 (Recommended Changes to Publications and Blank Forms)* directly to Commandant, United States Army Military Police School, ATTN: ATSJ-DD, 401 MANSCEN Loop, Suite 2060, Fort Leonard Wood, Missouri 65473-8926.

Unless otherwise stated, masculine pronouns do not refer exclusively to men.

Note: For the purposes of this manual, a dog team refers to one handler and one MWD unless specified otherwise.

Chapter 1

The Military Working Dog Program

Dogs have been used for the protection of life and property since ancient times. From these beginnings, dog training and employment has been continuously refined to produce a highly sophisticated and versatile extension of the soldier's own senses. Even the most complex machines remain unable to duplicate the operational effectiveness of a properly trained MWD team.

THE ROLE OF THE MILITARY WORKING DOG

1-1. MWDs provide a valuable asset to military police, infantry, special forces, the Department of Defense (DOD), and other government agencies. The MWD's senses of sight, smell, and hearing enhance his detection capabilities and provide commanders with a physical and psychological deterrent to criminal activity. Properly trained MWDs can prevent an intruder or suspect from escaping. When necessary, the MWD provides an added dimension of physical force as an alternative to the use of deadly force. Public knowledge of MWD team capabilities provides military police and various security forces with a formidable deterrent wherever the MWD team is employed.

1-2. MWDs are key resources for use in the military police combat support role. MWDs are trained for scouting, patrolling, and performing building and area searches. Commanders can use *Table 1-1* to determine what MWD type best suits their needs. Some MWDs have also been trained to track although this is not a required area of expertise. All of these skills contribute to the successful completion of the five military police functions across the full spectrum of military operations.

Table 1-1. MWD Mission-to-Type Matrix

Mission	*MWD Type*		
	PEDD	**PNDD**	**PD**
A. Area Security Operations			
(1) Area and reconnaissance	X	X	X
(2) Screening and surveillance	X	X	X
(3) Base/air-base defense	X	X	X
(4) Cordon and search	X		
(5) Checkpoints	X	X	X
(6) Roadblocks	X	X	X
(7) Response force	X	X	X
(8) Critical site, asset, HRP security	X[1]	X	X
B. Maneuver and Mobility Support Operations			
(1) Maneuver support	X	X	X
(2) Mobility support	X	X	X
C. I/R Operations			
(1) EPW and CI	X[2]	X[2]	X[2]
(2) Evacuation	X	X	X
D. Law and Order Operations			
(1) Force protection	X	X	X
(2) Military police investigations		X	

Table 1-1. MWD Mission-to-Type Matrix (Continued)

Mission	*MWD Type*		
	PEDD	**PNDD**	**PD**
(3) Customs support	X	X	
(4) Redeployment operations	X	X	
(5) Suspicious/unattended packages	X[3]		
(6) Health and welfare inspections	X	X	
(7) Crowd control	X[4]	X[4]	X[4]
(8) Alarm responses	X	X	X
(9) Bomb threats	X		
(10) Public MWD demonstrations	X	X	X
E. PIO	X	X	X

[1] PEDDs will not be used to provide security in ASPs, as the ammunition and explosives contained at the ASP will detract from the MWD's detection ability and may distract the MWD from his patrol function.
[2] MWDs will not be used in a correction/detention facility to ensure custody of prisoners.
[3] PEDDs can be used to search the area around a suspicious/unattended package for secondary devices. At no time will a PEDD or handler be used to search the package itself.
[4] Direct confrontation with demonstrators is not recommended, but is authorized with the commander's approval when lesser means of force have been unsuccessful.

1-3. MWDs are a unique item; they are the only living item in the Army supply system. Like other highly specialized equipment, MWDs complement and enhance the capabilities of the military police. MWD teams enable the military police to perform its mission more effectively and with significant savings of manpower, time, and money.

HISTORICAL OVERVIEW OF THE MILITARY WORKING DOG

1-4. Dogs have served in active service at the sides of their handlers for decades. They have been heroes, showing bravery under fire, saving lives (often losing their own), and bringing comfort to the injured and infirmed. The first recorded American use of military dogs was during the Seminole War of 1835 and again in 1842. In Florida and Louisiana, the Army used Cuban bred bloodhounds for tracking. During the US Civil War, dogs were used as messengers, guards, and unit mascots.

1-5. The Army Quartermaster Corps began the US Armed Forces' first war dog training during World War II. By 1945, they had trained almost 10,000 war dogs for the Army, Navy, Marine Corps, and Coast Guard. Fifteen war dog platoons served overseas in World War II. Seven platoons saw service in Europe and eight in the Pacific.

1-6. MWDs were trained at Fort Carson, Colorado, organized into scout dog platoons, and used in the Korean conflict for sentry duty and support of combat patrols. In 1957, MWD training moved to Lackland Air Force base (LAFB), Texas, with the Air Force managing the program.

1-7. Throughout the Vietnam Conflict, the Military Police Corps used dogs with considerable success. Most of these were sentry dogs used to safeguard critical installations such as ports and airfields. A new dimension in canine utilization was realized when marijuana detector dog teams were trained and deployed to assist military police in suppressing illicit drug traffic. Sentry and marijuana detector dog teams were then deployed worldwide in support of military police. An important outgrowth of the conflict was the development of canine research and development efforts. These ongoing efforts were able to initiate the first steps toward developing a more intelligent and stronger military dog, training dogs to detect specific drugs and explosives, developing multiple-purpose dogs, and employing tactical dogs by electronic remote control.

1-8. In the 1990s and early 2000s, MWDs were deployed around the globe in military operations such as Just Cause, Desert Shield and Desert Storm, Uphold Democracy, and Enduring Freedom and Iraqi Freedom. These teams were effectively utilized to enhance the security of critical facilities and areas, as well as bolster force protection and antiterrorism missions, allowing commanders to use military police soldiers and other assets more effectively elsewhere. MWD operations in Haiti showed the first large-scale utilization of MWDs since Vietnam. In the Balkans, military police deployed with MWDs to provide force protection and antiterrorism support and area security throughout the US base camps. MWDs trained as explosive detector dogs conducted vehicle sweeps and building searches to ensure the safety of US soldiers. US Army MWD teams operated beside Iraqi police officials to help in law and order missions and restore the war torn country into a peaceful region.

1-9. MWDs are force multipliers. Installation commanders should include MWDs when planning for force protection and antiterrorism countermeasures. They expand the individual soldier's effectiveness in the face of those that would come against him. They provide an immense physical and psychological effect when used as a show of force in day-to-day operations. The various uses of MWDs have been effectively employed in many aspects of military police missions. MWDs are utilized effectively at gates, camps and bases, and checkpoints and for random searches for narcotics and explosive devices. MWDs are also utilized for other missions in support of combat, combat support, and combat service support units.

1-10. MWD teams are assigned to perform high-risk personnel (HRP) security missions. MWD teams perform narcotic detection missions on the Mexico and US border to assist US Customs. They have been deployed on the United States and Canada border to perform law and order operations. The canine can easily detect what his handler cannot see, hear, or smell. The MWD's controllable aggressive behavior, coupled with the physical and psychological effects he creates, makes him and his handler an impressive and unmatchable team.

ORGANIZATION AND STRUCTURE OF MILITARY WORKING DOG UNITS

1-11. The US Army MWD program is an organization with structure and responsibilities that reach from the DOD level down to the handler level. For information concerning organization and responsibilities above the local level, see *Army Regulation (AR) 190-12.*

KENNEL MASTER TEAM

1-12. The kennel master (KM) team is comprised of two noncommissioned officers (NCOs) (one sergeant first class [SFC], who is the KM and one staff sergeant [SSG], who is the plans NCO). The KM team is modular and provides technical supervision of MWD teams assigned to the military police brigade. This team may be assigned to the corps military police brigade, echelon above corps (EAC) military police brigade, or the internment and resettlement (I/R) brigade. In addition to technical supervision, this team plans and coordinates MWD operations and support requirements. The KM team is also responsible for MWD team proficiency training. This team supervises the storage and accountability of the narcotic and explosive training aids as well as all assigned equipment.

PATROL EXPLOSIVE DETECTOR DOG TEAM

1-13. The patrol explosive detector dog (PEDD) team is comprised of three NCO MWD handlers (one SSG, and two sergeants [SGTs]) and three PEDDs. This team detects explosives and contraband in support of combat, combat support, and combat service support units and the five military police functions. It provides MWD patrol support (including I/R operations) when not employed with explosive detection. This team is normally assigned to the corps military police brigade, EAC military police brigade, military police I/R brigade, or the I/R command and control (C2) detachment. This entire modular team can be deployed together as a unit with all assigned equipment or, based on the specific mission, a handler and an assigned MWD may be deployed apart from the team.

PATROL NARCOTIC DETECTOR DOG TEAM

1-14. The patrol narcotic detector dog (PNDD) team is comprised of three NCO MWD handlers (one SGT and two specialists [SPC]) and three PNDDs. This team detects narcotics and other drugs in support of combat, combat support, and combat service support units and the five military police functions. It provides MWD patrol support (including I/R operations) when not employed with narcotics detection. This team is normally assigned to the corps military police brigade, EAC military police brigade, military police I/R brigade, or the I/R C2 detachment. This entire modular team can be deployed together as a unit with all assigned equipment or, based on the specific mission, a handler and an assigned MWD can be deployed apart from the team.

INDIVIDUAL DUTIES AND RESPONSIBILITIES

1-15. There are five levels of responsibility in the MWD program. The terminal level is the program manager and the beginning level, the handler. This is not to be confused with the operational chain of command or standards of grade. These individuals have specific duties in managing and performing the MWD program.

PROGRAM MANAGER

1-16. The program manager executes major Army command (MACOM) level responsibilities for the MWD program. He is the equivalent of an operations NCO and is an appointed certification authority whose duties include—

- Conducting MWD team patrol and detector certifications.
- Requisitioning personnel and MWDs.
- Providing policy and guidelines for the MACOM.
- Forecasting annual operational budgets.
- Coordinating MWD support for the office of the Secretary of Defense, US Secret Service (USSS), US Customs Service (USCS), and Department of State missions.
- Serving as the MACOM provost marshal's (PM's) advisor for the employment of MWDs in contingency operations, force protection and antiterrorism support, and homeland defense.

KENNEL MASTER

1-17. The KM is the senior NCO in charge of the installation's MWD program, consisting of 6 to 30 handlers and respective MWDs. KM duties are the equivalent of a platoon sergeant's duties when 9 or more handlers and MWDs are assigned. When 8 or fewer handlers and MWDs are assigned, the duties are equal to those of a squad leader. In addition to traditional leadership responsibilities, KM duties include—

- Monitoring all training to ensure that MWD teams are ready for annual certifications and worldwide deployments in support of various operational missions.
- Advising the commander on the employment of MWD teams.
- Ensuring that team proficiency is maintained.
- Preparing for MACOM and veterinary inspections.
- Forecasting annual operational budgets.
- Assisting the MACOM program manager in requisitioning personnel, MWDs, and equipment.
- Managing the daily maintenance and upkeep of kennel facilities.
- Monitoring MWDs' health and welfare.
- Managing procurements and accounts for explosive and narcotic training aids.
- Advising installation and unit force protection and antiterrorism officers on the use of MWDs.
- Coordinating MWD support for the USSS, the USCS, DOD protective services units, combat support operations, the Army Recruiting Command, and other civilian agencies.

MILITARY WORKING DOG PLANS NONCOMMISSIONED OFFICER

1-18. The MWD plans NCO is part of the KM team and performs the duties of the maneuver mobility support operations (MMSO) and training NCOs. The plans NCO is selected for his ability and overall knowledge of training, mentoring, and deploying MWD assets. This NCO is responsible for—

- Developing training plans.
- Coordinating and planning the unit mission-essential task list (METL).
- Supervising the execution and tracking of the deployment plan.
- Tracking the movement of MWD teams while they are deployed.
- Ensuring that deployed soldier's issues are resolved.
- Assisting in the supervision of the installation MWD program.
- Acting as the KM when the KM is absent.
- Preparing short- and long-term training outlines and risk assessments.
- Validating and supervising proficiency training.
- Overwatching preparation for the deployment of MWD teams.

SENIOR MILITARY WORKING DOG HANDLER

1-19. The senior MWD handler is the senior NCO of the PEDD or PNDD teams. His duties are the equivalent of the squad leader of the PEDD team or the team leader of the PNDD team. In addition to traditional leadership responsibilities, he is responsible for ensuring that—

- Annual certification training is conducted.
- MWD teams are ready for deployments.
- Team proficiency is maintained.
- Handlers are proficient with their assigned MWDs.
- Coordination for MWD team support for the USSS, the USCS, combat support operations, the Army Recruiting Command, and other civilian agencies is completed.

MILITARY WORKING DOG HANDLER

1-20. The handler provides the daily care and grooming for his assigned MWD. He ensures that proficiency skills are maintained in his assigned MWD. The handler and MWD perform either PNDD operations or PEDD operations. The handler and MWD are all trained for patrol but the variant for narcotics detector or explosive detector depends on the MWD team's training. The handler also—

- Fills out training records daily.
- Grooms the assigned MWD daily.
- Trains the assigned MWD daily.
- Maintains MWD kennels daily.
- Performs additional kennels duties.

Note: For more information about additional skill identifier (ASI) Z6 standards of grade, refer to *DA Pamphlet (Pam) 611-21*.

FUNDAMENTALS FOR THE USE OF MILITARY WORKING DOGS

1-21. An MWD can be trained to respond consistently to certain sensory stimuli (sights, sounds, odors, or scents) and alert the handler. Under most circumstances, a properly trained MWD can smell, hear, and visually detect movement better than a person.

1-22. Commanders and supervisors should consider the capabilities of the MWD team when identifying and evaluating possible MWD employment. Not all MWD teams are able or equipped to perform all missions equally (see *Table 1-1,* page 1-1).

USE OF PATROL DOGS

1-23. All MWD teams are trained to perform patrol missions. One benefit of employment is the deterrence of criminal activity. The obvious presence and well-published activities of MWD teams can successfully deter trespassers, vandals, violent persons, and other would-be criminals. When the desired task is to deter unauthorized or suspicious individuals, an MWD team should be assigned to a location during day or night.

1-24. The patrol dog's (PD's) contribution to the law and order effort is most effective when the team is utilized as a walking patrol. The law enforcement duties that a PD team can perform as a walking patrol include checking or clearing buildings and patrolling parking lots, family housing areas, and troop billet areas. Mobility significantly increases the potential area of coverage. The best and most effective method is the ride awhile, walk awhile method. This concept helps cover large patrol areas, and the periodic exercise helps keep the team alert. All PD teams can be utilized for—

- Antiterrorism operations.
- Force protection and antiterrorism efforts.
- Special-reaction team (SRT) operations.
- Assistance in military police investigations.
- Walking and mobile patrols.
- Alarm responses.
- Building checks and searches.
- The identification and apprehension of individuals.
- The protection of HRPs.
- The protection of funds.
- Open-area searches.
- Civil disturbances.
- Perimeter security.
- External intrusion detection.
- Response forces.
- Access control points (ACPs) support.
- High-risk targets (HRT) support.
- Mission-essential vulnerable areas (MEVA) support.
- I/R operations.

1-25. Many of the tasks above can be performed in combat support or force protection and antiterrorism operations. In addition, PD teams can and should be utilized as a tool for area defense, providing warning and response. PD teams are effective on combat patrols and as listening posts and reconnaissance team members. The employment of PD teams should not be overlooked during I/R or enemy prisoner of war (EPW) operations. PDs are trained to apprehend suspects at or near a crime scene, stop those who may attempt to escape, and protect their handlers from harm. Some PDs may be able to locate suspects who have left the scene of a crime.

1-26. The PD team should be utilized to provide a psychological deterrent and a show of additional force protection and antiterrorism measures. PD teams should be placed on roving patrols at ACPs, vulnerable areas, and the location of HRP. These missions should increase as force protection and antiterrorism levels increase.

USE OF PATROL NARCOTIC DETECTOR DOGS

1-27. Military police commanders, planners, and supervisors have a very effective means of detecting the possession and/or use of controlled substances with the employment of a PNDD team. PNDDs are trained to recognize the scent of certain illegal substances through a program of practice and reward.

1-28. In addition to PD missions, PNDD dogs can be utilized for the following:

- Criminal Investigation Division (CID) and military police investigator drug investigations.
- USCS inspections.
- Health and welfare inspections.
- Barracks searches.
- Workplace searches.
- Random gate inspections.
- Area searches.
- Vehicle searches.
- Postal inspections.
- I/R operations.
- Aircraft and luggage searches.
- Predeployment and redeployment searches.
- Other US government agency support.

1-29. When the PNDD responds on any of the substances he is trained to detect, his handler takes the appropriate action. PNDD teams perform a valuable service by helping to rid the military community of illegal drugs and the problems associated with drug and controlled substance abuse.

USE OF PATROL EXPLOSIVE DETECTOR DOGS

1-30. PEDD teams are very useful in many searches or investigations involving explosives. The search for explosives is critically important to ensure that the community is safe and protected from would-be terrorists. PEDD teams are particularly useful if there is a need to locate one or more of these explosive items, which may have been hidden in an area.

1-31. The deterrent value and the detection capabilities of PEDD teams are a very effective countermeasure to terrorism. Public knowledge that PEDD teams are in use is a deterrent to persons who may try to use explosives illegally. The knowledge that explosives can be detected by PEDD teams at entry points or in places where explosives have been hidden can prevent a person from attempting to commit terrorist acts.

1-32. In addition to PD missions, PEDD dogs can be utilized for—

- Bomb threat incidents.
- Suspicious/unattended package incidents.
- ACP searches.
- Checkpoint searches.
- Random gate searches.
- CID and military police investigations section.
- Health and welfare inspections.
- Building and area searches.
- Postal inspections.
- Aircraft and luggage searches.
- Predeployment and redeployment searches.
- Protective service missions.
- Other US government agency support.

1-33. A well–trained PEDD team can conduct a significantly more effective search of an area or facility in a much shorter time than a number of people can. Using PEDD teams helps reduce the potential risk to persons who would otherwise have to do the search without the benefit of the dog's superior sense of smell. More detailed information concerning the employment of PD, PNDD, and PEDD teams is discussed in *Chapters 5, 6,* and *7.*

Chapter 2
Administrative Requirements

This chapter contains information that is useful in planning an MWD program or adding MWD teams to an existing program. The guidelines in this section are not meant to be absolutes for utilization, training, or maintenance. They give general guidance and information that should be considered when planning a program. Leaders will find the following information helpful for writing authorization documentation statements and for planning the local MWD program. It is important to remember that all MWDs are trained as PDs in addition to the specific type of detector dog (narcotic or explosive).

DETERMINATION OF A NEED FOR MILITARY WORKING DOGS

2-1. The installation provost marshal (PM), military police unit commander, or task force commander must take the initiative to establish the local MWD program if it does not already exist. The specific needs must be determined and the costs of the program must be justified. The decision process involves a thorough risk and crime analysis, an accurate evaluation of the requirements of the entire military police mission, and the use of the working dog yardsticks located in this chapter.

2-2. Some of the factors to consider when determining the need for MWD teams include—

- The mission of the team.
- The size of the area of responsibility of the team.
- The size of the installation or the area population.
- The number of personnel to be served by the team.
- Incident rates for all crimes.
- The present capability and the commitment of a portion of the manpower resources as handlers.
- The types of terrain on the installation or in the probable areas of deployment.
- The types of combat support missions for which MWD teams can be used.
- The number of critical facilities or areas.
- The capabilities of each type of team.
- The number of HRP on the installation or in the area of responsibility.

VERIFICATION OF AUTHORIZATIONS

2-3. Initial MWD authorizations should be probationary using the following criteria:

- Adjustments must be made based on actual use data.
- A baseline ratio of 24 hours of MWD utilization to 8 hours training per week (4 hours patrol and 4 hours detection) must be maintained.
  - Utilization includes all missions (force protection/law enforcement, security, and combat support), whether patrolling, detecting drugs or explosives, or performing other functions (in combination), when a handler and dog are being employed together as a team.
  - Training includes all training activities to maintain, improve, regain, or develop dog skills such as obedience, controlled aggression, patrolling, scouting, detection of persons, and detection of drugs or explosives.
- An automatic review must be made by the respective MACOM after two years.

2-4. There is a trade-off when adjusting between the training and utilization necessary to reinforce basic skills or add new skills. For example, four hours per week may be adequate to maintain detector dog proficiency or to train in new skills, but additional training may be justified at times and training hours may be increased accordingly. This increase in training time should be accompanied by an equivalent reduction in utilization time. However, training hours should not be increased merely as a substitute for lack of utilization.

MISSION REQUIREMENTS

2-5. The mission requirements for the following support missions should be considered. These requirements should be used only as guidelines.

INSTALLATION FORCE PROTECTION AND LAW ENFORCEMENT SUPPORT

2-6. In order to provide continuous 24-hour, year-round MWD protection of the installation, nine MWDs are needed. Six of these dogs will be PEDD and three will be PNDD. This requirement is based on each MWD team being used a minimum of 24 hours per week. This does not include hours needed for training, grooming, or health care requirements. This does allow for temporary duty (TDY), handler leave, and illness.

2-7. The goal is to have one PEDD per installation available at all times for bomb threats or explosive type missions (not counted as part of the working force protection requirement). Installations with a three- or four-star headquarters require one additional PEDD authorization. This is due to the increased workload, facility criticality, facility vulnerability, location, past history of terrorist target selections, and present threat assessments.

HIGH-RISK PERSONNEL SECURITY MISSIONS

2-8. DOD tasks the Army to provide PEDD support for presidential, vice presidential, Secretary of State, and Secretary of Defense visits throughout the world. Currently, there is not a published formula to predetermine mission frequencies or requirements for these missions. On average, 40 teams inside the MACOM support the various missions above. These figures do not include travel or mission preparation time.

CUSTOMS SUPPORT

2-9. While no formula exists to quantify requirements for military customs support, this remains an ongoing mission that must be viewed as a requirement for additional MWD authorizations. MWDs are used to screen household goods shipments, privately owned vehicles (POVs), container express (CONEX) containers, military equipment, US mail, and port operations. Workload varies by installation population, location, unit missions and deployments, and the time of year.

CONTINGENCY OPERATIONS SUPPORT

2-10. Currently, approximately 50 MWD teams are supporting various contingency operations. This requirement will continue until mission completion. This figure represents 25 percent of Army MWD authorizations. Support for contingency operations has continued to rise since 1990 and will continue to do so. This increase must be factored into planning for MWD authorizations. A conservative average of 100 MWDs is needed for contingency operations on a yearly basis. This number is to ensure that MWD teams do not perform back-to-back deployments and that they receive proper proficiency training according to *AR 190-12* (4 hours per week in each specialty and 95 percent accuracy for PEDDs or 90 percent accuracy for PNDDs).

KENNEL CONSTRUCTION

2-11. When new MWD programs are started at locations where there are no facilities in place, kennel construction must be complete before the dogs arrive. The MWDs should be ordered through the military standard transportation and issue procedures (MILSTRIP) process approximately one year before the expected completion date of the kennel facility. In operations where MWD teams will be deployed for long periods of time and will be rotated, a permanent facility must be constructed within one year. Deployable field kennels may be used in the interim. The joint service MWD committee states that the DOD's design guide for MWD facilities must be referenced in planning the construction or renovation of a kennel facility.

2-12. Once the need and the justification for an MWD program are established, the PM, military police commander, or task force commander initiates a request to the installation facilities engineer to develop the kennel facility design. Before the kennel design has been finalized, the activity or installation veterinary corps officer (VCO), the physical security inspector, safety personnel, and the KM must be consulted to ensure that the minimum requirements are included in the final design. After kennel designs are completed and construction costs are estimated, the kennel project is submitted to the installation commander for approval and funding. The request for authorization of MWD teams must include the statement that the commitment has been made to fund and build kennels within one year prior to the arrival of the MWDs at the installation. Refer to *Appendix B* and *DA Pam 190-12* for details on kennel construction.

PROCUREMENT

2-13. All MWDs must be procured from the procurement section at the 341st Training Squadron, Lackland AFB, Texas. Local procurement of an MWD is prohibited. Requisition of an MWD team (one dog and one handler) requires two separate transactions. Handlers are obtained as a result of a personnel action requesting military police soldiers with the ASI Z6 in the required pay grades. MWDs are obtained as a result of a supply action. Commanders will assign an appropriately qualified handler to every MWD on hand. Assignment will be consistent with the policy of "one dog and one handler" as outlined in *AR 190-12*.

2-14. When the replacement of an MWD and not a handler is needed, requisitions are done according to *AR 700-81*. Initial requisitions for dogs should be submitted when positions for handlers are authorized. The MWDs in the Army supply system are—

- Dog, patrol: narcotic detector; national stock number (NSN) 8820-00-435-7542, line item number (LIN) D33800.
- Dog, patrol: explosives detector; NSN 8820-00-188-3880, LIN G33732.

2-15. The method of how to replace an MWD that has died, been adopted, or has undergone euthanasia is to submit a requisition in the MILSTRIP format that includes a statement that the unit or installation already has a trained handler on station. The MILSTRIP should also be submitted when the VCO has stated in writing that an MWD only has a 12-month life or work expectancy and/or is placed in deployment category IV (permanently nondeployable).

2-16. KMs should send requisitions for new or replacement dogs, in the MILSTRIP format, to their local property book office and forward a copy to their respective MACOM. The headquarters of the Air Force Security Police Agency will review all requisitions for dogs prior to processing by the 341st Training Squadron, Procurement Section.

TRAINING REQUIREMENTS

2-17. Training is needed at each installation to ensure that teams remain proficient. This realistic and challenging training should be scenario driven and relate to all of the possible operations in which MWD teams will operate according to the unit METL. Properly trained MWD teams are better prepared to accomplish all facets of MWD missions.

INITIAL TRAINING

2-18. MWD teams are initially trained at the 341st Training Squadron, Lackland AFB, Texas. This training is done at a basic level for both the handler and the dog and is not intended to prepare either the handler or the dog for immediate duty in an MWD team. The training is only intended to give a basic overview of the MWD program, the health and care of MWDs, and proper training techniques. It is the responsibility of the gaining unit to train a new handler or dog on all missions conducted by Army MWD teams. Proficiency training is continuously required to ensure that teams are reliable for use in patrol, narcotic detection, and explosive detection.

PROFICIENCY TRAINING

2-19. The effectiveness of MWDs trained in patrol, narcotic detection, or explosive detection depends on the continual reinforcement of the abilities of the dog team through proficiency training. This training is mandatory and if it is not conducted properly, an appointed certification authority can decertify the MWD team. Failure to meet the minimum training standard for at least the 35 days prior to a certification action will result in MWD teams not being evaluated for certification.

2-20. The minimum training standard is at least four hours of training each week in patrol and at least four hours of training each week in narcotic or explosive detection. The minimum standard of proficiency to maintain certification as a detector team is 90 percent or higher (with no more than a 10 percent false response rate) for narcotic detector dogs, and 95 percent or higher (with no more than a 10 percent false response rate) for explosive detector dogs.

2-21. Failure to maintain an average that meets or exceeds the minimum standard for three consecutive training events will result in the automatic decertification of the MWD team. Decertified teams may not be used in any way, and must undergo intensive retraining. These teams may be recertified only after retraining and a consistent demonstration to an appointed certification authority. Proficiency training should be conducted both on and off leash.

PROFICIENCY TRAINING OF PATROL DOG TEAMS

2-22. The effectiveness of PD teams depends on the continual reinforcement of mandatory proficiency training. PD teams must receive a minimum of 4 hours of patrol proficiency training each week. The training environment should approximate the military police working environment as closely as possible. Training should occur at varying times throughout the day and night and on varying days of the week, including weekends.

2-23. The emphasis in training is to develop the skills of both the handler and the dog so that they complement each other and the team becomes a working unit. As a proficiency evaluation criterion, the correct performance of the required task verifies that the handler understands how to control his PD and is able to do so.

2-24. Train the PD to protect with or without command based on any perceived threat to the handler. Because of the potential of serious bodily injury in this specialty, the certification of the handler and the dog as a team is very stringent. Ensure that the proficiency training of PDs includes an emphasis on critical tasks. There are no waivers for failure to accomplish a critical task by a PD team.

WEAPON FIRE TRAINING

2-25. Training on weapon firing ranges is essential for the PD to become proficient and not be deterred from attacking agitators during gunfire. The dog must not attack the handler during gunfire. The firing of all weapons assigned to the handler should be done with the MWD present whenever possible. PDs can be desensitized with the firing of many different types of weapons. This can often be accomplished by arranging for the handlers to take the dogs to weapon ranges of different units.

2-26. Determine the dog's reaction to the sound of gunfire. It may be necessary for the handler to use counterconditioning techniques until the desired proficiency is achieved. Counterconditioning techniques include starting at distances of 300 meters and slowly bringing the gunfire closer to the dog (or as safety allows, bringing the dog closer to the weapon). The goal is for the dog not to bark or show any signs of aggression when the handler fires all assigned weapons. This can be a slow process over the course of several days.

Obedience Course Training

2-27. The obedience course exposes the PD to various obstacles (see *Figure 2-1*) that simulate walls, open windows, tunnels, ramps, or steps. The dog's exposure to these obstacles reduces the amount of time required to adapt dogs to different environments. The dog learns to negotiate each of the obstacles. When confronted with a similar obstacle in the working environment, the dog is not deterred from completing his mission. The obedience course also develops the handler's ability to control the dog's behavior both on and off leash. The obedience course is not a substitute for exercise. A dog should never be required to negotiate the obedience course until he has been warmed up by proper exercise.

Figure 2-1. Obedience Course Obstacles

Proficiency Training of Detector Dog Teams

2-28. Continuous training of detector dogs is required so they maintain proficiency in the ability to seek and find drugs or explosives, depending on their specialty. Ensure that the detector training is always documented properly. Documentation and training are critical to both maintaining team certification and to establishing probable cause for a search based on a detector dog's number of correct responses.

2-29. There are several differences in the training requirements for PNDD and PEDD. Proficiency training scenarios for both types of detector dogs must be varied to avoid conditioning the dogs to a repetitious training pattern. The following factors will be varied for each proficiency training exercise:

- The general training exercise area.
- The number of training aids planted.
- The specific locations of the training aids in the area being used for training.
- The amount of explosives or drugs used in the training aid plant.
- The type of explosives or drugs used.
- The time of day or night of the training.
- The type of training aid container.
- The type of distraction material planted with the training aid.
- The length of time the training aid is left in place before the detector dog team search.
- The person used to handle and/or plant the training aid.
- The height above or below the floor level of the training aid plants.
- The size of the room or area in which the training aids are planted.

Decoys

2-30. The decoy is the person who role-plays the primary adversary for training and evaluating the MWD team. The decoy may be a suspect, a subject, an attacker, an agitator, a drunk, an escapee, an enemy, or any of a number of other persons an MWD team may expect to encounter while performing military police duties. A decoy may also be neutral or an ally, such as another military policeman, a supervisor, a lost juvenile, or an innocent person passing through an identification check. Use both males and females for decoys, agitators, and suspect role-players. Vary the clothing the role-players are dressed in to expose the dogs to civilian, military, and the ethnic dress of different countries, if possible.

2-31. Decoys must be taught how to handle the arm protector sleeve or wrap and other aggression tools such as the bite suit (*Figure 2-2*), the hidden wrap, and leg wraps for self–protection when the dog attacks. The sleeve or wrap used should be a hidden sleeve rather than the full, heavy padded sleeve. The hidden sleeve can be concealed under a sweater or field jacket to achieve more realistic conditions for training the MWD.

Figure 2-2. Bite Suit

2-32. KMs and plans NCOs teach decoys how to understand and use the different drives the MWD shows. Once understood, these drives or needs of the MWD can be exploited to cause the desired result in training. Decoys must learn the type of pattern to take to and from the MWD when training on aggression tasks or the bite. For safety and optimum training effect, decoys will learn how to target the MWD into the bite gear and what actions to take during a partial and full bite.

VALIDATION

2-33. The KM or plans NCO will conduct a validation prior to requesting certification. Validations will then be conducted quarterly to verify the patrol and detection accuracy rates recorded on *DA Form 2807-R (Military Working Dog Training and Utilization Record)* and/or *DA Form 3992–(R) (Narcotics or Explosives Detector Dog Training and Utilization Record)*. Additional validations will be given to any patrol or detector MWD whose demonstrated proficiency appears to differ significantly from the rating recorded on his *DA Form 2807-R* and/or *DA Form 3992–R*. The validation will verify proficiency or identify the specific patrol/detector weaknesses that need corrective training.

2-34. Validations include at least 30 trials for a narcotic detector dog and at least 40 trials for an explosive detector dog. These trials will be divided equally with each substance the dog is trained to detect. Training aids will be distributed in at least five different areas over a five–day period. These areas should include (if available) vehicles, barracks, open areas, warehouses, theaters, planes, and luggage. A written record of the evaluation and validation will be maintained on record for a minimum of two years.

CERTIFICATION OF MILITARY WORKING DOG TEAMS

2-35. *AR 190-12* defines the standards and specific requirements for all certification of MWD teams. Certification of an MWD and handler is valid for one year and is immediately nullified when—

- The dog is assigned to another handler.
- The handler and MWD have been separated or have not conducted training for thirty-five or more consecutive days.
- The team fails to meet the minimum detector standard of 95 percent accuracy (explosives) and 90 percent accuracy (narcotics) with no more than a 10 percent false response rate for three consecutive months.

SCOUTING OR PATROLLING

2-36. The PD must be able to detect a person by scent, sound, and sight (on leash only). This task is rated critical. The following tasks must be accomplished to demonstrate proficiency in scouting or patrolling:

- Daytime standards.
  - Detect and respond to the scent of a person who is hidden within 75 yards, and follow the scent to the person's location.
  - Detect and respond to a sound made by a person 100 yards away, and follow this sound to the source.
  - Visually detect and respond to a person in view 100 yards away, and pursue the person on command.
  - Demonstrate proficiency in pursuit, attack, and release.
- Nighttime standards:
  - Detect and respond to the scent of a person who is hidden within 75 yards, and follow the scent to the person's location.
  - Detect and respond to a sound made by a person 100 feet away, and follow this sound to the source.
  - Demonstrate proficiency in pursuit, attack, and release.

Note. There is no nighttime standard for sight due to the dog's natural ability to revert to other senses in the dark.

VEHICLE PATROL

2-37. The PD must be able to ride quietly and calmly inside a vehicle with the handler until provoked or commanded. This is a semicritical task.

TRACKING

2-38. Tracking is the only noncritical task on which PDs are trained. Tracking is noncritical because not all PDs have tracking potential. The local KM should know which dogs have been trained to track. PDs that demonstrate a potential for tracking are identified by the KM or plans NCO at each unit, and the dogs are then further developed in this task. Although tracking is a valuable skill for the PD team, improper performance of the task does not degrade the overall effectiveness of the PD team to perform the military police mission.

2-39. PDs selected for tracking training should be trained and employed for tracking on a tracking harness. The harness serves as a useful cue to the PD that it should focus his senses on tracking. At the start of the track, the handler will give the command "Track" and begin casting for the scent. Initially, scent pads are identified and then an item with the tracklayer's scent on it will be used. The dog is required to track and alert on any items left by the tracklayer.

2-40. The minimum level of proficiency a PD trained for tracking must attain is to be able to follow a human scent at least one hour old, for 1 mile over varied terrain, and on a course with several turns. Given suitable tracking conditions, a skilled tracker dog team can follow the natural wanderings of individuals or groups of persons for at least 3.1 miles over rugged and varied terrain and on a scent track that is at least 12 hours old. While tracking, the dog is also capable of alerting his handler to the presence of tripwires and ambushes.

2-41. Evaluation of tracking tasks must include various ages and lengths of the track. The areas used to conduct tracking training and evaluations will be rotated frequently. Tracks will be at least one-third of a mile long but no longer than one mile. The age should vary from ten minutes to one hour. Tracks will include turns, but evaluators should ensure that the person laying the tracks is not trying to deceive the dog with undue tracklaying. Articles will be placed on the track for the dog to find.

INSPECTIONS

2-42. Inspections help ensure that the MWD section is in compliance with Army standards. *Appendix C* contains an example inspection memorandum to assist the inspection officials on what areas to inspect at an MWD section.

DEPARTMENT OF THE ARMY INSPECTIONS

2-43. Personnel from headquarters, Department of the Army (HQDA), conduct annual inspections of the MACOM MWD program. These inspections are used to assist the MACOM PM and program manager in moving the MACOM MWD program forward. This inspection will be on all aspects of the MACOM MWD program (see *Appendix C*).

MAJOR ARMY COMMAND INSPECTIONS

2-44. MACOM program managers conduct annual inspections of handlers, MWDs, training, team use, team proficiency, equipment, and kennel facilities. This ensures compliance with *AR 190-12*.

PROVOST MARSHAL AND COMMANDER INSPECTIONS

2-45. PMs and/or commanders conduct monthly inspections of handlers, MWDs, training, team use, team proficiency, equipment, and kennel facilities. These inspections are conducted at least monthly on one or more areas of the kennels and quarterly on all aspects of the installation MWD section to ensure compliance with *AR 190-12*.

2-46. A written record of the monthly and quarterly inspection is maintained on file with the appropriate corrective action (both in memorandum format), stating the corrective action taken for each deficiency. This written record is maintained for one year past the inspection date.

2-47. The responsible VCO will conduct an inspection of the kennel facility at least quarterly to ensure health and welfare for both the handlers and dogs assigned to the MWD section. A copy of this inspection is maintained on file for a minimum of two years. The responsible PM and/or commander takes all actions necessary to correct any deficiencies noted by the VCO. Veterinary inspections are to be conducted monthly, if possible.

TRAINING RECORDS

2-48. There are two forms that must be used to annotate MWD training, *DA Forms 2807-R* and *3992-R*. These forms are official military documents and must be maintained daily with factual information. Many times these documents are used as court documents, so the importance of the completion and accuracy of these documents cannot be stressed enough. Example forms and the procedures for filling these forms out are in *Appendix D*.

PROBABLE CAUSE FOLDERS

2-49. Probable cause folders are one of the most important methods of proving an MWD team's reliability. Legal requirements to conduct a search of private property based on the response of an MWD do not allow the probable cause folder a margin for error. All the items contained in the probable cause folder are copies. The following original records must be kept on file at the kennel facility:

- ***Department of Defense (DD) Form 1834 (Military Working Dog Service Record).*** *DD Form 1834* must be maintained on the MWD and kept as the initial form in the probable cause folder. This form must be accurate and signed on a change of a handler. When the MWD is trained on new substances, ensure that they are listed on the *DD Form 1834.* The 341st Training Squadron at Lackland AFB, Texas, must approve new substances.
- ***Lackland AFB Form 375 (Patrol Dog Certification).*** *Lackland AFB Form 375* follows the *DD Form 1834* in the probable cause folder and lists patrol certification information for the MWD.
- ***Lackland AFB Form 375a (Detector Dog Certification).*** *Lackland AFB Form 375a* follows the *Lackland AFB 375* (if dual certified), and lists postgraduation certification information for the MWD.
- ***Air Force (AF) Form 1256 (Certificate of Training).*** A copy of the handler's *AF Form 1256* follows *Lackland AFB Form 375a.* The handler provides a copy of the *AF Form 1256* for the probably cause folder.
- **Certification Letter**. This letter attests to the fact that the appointed certification authority has witnessed a demonstration of the capabilities of the dog team. This letter, at a minimum, should list the handler and dog, the whelp, the type of MWD (PEDD or PNDD), the social security number (SSN) of the handler, and the date of certification. This letter should be maintained in the probable cause folder after the handler's graduation certificate until the next certification of the team or a change of handler.
- **Quarterly Validation and Evaluation by the KM**. This document is used to show that the KM has conducted a quarterly evaluation on the MWD team. If the KM is not available, the plans NCO can also conduct this evaluation. This document follows the certification letter in the probable cause folder.
- **Training and Utilization Records**. The last items that should be maintained in the probable cause folder are the last three months of training and utilization records for the MWD team. These records include *DA Forms 2807-R* and *3992-R.* Once an additional month has passed, the latest month will be transferred to the MWD team folder in the filing system.

TRANSPORTATION REQUIREMENTS

2-50. Every unit, installation, and activity with MWD teams should assign enough support vehicles so that all missions and kennel support activities can be properly conducted and maintained. The minimum standard for transportation is one support vehicle per two MWD teams (two dogs and two handlers). Vehicle requirements should be determined by mission requirements and the overall number of MWD teams. Some modifications are necessary so that the MWD can be seen and can provide a visual deterrence during law and order or force protection and antiterrorism missions. These modifications will maximize the mission effectiveness of the dog while providing the dog with a comfortable, supportive, and protective area inside the vehicle. This type of modification is normally a kennel crate or cage secured in the rear of the vehicle; the crate or cage should have a nonslippery surface.

2-51. MWD handlers perform complex missions. Four-by-four type passenger vehicles should be used to ensure that MWD teams can get to any mission during inclement weather or on rough terrain.

2-52. Pickup trucks are essential for the transportation of explosive training aids. *AR 190-12* lists the requirements for explosives transported in government vehicles. However, pickup trucks should not be used for transporting MWDs for training or during road patrol type duties. This vehicle offers limited protection for the MWD from inclement weather and innocent bystanders.

2-53. Warning signs stating "CAUTION—MILITARY WORKING DOGS" should be placed on all sides and the rear of any and all vehicles used to transport MWDs. The signs should also be printed in the host nation language. The only exception to this is the use of unmarked vehicles for President of the United States and Vice-President of the United States missions.

NARCOTICS AND EXPLOSIVE TRAINING AIDS

2-54. Detector dog training and evaluations require the frequent use of narcotic and explosive training aids. Due to the potential safety hazards and risks involved with these aids, it is absolutely mandatory for all MWD personnel and commanders to know and follow the regulatory guidance. MWD leaders and handlers must participate in periodic review and training on the requirements noted in *AR 190-12*. These requirements are clearly stated in *AR 190-12*.

Chapter 3
Managing Military Working Dog Operations

Managing an MWD program requires attention from the program manager, the KM, and the plans NCO. This chapter will help these individuals better understand their specific requirements and concerns regarding the scheduling of various missions.

SCHEDULING AND EMPLOYMENT

3-1. The KM or plans NCO assists and advises the commander or operations officer in scheduling and employing MWD teams. The KM and plans NCO are the most qualified to determine the duty schedules, work cycles, and hours of operation to support each mission. All MWD teams should be utilized at least 24 hours per week to perform nontraining related duties.

3-2. Scheduling must allow for a minimum of four hours of patrol proficiency training per week and four hours of detector training per week. Details concerning proficiency training requirements are discussed in Chapter 2. Daily scheduling should include enough time for feeding, grooming, and exercising the MWDs as well as maintaining equipment and kennel facilities.

PREDEPLOYMENT OPERATIONS

3-3. MWD leaders and planners need to ensure that handlers are prepared for all possible missions before the moment of execution. Careful planning and concurrent training is an ongoing process that involves all soldiers in every kennel section. No MWD team should ever be sent on a mission without the proper preparation.

PROGRAM MANAGERS PREDEPLOYMENT RESPONSIBILITIES

3-4. MACOM program managers play a key role in MWD team readiness. The careful management of information concerning each kennel enables leaders and planners to keep the unit in a constant state of readiness for any possible deployment or no-notice mission.

3-5. Program managers keep track of the—

- Total number of personnel at each kennel in the MACOM.
- Certification status of all teams at each kennel in the MACOM.
- Deployability status of all MWDs and handlers (see Chapter 8).
- Total number and location of deployed MWD teams.
- Status of kennel facilities at deployment locations.
- Status of VCO support at deployment locations.

KENNEL MASTER PREDEPLOYMENT RESPONSIBILITIES

3-6. The KM's responsibility is to ensure the daily accomplishment of all missions. The readiness of all assigned MWD teams is a process that requires continuous effort from all leaders. KMs must maintain statistics and make the necessary corrections to have all personnel and equipment available for future deployments and missions.

3-7. The KM is responsible for—

- Managing the number of personnel assigned to each kennel.
- Verifying the status of certified teams and the dates of certification.
- Maintaining the deployability status of MWDs and handlers.
- Ensuring that soldiers have soldier readiness packets prepared according to *AR 600-8-101*.
- Conducting monthly and quarterly counseling.
- Maintaining accurate and detailed training records that will give the forward deployed KM an understanding of strengths and weaknesses.
- Ensuring that MWD handler equipment for short-notice deployments is prepared at all times.
- Conducting annual 9-millimeter handgun qualification. More than one qualification should be conducted (if possible) to better train MWD handlers on the use of the 9-millimeter handgun and to ensure that handlers are always qualified.
- Ensuring that handlers meet the standards outlined in *AR 600-9* and *FM 21-20* by conducting height and weight testing and physical fitness tests.
- Ensuring that the mandatory common military training and common task tests are conducted according to *AR 350-1*.
- Training soldiers in tasks associated with deployments (for example, kennel setup, utilization, and missions).
- Ensuring that MWD handlers have received any needed driver's licenses or transportation motor pool (TMP) licenses before a deployment order arrives.
- Ensuring that soldiers have received the needed individual readiness training in preparation for deployment in a combat zone.
- Conducting mine awareness training.

Plans Noncommissioned Officer Predeployment Responsibilities

3-8. Deployment planning must include coordination with the supporting military VCO. The plans NCO must consider the expected duration of the deployment and the types of facilities anticipated at the destination.

3-9. The plans NCO is responsible for ensuring that the handlers and MWDs in the MWD team or kennel section are technically proficient. The plans NCO is best able to evaluate and recommend to the KM which MWD team is most qualified and prepared for deployments and missions. Through continuous and realistic predeployment training, the plans NCO ensures that the MWD teams are proficient in—

- Crowd control training.
- Building searches.
- Tactical movements (through obstacles and from point to point).
- Gunfire training (handler and decoy).
- Defense training.
- Controlled aggression.
- Explosive detection (area searches, bunker searches, and tactical vehicle searches).
- Narcotics detection (area searches, bunker searches, and tactical vehicle searches).

Additional Predeployment Considerations

3-10. Army mobility requirements necessitate that each MWD has a shipping crate. Shipping crates are used to transport MWDs, simplify handling and loading, and give MWDs adequate space and ventilation during transport. Shipping crates may also be used as temporary kennels at intermediate sites or at the deployment destination.

3-11. The average water requirement for a single MWD is 10 gallons a day. Enough dog food must be taken to last 90 days or until resupply can be established. MWDs must be fed the standard high-performance dog food, contracted and supplied by the General Services Administration (GSA), unless otherwise directed by the supporting VCO. MWDs are fed twice daily unless the supporting VCO directs another meal frequency.

3-12. First aid kits must accompany the MWD team on every mission. Information on the deployment destination, climate, and terrain help determine the specific contents of the first aid kits. For example, if MWDs are employed on rough terrain, a pad toughener is included in the first aid kit.

3-13. All mission-related equipment must travel and arrive with MWD teams so that they can be fully operational as quickly as possible. Any equipment that is not sent with the MWD teams should be sent by other means to arrive at the location within 60 days of the MWD team's arrival. An example equipment list is in *Appendix E*.

DEPLOYMENT CONSIDERATIONS

3-14. There are many important factors to consider when leaders and MWD planners are preparing for deployments. To ensure that every deployment results in mission success, every aspect from notification to redeployment must be considered. These aspects include, but are not limited to, field kennels, training, and operations.

FIELD KENNELS AND SUPPORT FACILITIES

3-15. A suitable kennel site must be carefully selected. Before deployment, some possible locations may be selected on a map from a leader's reconnaissance. As soon as possible after arrival, the sites should be inspected to choose the most suitable location. When selecting a site, consult a VCO (whenever possible) about possible health hazards.

3-16. Locating the kennel in a congested area may be necessary to protect it from enemy attack. A temporary screen or physical barrier may be needed to block the dog's sight so he will be able to rest.

3-17. When shipping crates are used as temporary kennels, the interior, exterior, and ground beneath the crate must be cleaned daily to prevent the accumulation of moisture and waste or insect infestation. It must be raised four to six inches off the ground to allow for adequate drainage and to reduce parasite breeding places. In hot climates, place crates under trees, a tarpaulin, or plywood to provide shade and ventilation. Spread gravel (if available) around and under the crates to allow for drainage and the easy removal of solid waste. A large tent or prefabricated building can serve as a temporary storage location for equipment, rations, and other supplies.

3-18. A second tent or building erected at the kennel site to house handlers and the KM provides the kennel with sufficient manpower for security of the kennel area. Although conditions around a temporary kennel may be primitive, a high standard of sanitation is essential to prevent diseases and parasite infections. Permanent kennel facilities must be erected within one year of the arrival of MWD teams to the location of operational use. Modular expandable MWD transport containers (NSN 961120, 961130, or an equivalent) (*Figure 3-1*, page 3-4) with integrated air conditioning, heating, and equipment storage are recommended for use as temporary kennels in field environments.

DEPLOYMENT TRAINING

3-19. As soon as possible after arriving at the deployment destination and while making sure that site security is established, deployment training must begin. Deployment training allows the MWD teams to quickly develop familiarity with their new location. This is also known as right-seat ride time. The KM or plans NCO needs to evaluate or reevaluate the degree of skill of each of the MWD teams so that the KM and NCO are fully prepared to support the deployed force. Initial evaluation and training should consist primarily of field problems such as attack, search, re-attack, and scouting and basic obedience. Training should also include the handler firing his assigned weapon while maintaining control of the MWD. Some training should be done at dusk, and then in darkness, so that the handler may see (for safety) and the MWD may adjust to the noise and flash of weapons.

Figure 3-1. Modular, Expandable MWD Transport Container

Maneuver and Mobility Support Operations

3-20. MMSO consist of measures necessary to enhance combat movement and the ability to conduct movement of friendly resources in all environments. These measures ensure that commanders receive personnel, equipment, and supplies as needed. MMSO is conducted across the full spectrum of military operations. The primary focus of military police during MMSO is to ensure the swift and uninterrupted movement of combat power and logistical support. KM teams will coordinate MWD operations with the plans and operations officer or task force commander.

Area Security Operations

3-21. When setting up listening posts or conducting combat patrols, consider the availability and suitability of PDs for the particular mission. The KM or plans NCO must be briefed early regarding the mission. This will ensure that the most suitable PD team can be selected and that the handler can prepare for the mission. The handler should check the MWD for any limiting ailments and conduct training rehearsals to become familiar with the mission. The KM or plans NCO and selected handlers must be involved in planning conferences and briefings involving the utilization of an MWD team.

3-22. In defensive operations, combat patrols are used to provide early warning, to confirm intelligence information, and to detect or deter enemy action. PD teams greatly enhance the security of reconnaissance and combat patrols.

3-23. On combat patrols, a PD works at maximum efficiency for only two or three hours. During inclement or extreme hot or cold weather, this time will vary. The KM, plans NCO, and VCO should advise the commander on the MWDs' particular capabilities. The team is most effective in uninhabited areas. If an MWD frequently alerts on friendly forces and is continuously taken off his alert, he soon loses interest and reliability. Only the best teams should be selected.

Defiles and Route Reconnaissance/Surveillance Operations

3-24. MWD teams can be used to enhance the security at defile and route reconnaissance/surveillance operations. Consider the following to ensure mission success:

- The mission at hand (the possible threat scenario).
- The size of the area to be covered.

- The type and number of MWD teams.
- The duration of the mission.
- Sustainment supplies.
- Communications.

3-25. Operational planning must ensure that the proper numbers and types of teams are requested to perform the mission. Planning considerations must allow for proper rotations and rest-to-work ratios for the MWD teams.

Checkpoints

3-26. MWD teams are often called on to enhance the effectiveness of checkpoints. After receiving the mission to perform checkpoint duty, planners and handlers need to consider the following:

- The force protection condition (FPCON) (see *Appendix F* for details).
- The type and number of teams required versus those available.
- Sustainment equipment.
- Communications.
- Available support elements.

Area Defense

3-27. Areas that cannot be covered by static defensive posts because of vegetation, terrain, or some other peculiar feature can be secured using PD teams. Prior to positioning any MWD team in a defensive posture, ensure that intelligence preparation of the battlefield (IPB) details any possible placement of friendly or enemy mines. When relieving others from established defensive positions, use range cards to familiarize the team with the area.

3-28. A PD team can detect personnel using concealed avenues of approach and provide an early warning to defenders. Each MWD team's area of responsibility should be large enough so that the team can move to take advantage of the prevailing wind directions. A fixed position should be established to provide covering fire for each MWD team in case the team must withdraw closer to the defensive perimeter after the MWD has alerted on an attempted hostile probe or penetration.

Warning and Response Procedures

3-29. Several procedures should be developed for warning friendly forces of a PD's response because circumstances may prevent the use of a radio. Also, different types of responses tailored to fit a variety of potential situations should be developed. For example, it may be desirable to have the MWD team follow the alert and locate the cause. At other times, combat intelligence may require the team to maintain the alert position until assistance arrives or to withdraw to a more advantageous position. Whatever action is taken, the handler should not release an MWD unless it is necessary to defend—

- Himself.
- Other personnel.
- Protected resources.

Internment and Resettlement Operations

3-30. I/R operations consist of those measures necessary to guard, protect, and account for people that are captured, detained, confined, or evacuated by US forces. In any military operation involving US forces, accountability and the safe and humane treatment of detainees is essential. US policy demands that all persons who are captured, interned, evacuated, or held by US forces are treated humanely. This policy applies from the moment detainees become the responsibility of US forces and continues until the time they are released or repatriated. (See *ARs 190-8* and *190-47* and *FMs 3-19.40* and *27-10*.)

3-31. MWDs offer a psychological and actual deterrent against physical threats presented by personnel housed in an I/R facility. They cannot be used as security measures against US military prisoners. MWDs reinforce security measures against penetration and attack by small enemy forces that may be operating in the area. MWDs also provide a positive, effective alternative to using firearms when preventing disturbances or escapes by personnel.

3-32. MWD handlers will not use their MWDs to guard prisoners inside prisons or detainee holding facilities and confinement facilities. In addition, MWD handlers will not use their MWDs to degrade, torture, injure, or mistreat EPWs, detained personnel, civilian internees (CIs), or other detainees in US custody.

Note: MWDs are not used during interrogation operations.

Demonstrations

3-33. Conduct periodic demonstrations in full view of personnel housed in an I/R facility to increase the psychological deterrent of an MWD. Emphasize how easily and quickly an MWD can overtake a fleeing individual. Highlight the MWD's ability to attack and overcome physical resistance. Demonstrate an MWD's scouting ability. To ensure a successful demonstration, use only the most qualified MWD teams.

Perimeter Security

3-34. Use an MWD team as an addition to perimeter security by making periodic, unscheduled patrols around the perimeter fence during periods of darkness. During inclement weather, a temporary blackout or an electrical failure increases the number and frequency of patrols.

WARNING

Never set patterns or establish routines when conducting security patrols. Enemy observing security measures can exploit set times by attacking when the patrol ends or before it begins.

Narcotic and/or Explosives Detection

3-35. Walk an MWD team through living areas to search for contraband. PEDD teams can be utilized to detect explosive devices and components in the I/R facility and PNDD teams can be utilized to detect narcotics.

Work Details

3-36. Position an MWD team between the work detail and the area offering the greatest avenue of escape. MWD teams provide a valuable addition to work detail guards, particularly those employed in areas offering the greatest potential for escape. However, MWD teams are deterrents and must not be used as guards.

Police Intelligence Operations

3-37. *AR 525-13* directs all military law enforcement to develop foreign and domestic criminal intelligence. MWD teams are an integral part of the police intelligence operations (PIO) process. PIO are conducted in conjunction with all functions in which the MWD teams perform. For example, a PEDD team working at a checkpoint to detect explosives will also be gathering intelligence. This holds true for all MWD functions.

3-38. Leaders and planners of MWD operations must develop procedures to ensure that the MWD teams gather and submit all intelligence in a timely manner. Standardized reports are an effective means that allow for the accurate and detailed submission of intelligence. Contact the supporting intelligence section (S2), provost marshal office (PMO), or CID office to ensure that all PIO coordination is in place prior to performing MWD functions.

LAW AND ORDER OPERATIONS

3-39. Law and order operations are those that MWD teams perform to enforce laws, orders, and regulations and to maintain discipline and safety. Law and order operations can be performed in both combat and peacetime environments

RANDOM ANTITERRORISM MEASURES PROGRAM

3-40. The random antiterrorism measures program (RAMP) is designed to be used at all FPCON levels to increase physical security awareness throughout an installation. RAMP involves the random use of measures, identified in *AR 525-13* and *TC 19-210*, for FPCON levels to make installation operations less predictable and therefore more secure.

3-41. The use of MWD teams during RAMP provides for more thorough detection of narcotics and/or explosives and for a psychological deterrent. An MWD presence can also increase confidence in installation security in general.

3-42. KMs coordinate with the PMO or S3 to ensure that RAMP checks are successful. Coordination will include—

- The reconnaissance of patrol areas (map reconnaissance as a minimum).
- Rehearsals and inspections before operations.
- Transportation.
- Equipment needed on site.
- A threat briefing.
- Rules of engagement.
- Use-of-force criteria.
- After-action review (AAR) requirements (see *Appendix D*).

TEMPORARY DUTY MISSIONS (UNITED STATES CUSTOMS SERVICE/HIGH-RISK PERSONNEL SECURITY MISSIONS)

3-43. USCS safeguards the internal security of the United States through border searches. Because of this, they have broad powers in conducting searches. By utilizing an MWD team, all vehicles, ships, luggage, and baggage can be checked at border clearance crossings or ports. MWD teams are called upon to support local or federal authorities in detection operations. The Joint Enforcement Military Customs Operation places MWDs at border clearance ports to conduct searches of vehicles and cargo entering the United States. The MWD team takes no law enforcement action. The MWD team only searches and reports. *AR 90-12* gives a more detailed explanation of civil support by MWD teams.

3-44. The Air Force has been designated to serve as the primary service point of contact for all very important person (VIP) missions received from the Office of the Secretary of Defense, USSS, and/or Department of State. The VIP coordination officer who (under authority from the Secretary of Defense) tasks and coordinates all missions performed by DOD PEDD team personnel, regardless of service affiliation, issues all mission tasks.

3-45. MWD leaders and planners ensure that the certification of the team selected to perform the mission is well documented and current. Other considerations in planning for USCS support include—

- The travel arrangements of the MWD team.
- Government travel credit cards.
- Lodging for the handler.

- The kennel and boarding.
- VCO support.

Military Working Dog Public Demonstrations

3-46. Public demonstrations are conducted to show the abilities of MWD teams. These demonstrations are coordinated through the KM to ensure that the appropriate team is available for the target audience. The goal of public demonstrations is to deter criminal activity and to promote public acceptance of the MWD.

3-47. At no time should demonstrations make an MWD a form of entertainment. The KM or plans NCO should ensure that the details of sensitive police training techniques are not explained to the public. Avoid lengthy or frequent public demonstrations that allow MWDs to tire easily or become bored.

Health and Welfare Demonstrations

3-48. Initially, the MWD team should provide the responsible commander with a basic demonstration of the MWD's ability. The KM or plans NCO places a training aid in an undisclosed location, and the handler enters the search area with the MWD. The handler presents and conducts a systematic search of the area. When the MWD alerts, the appropriate reward is given. This demonstrates the MWD team's proficiency.

3-49. After the demonstration has been conducted, the unit should ensure that the area being searched is free of soldiers and that all rooms are unlocked. MWD teams are a tool the commander uses to find any illicit narcotic or explosive devices. It is the responsibility of the unit to properly search the room after the MWD has cleared the room. If the MWD team responds, military police personnel are called to continue the search. The MWD handler conducting the search should not search for the illicit narcotics or explosives. Unit commanders should also ensure that no personnel enter the building or area until the search is complete. This is usually accomplished by placing NCOs at various entranceways to stop personnel from entering the area.

Competitive Events

3-50. MWD teams should be encouraged and allowed to participate in competitive events, seminars, or conferences conducted by outside agencies or police canine organizations. Exposure to competitive events can challenge MWD teams to develop skills or learn advanced employment and training techniques. These events may require new tasks or methods that the handler can use to identify and apply to military missions. In this way, the team's proficiency and effectiveness are enhanced.

Recovery Operations

3-51. After every deployment or extended off-site mission, MWD leaders need to ensure that vital actions are conducted upon the return of the MWD team. These actions include—

- Conducting an AAR after every MWD deployment or extended mission to assist in capturing lessons learned. See *Appendix G* for detailed information on AARs.
- Conducting veterinary inspections.
- Returning any explosive or narcotic training aids.
- Updating training requirements and records.
- Recertifying teams.

Chapter 4

Legal Considerations

The use of MWDs has the same legal considerations as the use of a military police soldier. Each situation and location may be subject to various Status of Forces Agreements (SOFAs), general orders, and other legal restrictions. The following chapter addresses some fundamental areas. It is important to consult with the Staff Judge Advocate whenever legal questions pertaining to an MWD arise.

USE OF FORCE

4-1. Procedures for the use of force are found in *AR 190-14*. This regulation states that personnel engaged in law enforcement or security duties will use the minimum amount of force necessary to control the situation. This regulation further details the levels of force and encourages commanders to substitute nonlethal measures for firearms when they are considered adequate for safely performing force protection, antiterrorism, and law and order functions. The regulation states that in evaluating the degree of force required for specific situations, the following options should be considered in the order listed:

- Verbal persuasion.
- Unarmed defense techniques.
- Chemical aerosol irritant projectors (subject to host nation or local restrictions).
- The military police baton.
- MWDs.
- The presentation of a deadly force capability.
- Deadly force.

4-2. Although high in the order of measures, it is clear that the release of the MWD as a use of force is a nonlethal option to be used before deadly force is considered. It must be understood that the release of an MWD to pursue, bite, and hold a person is likely to result in an injury to the suspect. This injury can be very minor such as scratches or bruises; however, serious bodily harm is possible. For this reason, it is required by *AR 190-14* that all lesser means of force must be used first whenever possible.

4-3. The use of force differs between MWD handlers and regular military police because of the uniqueness of the MWD. Positive control of the MWD must be maintained at all times throughout all of the use-of-force measures. Often, just the presence or arrival of a handler with his MWD can control a situation. If the desired level of control is not achieved, verbal persuasion is the first degree of force to be used. If verbal persuasion is not enough force to maintain or gain control, the next measure may be considered. If the situation allows, military police may use unarmed defense techniques, spray irritants, and the baton before MWD handlers use the MWD as a measure of force.

4-4. A handler is not likely to use unarmed defense techniques when holding an MWD on leash. The handler must remain in positive control of the MWD; therefore he will not be able to get close enough to a suspect to employ unarmed defense techniques. The use of chemical aerosol irritant projectors or the baton would also hinder the handler's ability to keep positive control of the MWD. Therefore, the options are not reasonable measures of force available to the handler.

4-5. Military police and task force commanders, along with PMs, establish clear policies regarding the release of MWDs according to *ARs 190-12* and *190-14*. If the minimum amount of force necessary to control the situation requires the use of MWD force off leash, the handler must ensure that the following steps are followed:

- Ensure that the MWD has identified the same target as the handler before releasing him to bite.
- Give the warning order, "Halt, halt, halt, or I will release my dog." If assigned in a foreign country, give the order in the primary language of the host nation.
- Warn all bystanders to cease all movement.
- Call the MWD off the pursuit and regain control of him if the suspect indicates surrender or stops moving.

4-6. Use extreme caution when removing an MWD from a suspect. Regain and maintain leash control of the MWD until he has become calm enough to obey the commands "Heel" and "Stay." Do not release an MWD if the suspect is out of sight, unless an off-leash building search is authorized. Do not release MWDs in areas where children are present.

4-7. MWD teams are not used for crowd control or direct confrontation with demonstrators unless the responsible commander determines it to be absolutely necessary. When it is necessary, dogs are kept on a short leash to minimize the danger to innocent persons. Dogs are not released into a crowd. Civil disturbance contingency plans include specific criteria for the use of MWD teams that are consistent with *FM 3-19.15, AR 190-12,* and *AR 190–14*.

SEARCHES

4-8. Rules 311 and 313 through 316, *Manual for Courts-Martial (MCM),* prescribe HQDA policies concerning the conduct of searches and inspections and the disposition of property seized in conjunction with those activities. Property refers to property of the United States or of nonappropriated fund activities of an armed force of the United States wherever it is located.

4-9. Commanders and law enforcement personnel must be aware of the distinction between searches, seizures, inspections, and inventories and the provisions for the execution of these actions as outlined in the *MCM*. Knowledge of this information is necessary to ensure the legal use of a PNDD or PEDD.

4-10. Probable cause is a reasonable belief that a crime has been committed and that the person, property, or evidence sought in connection with the crime is located in the place or on the person to be searched. Probable cause must be established prior to searching any persons, property, or residence. Probable cause can be established in the following ways:

- Personal observation.
- Information obtained from a reliable informant or from the witness who observed the crime.
- Use of a detector dog team.
- Any combination of these three.

4-11. The detector dog team is considered a "reliable informant," according to Rule 315 (f)(2), *MCM*. The reliability of a team is established through the initial and annual certifications of the detector dog team by the certification authority and through proper documentation of the MWD's training records.

4-12. Certification establishes the foundation for an officer to grant a search authorization or warrant. Detector dogs are first certified after finishing their basic training. To meet the legal requirements permitting their operational use, each detector dog team must undergo an annual certification and a validation.

4-13. Each detector dog team undergoes a certification when first assigned to duty. A designated certification authority conducts the certification. The detector dog team must perform a demonstration for the person who authorizes any search. This may be any commander from company level up to the post commander. The guideline is that the person who witnesses the demonstration can only authorize a search of units he has command authority over. This person must also review the probable cause folder of the dog and must validate the team as reliable, in writing. This validation is good as long as the commander is assigned to that command.

4-14. The certification authority should witness a demonstration of the dog's capabilities. All substances the dog has been taught to respond on should be used. Additionally, the dog's capability to detect past/present odor should be demonstrated. At the completion of the demonstration, the results are documented in letter form, signed by the commander, and entered into the dog's probable cause folder. The MWD handler is responsible for preparing and maintaining the probable cause folder and the contents (see *paragraph 2-49*, page 2-9). The probable cause folder accompanies the handler each time the dog is used on an actual search. Each document in the folder should be explained to the commander requesting a detector dog search. The commander with command authority should witness a simple demonstration of the dog's capabilities prior to a search. The above procedure should be followed whenever possible. This procedure establishes the dog's reliability on a continuing basis and is essential in establishing probable cause.

4-15. PEDD and PNDD teams must maintain high standards of accuracy to provide the desired level of confidence in their searches. PNDD teams must maintain a 90 percent accuracy rate for the number of trials conducted, while PEDD teams must maintain a 95 percent accuracy rate for the number of trials conducted. The installation commander authorizes the support of local and federal law enforcement agencies. *DOD Instruction 3025.12* governs the use of all PEDD teams. It governs the use of PNDD teams and ensures compliance with *Section 1385, Title 18, United States Code (USC) (The Posse Comitatus Act) (18 USC 1385 [The Posse Comitatus Act]).* Complete certification, recertification, and decertification details for narcotic and explosive detector MWD teams can be found in *AR 190-12*.

Chapter 5

Patrol Dogs and Characteristics

PDs can be used in traditional law and order functions, but commanders and planners should not overlook the tactical enhancements that a PD team offers to combat support operations. All aspects of military police functions can be enhanced by the employment of MWD teams. This chapter discusses many practical applications of the PD team.

EMPLOYMENT TECHNIQUES DURING COMBAT SUPPORT OPERATIONS

5-1. PD teams can perform many combat support operations. These teams enable the tactical commander to free up soldiers and employ their resources in other areas. This force multiplier is especially valuable in area security and force protection and antiterrorism operations.

PERIMETER SECURITY

5-2. PD teams are useful for perimeter (*Figure 5-1,* page 5-2) and distant support posts because they are usually located away from normal activity and large numbers of people. These posts also usually enclose large areas, and it would take a large number of single sentries to secure the area effectively. The large size of these areas allows the PD team to change positions to take advantage of the prevailing wind direction. These posts may be secured only during periods of advanced security and high threat. Occasional random posting of these areas is recommended, especially during periods of limited visibility (rain, snow, dust, smoke, and fog) or night conditions. Barriers and obstacles (such as fences, buildings, gullies, and streams) must also be identified and considered in security post planning. Place the PD team so that these obstacles offer the least interference to security.

5-3. Perimeter security, depending on the area to be covered, should use a two-team concept. The teams overlap each other while walking around the perimeter. PD teams should not come into direct contact with each other. Considerations include—

- The commander's intent.
- Priority intelligence requirements.
- Rules of engagement.
- Use-of-force criteria.
- The wind direction and distracters in the area (see *Chapter 8*).
- Friendly and enemy forces around the area.
- Rest areas within the perimeter to allow the MWD to relieve himself.
- Watchtowers or other security forces in the area.

5-4. The PD team should initiate the perimeter security operation by walking in a zigzag pattern around the exterior perimeter. The team should be allowed to go outside of the security area to ensure that no unwanted intruders are present.

Figure 5-1. Perimeter Security

INTRUSION DETECTION SYSTEM AUGMENTATION

5-5. Continuing development and use of Intrusion Detection Systems (IDSs) for security make it necessary to reassess the role of PD teams in the security of priority resources where IDSs are used. Although IDSs are designed to greatly reduce the probability of a successful area penetration, they are not designed to replace PD teams. IDSs cannot see beyond the outer clear zones, counter intrusions, or scout the progress of an intruder force after penetration. They are also subject to malfunction and breakdown. PD teams cannot counter all potential problems, but they can be used selectively to counter some of the IDS limitations. With the advice of the KM, security planners should develop plans that set an effective balance between sensor systems and PD teams. Some suggested functions for integrating PD teams into IDS-augmented security systems are as follows:

- PD teams may be used inside or outside critical sites or locations of HRP type personnel areas to give immediate response to areas not clearly seen by closed circuit television (CCTV) or to support security response teams in their role of alarm assessment
- PD teams may be used to scout hidden intruders from the point of penetration to the hiding location inside protected areas.
- PD teams may be posted outside critical sites during periods of increased threat to expand the protected area beyond the range of the IDS. This use is particularly desirable when there is heavy vegetation beyond the perimeter clear zone or when protected areas are located on or near installation perimeters and may, therefore, be easily accessible to the public.
- PD teams may be used as an element of the security response team or as an element of other designated response forces.
- PD teams may be used to compensate for decreased IDS effectiveness when poor visibility or other environmental conditions adversely affect the IDS equipment or when the IDS is not functioning.

Secondary Response Forces

5-6. Secondary response forces are typically used as blocking forces and for sweeps and counterattacks. PD teams can greatly improve the capabilities of the secondary response force. Some factors that must be considered are that—

- Blocking positions must be selected downwind from the PD team's position when PDs are used with blocking forces. In this type of mission, PDs should only be used to alert the forces of possible intruders.
- The secondary response force moves into the wind with the PD team in the lead position when the PDs are used for sweeps. It is imperative that security be placed on the PD teams to ensure that they are protected.
- A PD team can often pinpoint the exact location of individual intruders.

5-7. A PD team can help detect, alert, and catch hidden intruders while clearing a position that has been counterattacked successfully. PEDDs can be used to help clear the area of explosives. (See Chapter 7.)

Listening Posts

5-8. On a listening post, the PD team should be positioned forward of the tactical area to reduce distractions to the PD, yet close enough to maintain contact with friendly forces. In selecting a location, consider the primary detection senses of the PD. Whenever possible, PD team listening posts should be located downwind from any potential enemy position or dismount avenue of approach. Other locations force the PD to rely only on his hearing and sight for detection, reducing his effective use of scent.

Combat Patrols

5-9. In defensive operations, combat patrols are used to provide early warning, to confirm intelligence information, and to detect or deter enemy action. PD teams greatly enhance the security of reconnaissance and combat patrols.

5-10. On combat patrols, a PD works at maximum efficiency for only two or three hours. The team is most effective in uninhabited areas. If a PD frequently alerts on friendly forces and is continuously taken off his alert, he soon loses interest and reliability.

5-11. Patrol leaders and members must be briefed on actions to take when a dog handler is seriously wounded or killed. Dogs that have worked closely with patrol members and have developed a tolerance for one or more of them will usually allow one of the patrol members to return it to the kennel area. However, the dog may refuse to allow anyone near his handler, and other handlers may need to be called to help. The canine must be separated from the handler by—

- Coaxing the dog away with friendly words or food.
- Covering the dog with a poncho to immobilize it.
- Staking out the dog or leashing him.

5-12. A human will always take precedence over an animal. If no effort is effective, the dog may be destroyed. However, all efforts should be made to avoid this; an MWD is an expensive and valuable asset to replace.

5-13. If a situation arises where a canine is injured but the handler is not, the handler must be allowed to accompany the canine. If wounded or killed, an MWD should be evacuated using the same assets and should receive the same consideration as that given to a soldier under the same circumstances.

Tactical Patrols

5-14. PD teams can be used in tactical patrols to detect enemy presence, to help the patrol avoid discovery, and to locate enemy outposts. When the PD alerts, the handler should signal the patrol to halt until the cause of the alert can be identified and the patrol can proceed safely. If a firefight develops while the PD team is at the point position, the PD team should respond to fire team's directions and act as a regular member of the patrol.

5-15. Generally, the best locations for PD teams are directly in front of the patrol or on the flanks to enhance the flexibility and security of the formation. Prevailing wind direction should be used to improve the patrol's chances for early warning of enemy forces. If the wind direction is from the rear of the patrol, the PD is forced to rely on sight and sound and may not be as effective. The handler must concentrate on the PD to read his alert and will not be able to use a weapon rapidly. Therefore, members of the patrol should be assigned to protect the PD team.

5-16. PD teams can be used to enhance the capabilities of an ambush patrol. A PD team should be placed in front of the patrol to minimize distractions, yet close enough to maintain contact with other patrol members. PDs on ambush patrols must remain silent and not respond aggressively to approaching enemy forces.

Internment and Resettlement Operations

5-17. PD teams are used during I/R operations for camp perimeter security and to guard against escapes. The PD's keen senses of sight, smell, and hearing assist the handler and camp authorities in detecting unauthorized activity. A PD team's presence also acts as a deterrent to escapes. In the event of an escape, the PD is a valuable asset that can assist in the search and recovery of internees.

PD Duties and an Internment and Resettlement Camp in Iraq

On 13 April 2003, SGT Carey A. Ford and his MWD Rex responded to a riot at Camp Bucca, Iraq. It was believed that an Iraqi internee had taken a pistol belt and some personal equipment from a guard. The prisoner had taken this equipment back to his tent and the 600 other internees refused to cooperate with the guard force. SGT Ford and MWD Rex went into the I/R camp unarmed. The MWD team moved through the disruptive and unruly crowd providing a secure environment for the quick-reaction force to extract the prisoner and retrieve the stolen pistol belt and personal equipment. SGT Ford related that EPWs would not come anywhere near his MWD Rex. "We work that fear to our advantage."

Four days later, SGT Ford and Rex were called on to scout for and apprehend an escaped EPW. The fugitive had a substantial lead on them, having made his escape two hours prior to SGT Ford's notification. SGT Ford entered a possible unexploded ordnance and/or antipersonnel minefield with Rex, who was also trained to detect explosives. The MWD team was determined to locate the dangerous fugitive and prevent access to any possible weapons caches since many had been located in the area. After hours of relentless searching, SGT Ford and Rex located and apprehended the escapee. Their actions provided the psychological deterrent necessary to prevent future escape attempts.

Riot/Crowd Control

5-18. Extreme caution should be applied before committing PDs to a potential crowd control situation. Although the presence of a PD can be a powerful statement to an enemy or terrorist, it can elicit a reverse and potentially adverse effect on individuals in a riot/crowd control situation.

5-19. It is difficult to employ PDs properly during demonstrations or riots due to the high levels of confusion and excitement and the large number of potential antagonists. PDs should be trained to remain undistracted by loud noises and crowds. A crowd's reaction to a PD in a riot environment may be dependent on the location. In some areas of the world, the appearance of PDs at an angry crowd scene results in an escalation of violence. The crowd will often challenge soldiers to use the PDs as a measure of force. This is particularly true if a situation can be manufactured or provoked so as to be interpreted as an unreasonable use of force by the military. The presence of PDs can cause extreme fear (deterrence) in some cultures and regions. This fear can be advantageous and should be exploited when needed.

5-20. Committed PD teams may be used as an alternative to the use of deadly force to gain control of a riot situation. If the commander located on the ground directs the employment of PD teams for direct confrontation with demonstrators, PD teams should be held in reserve, out of the sight of the crowd. As the situation escalates, PD teams may be moved forward to within the sight of the crowd, but well away from the front lines.

5-21. When committed to the front lines in direct confrontation, PDs are allowed to bite only under the specific circumstances authorized by the responsible commander. PDs should never be released freely into a crowd. Other riot control force personnel should be positioned approximately three meters from any PD team when conducting any form of riot control. This is to prevent unintentional bite injuries to friendly, riot control force personnel that may happen to get between a PD and an instigator participating in the riot or demonstration.

5-22. PD teams must move a safe distance from the crowd to ensure the safety of the PD. Use of PD teams for direct confrontation with demonstrators is not recommended. According to the Army's use-of-force policy, the application of physical force against a person who cannot reasonably be suspected of committing an offense is not appropriate. Presence in a crowd does not necessarily constitute an offense or a situation where it would be reasonable to apply the degree of physical force of a PD. A PD would not be used unless the person is engaged in some other readily apparent criminal behavior such as theft, destruction of property, or assault.

EMPLOYMENT TECHNIQUES DURING FORCE PROTECTION AND ANTITERRORISM OPERATIONS

5-23. PD teams contribute significantly to the military police force protection and antiterrorism effort. PDs are directly responsible for executing specific security-related measures and for providing support to other government agencies that contribute to force protection.

5-24. MWD support is accomplished under the leadership of the respective command PM. Force protection is accomplished through an active role in physical security, HRP security, law enforcement, and antiterrorism efforts. Force protection and antiterrorism operations consist of those actions that prevent or mitigate hostile actions against DOD personnel (to include family members), resources, facilities, and critical information. During force protection operations, offensive and defensive measures are coordinated and synchronized to enable the joint force to perform while degrading opportunities for the enemy.

5-25. In operations, PD teams are used to enhance the detection capabilities of the combat support force and to provide a psychological deterrent to hostile intrusions. If the PD teams are properly located in a hostile environment, the PD's alert provides an initial warning to the presence of a hostile force. In past combat operations, PD teams often provided warning of attacks early enough to allow a response force time to deploy and to prevent enemy forces from reaching their objectives. PD teams have also helped clear protected areas of hostile persons, explosives, and weapons after attacks.

Crowd Control in Kosovo

During Task Force Falcon (2A) in Kosovo, LTC Cloy (the battalion commander of 1-36 INF, 1 Bde, and 1AD) requested MWD support to assist the Alpha company clear a main supply route (MSR) that had been blocked by 10,000 Kosvar Albanians. LTC Cloy had already deployed both quick-reaction forces to assist Company A because the crowd would not stay off the MSR. LTC Cloy called the battalion operations center and directed that the MWD teams be sent to the scene.

SFC Barnes, SSG Rogers, SSG Barters, and SGT McClintock responded with their dogs Queeny, Danny, Sandor, and Fedor and linked up with an infantry squad. The MWD teams and infantry squad moved dismounted to the town of Gjilani. The crowd was starting to become violent. LTC Cloy was in the middle of the crowd, standing in the center of the city at the critical MSR intersection. LTC Cloy attached the MWD teams to Company A. CPT Curtain, the company commander, told SFC Barnes to move the crowd away from his area and to establish a perimeter. The MWD teams moved into the diamond formation with the dogs on guard. Splitting the crowd in half, the MWD teams moved through the crowd and reached the city center MSR intersection in the center of the city. The handlers then used the dogs to move the crowd back out of the road. Company A and the four MWD teams established a perimeter line to keep the MSR open and traffic moving. They also apprehended troublemakers in the crowd.

Later that evening when the dog handlers entered into the DFAC, LTC Cloy stood up, gained everyone's attention, and stated, "SFC Barnes, you and your handlers were like Moses. Your dogs parted the Red Sea!"

Access Control Points

5-26. A PD team may be used to enhance force protection and antiterrorism efforts at ACPs. The PD team's function at an ACP is to deter unauthorized access, crime, and terrorism. The handler that is assigned to the ACP should use the PD so that the public can observe and hear the dog. Care must be taken to ensure that the handler constantly keeps the PD under control to prevent incident or injury to innocent personnel.

Building and Area Searches

5-27. A PD team is especially effective in checking and searching buildings such as commissaries, post exchanges, finance offices, banks, and warehouses. The PD team can check buildings visually while patrolling and can stop and dismount so that the handler can physically check doors and windows. These checks should be made with the PD on leash. To take maximum advantage of the PD's scenting ability, the handler should approach buildings from downwind.

5-28. If a building is open or forced entry is evident, PD teams may be used to track hidden intruders from the point of penetration to their location. PDs may be released to apprehend an intruder suspected of committing a serious offense when the only alternatives are escape or the use of deadly force.

5-29. Prior to searching a building, the handler will determine whether to make the search with the PD on or off leash. Some of the factors that must be considered include the following:

- Friendly forces or innocent bystanders in the area.
- The time of day or night.
- Evidence of forced entry.
- The type and size of the building or area to be searched.
- The danger to any innocent persons in the area.
- The PD's ability to work off leash.
- The threat level and IPB.

5-30. Prior to releasing the PD inside a building or enclosed area, the handler will give a clear voice warning to any persons inside the building or in the area. They will be told to come out immediately and that failure to comply with these instructions will result in the release of a trained MWD. An interpreter may be needed to assist the PD team in giving the warning in the host nation language. All persons will be warned that the PD may attack without warning and that they could receive physical injuries. The handler maintains voice control of his PD throughout the search.

Note: A warning similar to the one for off-leash searches will precede on-leash searches. The searching team should be followed by at least one other military policeman to protect the search team.

Apprehension of Subjects

5-31. The use of PDs to attack and hold an individual enemy or terrorist should not be overlooked. A PD can be sent into a building, confined space, or cave to find, attack, and hold an individual until the handler and friendly forces can arrive to place the subject into custody. Although disadvantages do exist (injury or death of the PD), the PD is better-suited than a human to immediately find and subdue a subject in an area where the interior structure is unknown to friendly forces.

Walking Patrols

5-32. The PD team's contribution to MWD team law enforcement efforts is most effective when the team is on foot. Some of the law enforcement duties that a PD team can do as a walking patrol include checking or clearing buildings and patrolling parking lots, dependent housing areas, and troop billet areas.

5-33. Since it is easily seen, a PD used during daylight in congested areas provides a good psychological deterrent to certain crimes. PDs are tolerant of people, and the presence of a large number of people does not significantly reduce the PD's effectiveness except when conducting detection missions.

5-34. Using the PD team during darkness or periods of reduced visibility and in areas where there are fewer distractions can enhance the PD's detection ability. Military police may not detect a person fleeing a crime scene at night, but a PD will normally be able to detect and locate a fleeing suspect. When necessary and appropriate, the PD is also much more effective at pursuing and holding the suspect.

5-35. Periodic use of PD teams around on–post schools, especially when school is starting and dismissing, may help deter potential vandalism, child molestation, exhibition, and illegal drug activities. PD teams can also provide effective security for communications facilities, military equipment, and command posts (CPs).

Alarm Responses

5-36. In responding to an alarm condition at facilities such as clubs, finance offices, or banks, the PD team should be among the first military police on the scene. The PD team may be used to search and clear the building and immediate area and help with any apprehensions. If the PD must be used to find a suspect, other persons should attempt to avoid contaminating the area with scents or tracks that may confuse him.

Vehicle Parking Lots

5-37. PD teams are effective in detecting and apprehending thieves and vandals in vehicle parking lots. The presence of the PD team may deter potential acts of theft and vandalism. The PD team can be most effective by alternating between vehicle and foot patrols of the parking lots. During hours of darkness, when there is no activity in a parking lot, the PD team should approach the lot from the downwind side. If the PD alerts, the handler should locate and challenge the suspect for identification.

5-38. If a suspect tries to escape or avoid apprehension and the handler is reasonably certain the suspect has committed or is attempting to commit a serious offense, the handler should give a clear warning to the suspect to halt. When lesser means of force have failed to stop the fleeing suspect, the PD may be released to pursue, attack, and hold. If it is not safe to release the PD, the handler may attempt to follow the suspect if the PD has the ability to track.

Chapter 6

Narcotics Detector Dogs and Characteristics

MWDs are trained to perform dual roles. In addition to the uses and characteristics of the PD, the PNDD is trained to detect the odor and presence of specific drugs. This specialized ability to detect certain illegal drugs makes the PNDD a valuable tool to help commanders and other government agencies maintain law and order. The PNDD can be employed in combat and/or force protection and antiterrorism operations worldwide.

EMPLOYMENT FOR COMBAT SUPPORT OPERATIONS

6-1. The PNDD provides commanders with unique capabilities in combat support environments. The narcotic detection capabilities of these dogs can be used to maintain the order and discipline of US soldiers as well as combatants and other persons involved in operations (EPWs, CIs, refugees, or other detained/interred persons [during time of war/conflict]). In addition to narcotic detection functions, PNDD teams should not be overlooked for their PD functions as discussed in *Chapter 5*.

INTERNMENT AND RESETTLEMENT OPERATIONS

6-2. In addition to perimeter and other security duties, PNDD teams assigned to I/R camps can be used to ensure that no illegal drugs are at the I/R camps by conducting searches of all areas. Commanders and planners should consider having all inbound supplies or equipment searched by PNDD teams. Inbound and outbound mail and vehicles should also be considered as a means for the movement of illegal drugs and, as such, subject to a search by PNDD teams.

POSTAL OPERATIONS

6-3. Postal inspections, including mail storage areas and drop boxes, can be conducted in conjunction with a military customs inspector (MCI) and according to *DOD Directive 5030.49*. The postmaster or MCI can request PNDD support from the area PM or military police brigade S3 with coordination through the KM team.

6-4. A PNDD team can be used to search mail that is received or shipped. Searches can be conducted as the mail comes off or before it goes on the truck prior to being shipped. The PNDD team can search up and down the conveyer belt as the mail is removed from or placed on the truck. If a conveyer belt is not being used, the handler can instruct postal personnel how to arrange the mail in a secure area so that the PNDD team can search it more effectively. When a PNDD team is used in the close quarters of a mailroom or delivery area, all nonessential personnel should remain clear of the area while the team is actively working.

6-5. When the PNDD responds to a parcel indicating the possible presence of illegal drugs, the CID or the military police investigations section will be notified for further action. At this point the PNDD team's actions are finished.

AIRCRAFT AND LUGGAGE SEARCHES

6-6. Before beginning the search, ensure that all personnel have exited the aircraft and all compartments are opened. The PNDD team searches the aircraft before searching any luggage or cargo. When approaching the aircraft, conduct a fresh-air perimeter search from the downwind side. Once a perimeter search has been conducted, the team boards the vessel at either the front or rear. Once the area has been cleared to search, the handler ensures that only one door is open to limit draft and wind current. A systematic search of all areas within the aircraft is conducted from the top to the bottom. The handler advises other military police on the scene where the PNDD displayed a positive response, and the augmenting military police conduct a detailed search of the area surrounding the response.

6-7. When conducting inbound baggage inspections, the items should be aligned in a sterile area in rows with at least six feet between rows. The handler conducts an on-leash search, presenting each piece of baggage. When searching cargo, every effort should be made to search the cargo before it is palletized. The cargo should be aligned in rows, with at lest six feet between rows, in a sterile area (same as baggage). The handler conducts an on-leash search, presenting cargo as required. Once the PNDD team has searched the cargo, it may be palletized while remaining in the sterile area prior to embarkation.

6-8. Conducting debarkation operations is the same as inbound operations. The host unit (to which the cargo belongs or is being delivered) provides a representative (E6 or above) and a work detail to assist the PNDD team. When pallets are removed from the aircraft, they are placed in a sterile area. Once in the sterile area, the work detail unloads the pallets and arranges the cargo, as required by the PNDD handler.

Force Protection and Antiterrorism Operations

6-9. PNDD teams are extremely useful in force protection and antiterrorism operations. Commanders and other government agencies use PNDD teams to protect the community from the harmful effects of illegal drugs. PNDD teams protect the force and surrounding communities by deterring criminal activity and enhancing law enforcement capabilities.

Health and Welfare Inspections

6-10. PNDD teams conduct health and welfare inspections when requested by the commander and coordinated through the KM, PMO, and S3. The commander contacts the KM directly to schedule a health and welfare inspection. The KM, in conjunction with the installation PM, the CID special agent in charge, and the military police investigations supervisor, briefs all levels of the command on the availability and capabilities of the PNDD teams available to them on the installation.

6-11. When scheduling the inspection, the commander should have the following information available:

- The requested time of inspection.
- The areas to be inspected.
- The number of living quarters.
- The number and size of common areas.
- Other pertinent information.

Appendix H provides a guide to assist handlers with health and welfare inspections.

6-12. Once the coordinating instructions have been established, the KM coordinates a time and location to brief the commander and demonstrate the PNDD team's capabilities. It is important that the unit requesting the health and welfare inspection practices operational security (OPSEC) and informs only those personnel who have an immediate need to know. This usually consists of the commander and first sergeant or sergeant major. The KM advises the PM and the military police investigation supervisor when the inspection is scheduled and if any additional support will be needed.

6-13. A systematic inspection of the living quarters and common areas is conducted. If the PNDD responds to a locked wall locker, the soldier assigned to that wall locker is summoned to open it. Once the wall locker has been opened, unit representatives conduct a detailed search of the wall locker and its contents.

Department of Defense Schools and Drug Abuse Resistance Education Program Inspections

6-14. KMs may receive requests from Drug Abuse Resistance Education (DARE) officers or DOD school principles to conduct a health and welfare inspection of a school. When conducting school inspections in conjunction with DARE, basic building inspection techniques apply. The primary concern when inspecting schools is the safety of the students and the secrecy of the search.

6-15. When possible, searches should be conducted in conjunction with a school activity, such as an assembly or fire drill. This will allow the handlers to conduct the search with a decreased risk of student involvement. Another option is to search the school after normal school hours or on a weekend or holiday. In any event, the use of the PNDD team on school grounds must be cleared by the school administration and the PM must be notified in advance.

Assistance to Military Police Investigations Section and/or the Drug Suppression Team

6-16. The military police investigation section or the drug suppression team (DST) may contact the KM to schedule an inspection for a number of reasons. PNDD teams may be requested to conduct an inspection of a residence or building where illegal drugs have already been found. In addition, unit commanders may request a PNDD team to inspect the quarters of a soldier who has shown a positive response on a urinalysis. The PNDD is not used to verify or test a substance that has already been found. PNDD handlers contact the military police investigations section or the DST in any circumstance that a PNDD has responded positively according to local standard operating procedures (SOPs).

Military Customs Missions

6-17. The US Army has been designated as the proponent service for the military customs program. According to current regulations, PNDD teams are used to inspect aircraft cargo areas and airframes when the military customs inspector program manager (MCIPM) directs these inspections. When conducting military customs missions, the handler should first inspect the vessel, then if deemed necessary by an MCI, continue with baggage.

6-18. When scheduling military customs missions, the MCIPM coordinates with the KM according to local SOPs. At a minimum, the MCIPM notifies the KM of—

- The requested time and date of the search.
- The number and types of vessels to be searched.
- The amount of cargo to be searched.
- The number of passengers and crew requiring the MCI's clearance.
- The determination on whether the search is a high-risk or selective search.

6-19. PNDD teams may inspect vessels such as military or civilian aircraft, boats or ships, vehicles, rail cars, or any other vehicle capable of carrying passengers and/or cargo according to the military customs program. These inspections are primarily conducted outside the continental United States. However, some missions in the continental US may be necessary for units returning from foreign theaters where the installation is their first stop within the customs territory of the United States. In addition, the military customs program may require preclearance for units or individuals deploying to areas outside the customs territory of the United States where an MCI has not yet been assigned or is not available. When conducting preclearance of mobilized or deploying soldiers' personal baggage while in formation, the MCI must ensure that the soldiers do not distract the PNDD team.

VESSELS, BAGGAGE, AND CARGO INSPECTIONS

6-20. When establishing sterile areas, the MCI coordinates with the PNDD handler and ensures that the area is of sufficient size with minimal distractions to ensure the optimal working environment for the PNDD. When scheduling working parties for embarkation or debarkation, the PNDD handler coordinates with the MCI to ensure that sufficient personnel will be present to accomplish the mission in a timely manner, but not so many soldiers that they might present a distraction to the PNDD. If an examination has been deemed necessary by the MCI, the handler presents each piece of baggage or equipment to the PNDD.

UNITED STATES CUSTOMS SERVICE MISSIONS

6-21. Missions in support of the USCS come from the MWD program manager via the KM. Handlers must have—

- Active government credit cards.
- Current passports.
- Appropriate civilian attire (or a civilian clothing allowance).
- Complete transportation requirements.

6-22. When temporarily assigned to the USCS, local USCS agents, policies, and procedures direct the PNDD team. Weapons are not authorized for handlers assigned to the USCS. The uniform of the day is determined by the USCS and is normally civilian attire.

6-23. Prior to deploying, every attempt should be made to conduct pretraining exercises with the USCS. These pretraining exercises will allow the MWD to acclimatize to larger quantities of contraband, which may be found while on the mission. The USCS will provide large training aids and determine the PNDD saturation point prior to deploying.

6-24. PNDD teams can expect to perform searches on aircraft, baggage, and tractor trailers as well as vehicle lines at US borders. When assigned to the USCS, the handler does not conduct physical searches of any kind. If the PNDD responds to an area, the handler advises a USCS agent and continues with the search, leaving the agent to conduct the actual physical search of the area.

6-25. Tractor trailers are searched in a USCS sterile area. Once the cargo has been offloaded and placed in standard lines, the PNDD team performs an exterior inspection of the vehicle before moving to all interior compartments and concluding with a systematic search of the cargo.

6-26. While assigned to a US border, the PNDD may be employed in vehicle lines. Vehicle lines are generally 10 to 20 vehicles in a row (placed bumper to bumper) that are inside a sterile area. The PNDD team travels up one side of the vehicle line and back down the other. The handler presents all productive areas. In the event that a USCS agent selects a vehicle for an individual vehicle inspection, that vehicle will be placed in a sterile area and the driver will open all the compartments of that vehicle. Once all occupants are clear of the vehicle, the PNDD will conduct a systematic inspection of the entire vehicle.

6-27. The PNDD will only be used in his narcotics detector role while deployed in support of the USCS or other federal agency. The PD capabilities of the PNDD will not be used for any reason. If the handler is instructed to use his PNDD in any method other than that of a narcotics detector dog, he must immediately notify his KM.

PNDD Support to US Customs Counterdrug Missions

In the summer of 2003, SGTs Payne and Mason from Fort Knox, KY, and SPC Johnson from Fort Sill, OK, were tasked to support a US Customs counterdrug mission along the southwest border between Mexico and the United States. After an overview and train up, the PNDD handlers and their assigned dogs went to work with the customs agents to detect illegal drugs at various points of entry into the country.

These teams became very proficient in detecting large quantities of illegal drugs. The PNDD teams detected narcotics in false floors, dashboards, gas tanks, seats, and hidden panels of many vehicles.

After three months of searching vehicles at checkpoints, the PNDD teams were responsible for detecting over 4,800 pounds of marijuana, over 95 pounds of cocaine, and other undisclosed contraband.

RANDOM VEHICLE SEARCHES

6-28. The use of a PNDD at ACPs and installation gates can increase installation antiterrorism capabilities by providing a visual deterrent to terrorist groups and other criminal elements. The installation commander, task force commander, or base camp commander may task the PM to conduct random vehicle inspections according to local SOPs and the FPCON. When this occurs, the PM submits a schedule of random vehicle inspections to be conducted within a time frame that includes times, dates, and locations of inspections to be conducted. The schedule will be approved or disapproved and the inspection parameters will be dictated (will every vehicle be inspected or will vehicles be inspected at random according to the guidance provided) by the commander and returned to the PM. The PM then notifies the KM accordingly and provides him with a copy of the schedule with the commander's signature and inspection parameters. The PNDD's patrol capabilities may be used at ACPs to search for stowaways and personnel trying to gain access to the installation or a controlled area by hiding in delivery vehicles or other large cargo type vehicles. Extreme caution must be used when using the team in this manner. The commander must be aware that once the PNDD is given the command to search for personnel the MWDs aggression level will increase, thus increasing the chances of the PNDD biting personnel.

ON-CALL DRUG RESPONSE

6-29. PNDD handlers will normally be placed in an on-call status. On call means the handler must remain within response time and distance according to the local SOP. When on call, PNDD handlers do not consume alcoholic beverages; however, they are in a duty position that allows them to wear appropriate civilian attire as well as leave the post according to the local SOPs.

6-30. When on call, PNDD handlers maintain positive control of a secondary communication device issued by the command. The secondary communication device may be a digital pager or cellular telephone as determined by the command. When recalled, the PNDD handler returns to the kennel in the most expedient legal manner to retrieve his PNDD. The PNDD will only be transported in an approved government owned vehicle (GOV). At no time will an MWD be transported in a POV. The handler may respond in appropriate civilian attire according to the local SOP.

6-31. KMs provide insight and guidance to the PM on technical matters of the SOP. The KM should provide recommendations as to acceptable distances which may be traveled while on call, required response times, the required secondary mode of communication, and any installation-specific items of interest, such as an alternate uniform based upon the customary civilian attire that is prominent in the area.

Chapter 7

Explosive Detector Dogs and Characteristics

In addition to the uses and characteristics of the PD, the PEDD is also trained to detect the odor and presence of explosives. This specialized ability to detect explosives makes the PEDD a valuable asset to help commanders and other government agencies maintain law and order, while enhancing safety procedures. PEDD teams can be employed in combat and/or force protection and antiterrorism operations worldwide. PEDD teams are valuable assets. The task force commander, the PM, and military police planners should consider the use of PEDD teams during combat support operations. This chapter details sound employment techniques and the capabilities of the PEDD teams that enable leaders to develop plans.

INTERNMENT AND RESETTLEMENT OPERATIONS

7-1. The Army is the DOD executive agent for all EPW/CI operations. The military police is tasked with coordinating shelter, protection, accountability, and sustainment for I/R operations. The I/R function addresses military police roles with US military prisoners, EPWs, CIs, refugees, or other detained persons (during time of war/conflict). Although the combat support military police unit initially handles EPWs/CIs, modular military police (I/R) battalions with assigned military police guard companies and supporting MWD teams are equipped and trained to perform I/R operations for sustained amounts of time (see *FM 3-19.40*).

7-2. PEDD teams may be used for searching internees' belongings or confinement cells. PEDD teams should not be used to search a person. A PEDD is trained to attack with or without command; any sudden movements the internee may make could be perceived by the PEDD as an aggressive act. Safety must remain a priority for both the handler and military personnel.

7-3. The commander can also use PEDD assets to search the property of attached military personnel when needed. This is especially helpful when new units are rotating or replacing old units. These types of searches ensure that no explosives are taken out or brought in by US forces.

POSTAL OPERATIONS

7-4. Postal inspections, including mail storage areas and drop boxes, can be conducted in conjunction with an MCI and according to *DOD Directive 5030.49*. The postmaster or MCI can request PEDD support from the area PM or military police brigade S3 with coordination through the KM team.

7-5. The PEDD team can be used to search mail that is received or shipped. Searches can be conducted as the mail comes off or before it goes on the truck prior to being shipped. The PEDD can search up and down the conveyer belt as the mail is removed from or placed on the truck. If a conveyer belt is not being used, the handler can instruct postal personnel how to arrange the mail in a secure area so that the dog can search the mail more effectively. When a PEDD team is used in the close quarters of a mailroom or delivery area, all nonessential personnel should remain clear of the area while the team is actively working.

7-6. When the dog responds to a parcel, indicating the possible presence of explosives, the area is evacuated immediately. Military police secure the area and set up a temporary CP a safe distance from the suspected explosives. Use the stand-off distance according to *Appendix I*, as applicable. Use the PEDD team to search the safe area for any possible secondary explosives. Explosive ordnance disposal (EOD) personnel are requested to dispose of the explosives or render them safe. EOD personnel may give explosives and improvised explosive device (IED) components to federal agencies for further investigation or turn over the scene to federal agencies after explosive hazards have been neutralized. The PEDD team remains at the temporary CP and resumes detection operations once EOD personnel remove the explosives.

CHECKPOINTS AND ROADBLOCKS

7-7. PEDD teams can be used at checkpoints and roadblocks to supplement the control and movement of vehicles, personnel, and materiel (*Figure 7-1*). PEDD teams are added to security forces to detect explosives and to prevent illegal actions that may aid the enemy. Using these control measures deters terrorist activities, saboteurs, and other threats.

Figure 7-1. Checkpoint Search

7-8. When using PEDD teams at hasty or deliberate checkpoints and roadblocks, a search area must be designated for vehicles and a separate area must be designated for the vehicle occupants or any dismounted personnel. The PEDD team conducts a search of the vehicle only after security forces at the checkpoint ensure that all of the occupants have exited the vehicle and that a threat to friendly forces does not exist. Security forces at the checkpoint provide overwatch while the PEDD team searches for explosives.

Note: At no time should a PEDD handler be used to search the interior of a vehicle. PEDD handlers are trained on searching techniques with a PEDD and should not be used for security force tasks.

7-9. In the event that the dog indicates the presence of explosives, the PEDD team withdraws immediately from the vehicle. The area is evacuated and EOD personnel and the chain of command are notified. The PEDD team stays behind a barrier at the safe-distance location in case the EOD team leader requests further detector assistance. (See *Appendix I* for stand-off distances.)

Note: Be sure that radio operations are stopped until the PEDD team and security forces reach a safe distance from the suspected explosives.

CORDON AND SEARCH OPERATIONS

7-10. Military police leaders and other task force commanders should consider using PEDD teams when conducting cordon and search operations. PEDDs are trained and able to greatly improve the combat force's success in uncovering hidden explosives and people.

7-11. Once the military police or other security forces have established the cordon, the PEDD teams should move in and operate with the search elements. During the search, the PEDD teams provide warning of hidden explosives or persons while the rest of the search element employs overwatch and search techniques.

7-12. The search element leader directs the actions of the team if people or explosives are detected. The PEDD handler should be prepared to advise the commander on the capabilities of the dog and actions required by the rest of the team to ensure the PEDD team's safety. It is important that all members of cordon and search teams rehearse together before conducting operations.

FORCE PROTECTION AND ANTITERRORISM OPERATIONS

7-13. PEDD support for force protection and antiterrorism operations can be requested anywhere in the world and in different circumstances. In forward deployed areas or hostile environments, PEDD teams are often requested to ensure that any suspected explosives are detected before harm can come to US Army, joint, and allied forces. In peacetime environments around the world, PEDD teams perform the detection operations much the same as in combat support operations. The environment surrounding explosive detection operations does not interfere with the explosive detection capabilities of the PEDD team.

SUSPICIOUS/UNATTENDED PACKAGE RESPONSES

7-14. Once a suspicious/unattended item that may contain explosives has been discovered, the military police or security forces on the scene request a PEDD team. Normally, military police evacuate the area and establish a CP a safe distance away from the threat using the stand-off distance listed in *Appendix I* of this manual. The PEDD team checks the perimeter and secures the area for secondary explosives. EOD personnel should respond to the scene and make contact with the PEDD team to be briefed on the location of the package. The PEDD team moves to within 50 meters of the package and begins an area search up to package. The PEDD team also scans the immediate area for secondary devices, when applicable.

Note: At no time should a PEDD team be used to search the package itself. EOD personnel have received special training in handling suspicious/unattended packages and determining if a suspicious/unattended package contains explosives.

7-15. A PEDD handler must not present, touch, open, or physically disturb a package that is suspected of concealing explosives. Once the MWD team has discovered a suspicious item, EOD personnel assume responsibility for the area and take control of the situation in order to render the package safe or dispose of it as necessary.

PEDD Operations in Support of Operation Iraqi Freedom (OIF)

In April of 2003, SGT Raymond Rivera was performing explosives detection missions with his PEDD Kevin in support of OIF near Baghdad, Iraq. SGT Rivera was requested to sweep an area that a support element was in the process of clearing. There was so much debris that a forklift was being used to move large amounts of furniture and building supplies.

The PEDD team would search a portion of the area and then the forklift would come in and haul off the debris searched. After the third load was hauled off, the dog responded to a mattress on the ground with an office table on it. Expecting an explosive devise, SGT Rivera notified the NCOIC of the detail and the area was cleared of all personnel (to a safe distance) and secured. EOD personnel were notified of the suspected explosives and responded to the scene.

SGT Rivera was standing by at the safe area while EOD personnel uncovered two hand grenades that were buried in the sand just below the surface. SGT Rivera said, "Even though it was only two grenades, I was amazed that Kevin found them under all the debris. I'm glad we found them and not our soldiers."

BOMB THREATS

7-16. A PEDD team should be requested for assistance in detecting explosives whenever bomb threats are made against persons or property for which the Army is responsible. The PEDD team's search for explosives begins only after the area or building has been evacuated of all personnel and security has been established. EOD personnel only respond when a suspect item is found; they do not conduct or participate in searches except in support of the USSS. EOD units place a team on standby once the unit is alerted of a bomb threat. PEDD teams coordinate this with the local EOD unit prior to any operation.

7-17. The PEDD team, EOD personnel, and site leadership collaborate to resolve the situation at a nearby CP (see *Appendix I* for stand-off distances). When PEDD teams are located on an installation, their use must be included in the bomb threat planning guidance.

7-18. A systematic search by the PEDD team should include the assistance of a spotter to ensure safety and accuracy. PEDD teams are to be used only to detect the possible presence of explosives.

WARNING:

At no time will the handler touch, open, or otherwise disturb a possible explosive device. IEDs are easily disguised and can be triggered to explode by motion, sound, light, heat, or remote devices.

7-19. Handlers and spotters must not change or otherwise disturb anything in the search. If the lights or any electric device are off or on, do not touch the switch until a thorough search is conducted.

7-20. Ensure that the handler and spotter are provided with available safety equipment including—

- A flack vest (in addition to body armor).
- Flashlights or helmet-mounted lights.
- Gloves.
- Safety goggles.

WARNING:

Always keep track of how much time has elapsed when using a PEDD team at bomb threat incidents. Sufficient time must be maintained to ensure that the team exits the building or area well before the announced or suspected detonation time.

7-21. If the PEDD responds to the presence of explosives, the PEDD handler and spotter note the location and leave the area. EOD personnel then begin measures to render the device safe or remove it. On the request of EOD personnel, the PEDD team may then continue the search until another suspected explosive is found or until the area is cleared by EOD personnel.

7-22. If the search is 30 minutes or longer in duration, explosive training aids should be planted periodically so that the dog can find them. This helps to keep the dog interested in the search. Training aids should only be planted in areas that have already been searched by the PEDD team.

HIGH-RISK PERSONNEL SEARCHES

7-23. Military police provide the task force commander with valuable area security capabilities to protect C2 headquarters, equipment, and services essential for mission success. HRP security is an important task within the area security function. PEDD teams may be employed to enhance explosive detection and protection capabilities.

7-24. The KM or plans NCO determines which PEDD teams will perform the mission. The number of PEDD teams for any mission is determined by the threat, the number of PEDD teams available, any conflicting missions, and other environmental considerations noted in *Chapter 8*.

7-25. PEDD teams may be used as part of a combined arms operation or as an element of a larger security force. When used as part of a combined arms team, the PEDD teams need to be involved will all troop-leading procedures, including rehearsals. All personnel must have a complete understanding of the mission in order to create a safe and secure environment for the HRP. The PEDD teams provide explosive detection and personnel detection capabilities. PEDDs are also a deterrent to hostile forces or terrorists operating within the area of operations.

7-26. Prior to an HRP visit to an area, the PEDD team assists with force protection actions by searching for explosives and IEDs. PEDD also conduct searches of and provide security for outer perimeter areas. During HRP operations, the handler—

- Briefs the on-scene commander of the dog's capabilities.
- Communicates with EOD personnel.
- Obtains the HRP's itinerary and determines locations to be searched.
- Requests a spotter or helper to assist.
- Ensures that areas are secure after searches are done.
- Informs the commander of results and continues follow-on missions.

ACCESS CONTROL POINTS

7-27. Using PEDD teams at ACPs can increase an installation's or base camp's force protection and antiterrorism capabilities by providing a visual deterrent to terrorist groups and other criminal elements. The PM notifies the KM of the requirement to perform vehicle searches at ACPs.

7-28. In the event that the PEDD detects the presence of suspected explosives, the military police forces at the ACP secure and evacuate the area (see *Appendix I* for stand-off distances). EOD personnel are then notified to render the item safe or remove it. PEDD handlers do not attempt to remove or otherwise disturb any suspected explosives.

7-29. PEDD team patrol capabilities may be used at ACPs to search for stowaways and personnel trying to gain access to the installation or a controlled area by hiding in delivery vehicles or other large cargo type vehicles. Extreme caution must be used when utilizing the team in this manner. The commander must be aware that once the PEDD is given the command to search for personnel the potential for injury exists.

HEALTH AND WELFARE INSPECTIONS

7-30. PEDD teams conduct health and welfare inspections when requested by the commander and coordinated through the KM, PMO, and S3. The commander contacts the KM directly to schedule a health and welfare inspection. The KM, in conjunction with the installation PM, the CID special agent in charge, and the military police investigations supervisor, briefs all levels of the installation command on the availability and capabilities of the PEDD teams available to them.

7-31. When scheduling the inspection, the commander should have the following information available:

- The requested time of the inspection.
- The areas to be inspected.
- The number of living quarters.
- The number and size of common areas.
- Other pertinent information.

Appendix H provides a guide to assist handlers with health and welfare inspections.

7-32. Once the coordinating instructions have been established, the KM coordinates a time and location to brief the commander and demonstrate the PEDD's capabilities. It is important that the unit requesting the health and welfare inspection practice OPSEC and inform only those personnel who have an immediate need to know. This usually consists of the commander and 1SG or sergeant major. The KM advises the PM and the military police investigations supervisor when the inspection is scheduled and if any additional support is needed.

7-33. A systematic inspection of the living quarters and common areas is conducted. If the PEDD responds to a locked wall locker, the soldier assigned to that wall locker is summoned to open it. Once the wall locker has been opened, unit representatives conduct a detailed search of the wall locker and its contents.

AIRCRAFT AND LUGGAGE SEARCHES

7-34. Before beginning the search, ensure that all personnel have exited the aircraft and that all compartments are opened. The PEDD team should search the aircraft before searching any luggage or cargo. The PEDD team conducts a fresh-air perimeter search from the downwind side when approaching the aircraft. Once a perimeter search has been conducted, the team boards the aircraft at either the front or rear. Once the area has been deemed clear, the handler ensures that only one door is open to limit draft and wind current. A systematic search of all areas within the aircraft is conducted from the top to the bottom.

7-35. If the dog alerts the handler to a possible presence of explosives, military police or security forces evacuate and secure the scene. The PEDD team moves a safe distance away from the suspected explosives but remains available to further assist EOD personnel, if requested. Once EOD personnel clear the area, the PEDD team may resume the search until another suspected explosive is found or until the area is cleared by EOD personnel.

7-36. When conducting inbound baggage inspections, the items should be aligned in a sterile area in rows with at least six feet between them. The handler should conduct an on-leash search, presenting each piece of baggage. When searching cargo, every effort should be made to search the cargo before it is palletized. The cargo should be aligned in a sterile area in rows with at least six feet between them (the same as baggage). The handler should conduct an on-leash search, presenting cargo as required. Once the PEDD team has searched the cargo, it may be palletized while remaining in the sterile area prior to embarkation.

7-37. Conducting debarkation operations is the same as inbound operations. The host unit (to which the cargo belongs or is being delivered) provides a representative (E6 or above) and work detail to assist the PEDD team. When pallets are removed from the aircraft, they are placed in a sterile area. Once in the sterile area, the work detail unloads the pallets and arranges the cargo as required by the PEDD handler.

Chapter 8
Veterinary Support

At most locations, veterinary support is provided by an assigned Army VCO. When there is no assigned VCO, an attending VCO from a nearby military installation is assigned responsibility for providing veterinary support. The Army provides veterinary support for all MWDs as authorized by *DOD Directive 6400.4* and *ARs 40-1, 40-3*, and *40-905*. Emergency, civilian veterinary support is authorized for MWDs when there is not a VCO available and/or care is beyond the capability of the veterinary treatment facility (VTF). The responsible VCO and Army Veterinary Service chain of command are responsible for establishing standard procedures for approval, provision, and payment for emergency, civilian veterinary support.

VETERINARY RESPONSIBILITIES

8-1. The US Army Surgeon General provides professional veterinary support for the entire MWD program through the Army Veterinary Corps. This responsibility includes—

- Medical and surgical care at training facilities, bases, and installations.
- Inspections to ensure that the kennel facilities are sanitary.
- A professional review of plans for new construction and modification of kennels, support buildings, and sites.
- Prescribing an adequate feeding program.
- Instructing handlers and supervisors in all matters related to the health of MWDs.
- Conducting research to improve the MWD program.

8-2. The responsible VCO provides treatment for the diseased or injured MWD at the kennel site or at the installation VTF. The Army Veterinary Corps is responsible for equipping this facility and for providing medical and surgical supplies. The VCO is responsible for the MWD veterinary treatment record (VTR) and related information about examinations, immunizations, and treatment.

8-3. The VCO instructs handlers about dog health, care, feeding, and first aid. This instruction helps the handler to develop a better understanding of the MWD's health needs and improves the handler's ability to care for the MWD. The VCO prescribes an appropriate feeding program based on the MWD's health, the climate, and working conditions.

8-4. The VCO is involved in the review of all plans for new kennel construction or kennel modifications. This ensures that potential health and safety hazards can be corrected before construction begins.

VETERINARY INSPECTIONS

8-5. At least quarterly, the VCO inspects the MWD kennel facility to make sure that proper sanitary standards are being maintained. It is recommended that an animal care specialist conduct a monthly check of the kennels to ensure that no major safety violations are present. The VCO ensures that insect and rodent control efforts are adequate and that the handlers are properly maintaining the general health of the MWD. The quarterly inspection also includes the following elements:

- An examination of the kennel facilities for safety hazards and distractions that may interfere with the rest and relaxation of the MWDs.

- A review of the adequacy of the kennel structure (particularly for environmental conditions) and feeding and watering schedules. The VCO should make recommendations to help prevent disease and injury.
- A review of kennel sanitation and cleaning procedures and MWD food storage procedures to minimize the risk of disease transmission to the MWD.

MANDATORY TRAINING

8-6. The local VCO is required to provide training to the MWD handlers upon initial assignment and at least annually thereafter. More frequent training of handlers is recommended but not required. The Department of Defense Military Working Dog Veterinary Services (DODMWDVS) recommends that handler training be conducted at least quarterly. Topics include the following:

- Procedures on when and how to obtain veterinary support (an SOP is appropriate for this).
- General animal health.
- Grooming.
- Prevention of environmental injury (such as heatstroke or cold injuries).
- Emergency first aid (for bloat [gastric dilatation-volvulus], heat stroke, snake bites, and cold injuries).
- Administration or application of routine medications.
- Proper MWD feeding.
- Feed sanitation (proper storage, preparation, and cleanup).
- MWD chemical, biological, radiological, nuclear, and high-yield explosives protective measures.

8-7. It is best to schedule mandatory training so that seasonal problems are addressed prior to their occurrence and appropriate refresher training is provided prior to a deployment. This training is not meant to turn handlers into animal health technicians, but to give them the tools needed to provide emergency care until they can acquire veterinary support.

WEIGHT CHECKS

8-8. Each month the MWDs body weight must be recorded and sent to the responsible VCO. The KM may communicate this to the VTF. Body weight is also measured and recorded monthly in order to monitor body condition and detect important trends in weight loss or gain that may indicate illness. The KM and/or handler documents data such as food intake, stool consistency, and vomiting in the monthly records.

RECORDS/HEALTH CERTIFICATES

8-9. VTRs are maintained by the local VTF. It is very important that MWD handlers ensure that all health information is annotated in these records.

8-10. An examination is given just before shipping an MWD from one installation to another. The MWDs are examined to detect injury or disease and to support the safe shipment and continued good health and performance of the MWD. A health certificate is issued for the state or country into which the MWD is being shipped for TDY and must accompany the MWD. When the TDY period exceeds 10 days, a health certificate may be required for the MWD's return to his home station. A VCO signs this certificate no more than 7 days prior to reembarkation. If the TDY location is in a different Army VCO's area of operation, the gaining VCO is notified prior to the MWD's departure from the home station. An examination is given as soon as possible, but not more than 10 days after the arrival of an MWD at a new assignment, TDY, or deployed location and within 3 days of the MWD's return to his home station. It is important that the handler who deployed with the MWD is present for the examination so that the VCO can get accurate and complete information on any health issues while the MWD was deployed.

Feeding and Food Control

8-11. MWDs require a diet that is significantly different from that of pet dogs. Their work demands much higher levels of energy and larger quantities of essential nutrients. The standard high-performance diet, contracted and supplied by the GSA for MWDs, meets these nutritional needs and is the only approved feed unless otherwise directed by the attending veterinarian.

8-12. Special diets may be procured and fed to individual MWDs when the VCO determines that other than the standard diet is required. Normally, these diets are available from the VTF, but the MWD unit is responsible for procurement costs.

8-13. Some MWDs have been trained using a food reward schedule that also requires special food. Food reward is used only if an MWD fails to respond to any other type of reward. Procure the special food through normal supply channels or authorized local purchase. The amount of food each MWD should be fed depends on his weight, the amount of activity, and the climate. Local VCOs determine the proper amounts to feed. The responsible VCO must be told when an MWD is trained using food.

8-14. The VCO also prescribes the time of day each MWD is to be fed. This depends on the MWD's duty schedule and the schedule of other kennel activities. MWDs should be allowed 2 hours to eat; leftover food is disposed of within another two hours and feeding pans are cleaned and put away. Never leave uneaten food in the kennel past the authorized 2-4 hour feeding period. If the MWD finishes his meal prior to the end of the feeding period the feed pan may be removed.

Parasite Control

8-15. MWDs are at risk of becoming infested with internal parasites (intestinal worms and heartworms), and external parasites (fleas, ticks, and mange). Some of these parasites may also transmit infectious diseases to the MWD. In order to prevent infestation and disease transmission, all MWDs must be on a routine parasite prevention program supervised by the responsible veterinarian. At a minimum, this must include the administration of a monthly oral (pill or chewable) heartworm preventative and internal parasite preventive and the administration of a topical, veterinary-prescribed external parasite preventative against fleas and ticks.

8-16. When deployed to areas with high parasite risk, additional preventive measures such as prescription flea and tick collars and the oral antibiotic doxycycline will be prescribed by the supporting veterinarian. Doxycycline must be given with food. The handler must ensure that adequate supplies of all preventives are issued by the VTF prior to departing for TDY or deploying. Flea and tick collars will only be used when prescribed by the veterinarian and must be removed when the MWD is not under direct physical control (when off the choke chain and leash) of the handler to prevent accidental ingestion.

Food Storage

8-17. Food should be stored in a sealed container that will not allow contamination or rodents to enter the food (for example, pail, utility 7 gallon, NSN 7240-01-411-0581). Bulk storage of food must be in an area separate from the storage of other equipment, cleaning supplies, or materials that may contaminate the food. Bags should be elevated 4-6 inches above the ground, and neatly stacked with a 4-inch space between the stacks and the walls. The sanitation standards for the storage of MWD rations are identical to those required for the storage of bulk rations for personnel in a troop issue subsistence activity, exchange, or commissary.

Deployability Categories

8-18. The purpose of reporting MWD deployability status is to assist the MWD program managers, PM, commanders, and KMs with the selection of MWD teams for deployment and TDY and to rapidly identify MWDs with limited duty requirements. These guidelines take medical and physical fitness factors into consideration including—

- The anticipated MWD duty stress at deployed locations compared to the home location.
- The availability of veterinary support at the deployed location.

- The climate and environment of the deployed location.
- Any diagnosed or suspected medical conditions that may affect the MWD's performance.
- The MWD's physical conditioning and stamina as observed by the KM and handler.

8-19. It should be noted that some determinations are and will remain subjective; thus the need for prudent judgment by all individuals involved in these assessments remains. This is particularly true regarding the effect of age on an individual MWD. Old age is not a disease and MWDs are not downgraded just due to age; however, with increased age comes increased incidence of illness or disability. These problems may warrant a change in deployability status.

8-20. Deployability status is initially determined after the VCO examines the MWD at the receipt physical or semiannual physical exam. The KM determines training and proficiency status. The VCO and KM should discuss their findings and then categorize each MWD's status and forward the listing through the KM to the PM and MWD program manager. Updates to deployability status lists are performed at least monthly or as needed as changes occur in an individual MWD's status.

Category 1—Unrestricted Deployment (Both Outside the Continental United States and in the Continental United States)

8-21. The MWD is medically fit for any contingency or exercise. There are no limiting or compromising factors. Medical factors may exist but will not limit performance. Special diets, controlled drugs, or special medications can be sent with the MWD (or be carried by the handler) in a quantity sufficient to last the full duration of the anticipated deployment.

Category 2—Limited Deployment (Continental United States or Home Theater of Operation)

8-22. The MWD is medically fit for regions and missions with a minimal requirement for acclimation to heat or physical stress where complete veterinary support is available. The MWD is fit for short duration deployments. There are no significant limiting or compromising factors. Medical problems may exist which slightly limit performances but are controllable. Special diets or medications are required; however they are available in the deployed location or can be sent with the MWD. The reason for limited deployability must be reported.

Category 3—Temporarily Nondeployable

8-23. A medical condition exists that impedes daily duty performance and is under diagnosis, observation, or treatment. The MWD may still be allowed to work at the home station but in a limited capacity.

Category 4—Permanently Nondeployable

8-24. An unresolved medical or physical problem exists that frequently or regularly impedes daily duty performance and an early retirement date cannot be given. The reason for nondeployability must be reported. Medical or physical conditions warrant replacement within one year. The responsible veterinarian and KM should initiate an MWD disposition process at the time an MWD is declared category 4.

Kennel Sanitation Procedures

8-25. Cleanliness is one of the most important factors for the good health of an MWD. The KM must enforce sanitary measures in and around the kennel area. A good standard of sanitation is the result of the cooperation of the handlers, KMs, supervisors, and the VCO.

8-26. The VCO and the PM or military police commander should set the standard of sanitation. This standard must be maintained by each of the handlers and the KM.

8-27. Kennels must be sanitary, in a good state of repair, and thoroughly cleaned every day. Kennels should be disinfected at least once every week using only those disinfecting products approved by the VCO. Kennels should also be disinfected whenever an animal is removed from a kennel so that the kennel will be ready to be occupied by another animal.

8-28. Stool must be removed from the runs as often as necessary as but no less than twice daily to prevent the MWD from becoming dirty. The method of disposing of stool depends on local conditions and the type of sewage system present. If stool must be carried from the area in cans, the cans must be cleaned and disinfected after each use.

8-29. The entire kennel area must be free of refuse and garbage that could attract rats and insects. Mosquito control measures must be used in ditches and swampy areas in the vicinity of the kennels. Disinfectants and disinfectant procedures should be used only with the approval of the VCO.

MEDICATION

8-30. At times, a VCO prescribes special medication (to be given separately or mixed with food) for a sick or injured MWD. The handler must know how to administer these medications in both pill and liquid form.

8-31. When any foreign substance is placed directly into a dog's mouth, his first reflex is to spit it out. The handler must learn how to administer medication properly so that the MWD is forced to swallow. With the VCO's approval, pills or liquids can be mixed with a small amount of canned dog food or other soft food and fed to the MWD. To administer capsules or tablets—

- Place the fingers of your left hand over the MWD's muzzle and insert your left thumb under the lip and between his upper and lower teeth right behind the canine tooth, pressing your left thumb against the roof of his mouth to open it.
- Place the capsule or tablet into the MWD's throat at the extreme rear of the tongue to prevent him from spitting it out. Quickly remove your hand, close the MWD's mouth, and gently massage his throat until he swallows. The entire procedure must be quick and smooth to ease the MWD's apprehension and resentment.

8-32. Liquid medication is best administered with the help of another person. The handler—

- Holds the MWD's upper and lower jaws together with his left hand. The assistant pulls the MWD's lips away from the teeth at one corner of the mouth with his right hand. The MWD's nose is then pointed slightly upward, forming a natural funnel by the lip, and the assistant pours the liquid into this funnel.
- Elevates the MWD's head only slightly above the horizontal when giving liquid medicine. If the MWD's head is raised higher, he has difficulty in swallowing. Give the MWD adequate time for swallowing to prevent the liquid from getting into the trachea, nose, or lungs. Use extreme caution in giving oily liquids.
- Allows the MWD to lower his head and rest before proceeding if any signs of distress, such as coughing or struggling, appear. Do not give oral medications or any liquids if the MWD is unconscious or cannot swallow.

FIRST AID

8-33. Normally, the handler's early recognition of symptoms of illness or injury allows sufficient time to get assistance from a VCO. However, situations may arise when medical help is not immediately available. The seriousness of the incident may require that the handler take emergency actions to protect the life or health of the MWD. The first aid instructions that follow are commonly used to save life, prevent further injury, and reduce suffering from pain. In all emergency situations, notify the VCO as soon as possible so that the MWD can receive professional medical attention.

8-34. First aid kits should always be available in the kennel area and at training sites. The handler should also carry a first aid kit as part of his or her equipment for the MWD on all operational missions. The VCO will determine the contents of first aid kits. When items are used from the first aid kit, they should be replaced with new items immediately.

MUZZLES

8-35. When an MWD has been injured, the first consideration is to calm and immobilize the animal. Pain and distress, however, may cause the MWD to respond to the handler in an unpredictable manner. The MWD may not respond to verbal commands and may attempt to bite the handler and anyone helping the handler. Whether to apply a muzzle (*Figure 8-1*) or not depends on the nature of the emergency. If the MWD is unconscious, shows any difficulty in breathing, or has suspected head injuries, a muzzle should not be used. Otherwise, a muzzle should be used for safety.

Figure 8-1. Muzzle Use

8-36. There are several types of muzzles, but the leather basket muzzle is the best and most comfortable. It allows for free breathing and causes the least alarm and apprehension. The MWD can still inflict a wound while wearing this type of muzzle, so caution is required.

8-37. An improvised muzzle can be made using the MWD's leash and is called a leash muzzle. To apply the leash muzzle, tighten the choke chain on the MWD's neck by pulling the leash tightly with the right hand. Place the left hand, palm up, under the choke chain on the neck. Grasp the leash tightly as it passes through the palm of the left hand, wrap the leash once around the MWD's neck, and bring it up and across the left side of his head. Finally, wrap the leash twice around the muzzle and grab it tightly with the left hand.

CAUTION

The leash muzzle may be used when the leather muzzle is not available or when it would not provide adequate safety. Do not use the leash muzzle when the MWD is overheated, is having difficulty breathing, or is indicating that he may vomit. Do not leave it on for long periods in hot weather.

FRACTURES

8-38. If a fracture occurs, restrain the MWD and encourage it to lie down and remain quiet. If the face and jaw are not injured, put the MWD in a loose muzzle. Notify the veterinarian immediately when a fracture occurs or is suspected. If there is any bleeding, treat the MWD according to *paragraph 8-40*, then proceed with splinting and moving the MWD to the veterinarian if he is not immediately available.

8-39. Prior to moving the MWD, the fracture area needs to be immobilized with a splint. This should be possible for most leg fractures. The leg should be splinted in the same position it is placed by the MWD when it is hurt. Do not try to straighten or reduce the fracture. Splinting an MWD's leg is similar to splinting a soldier's leg and may be done with the same materials: bandages, torn cloth, splints from the medic's bag, or other available material including sticks, boards, and rolled newspaper. After the leg is splinted, or if the fracture cannot be splinted, gently move the MWD onto a litter or improvised litter, such as a piece of plywood, or into a kennel and carry it to a vehicle for transport. It is best to slide the litter under the MWD rather than lifting the MWD onto the stretcher. The MWD may need to be strapped to the litter to allow safe movement. Make sure to support the injured leg or area during movement.

Bleeding Wounds

8-40. Bleeding must be quickly controlled, particularly wounds in the foot or leg that are prone to bleed freely. Applying pressure directly on the wound may control bleeding; use a sterile bandage or clean handkerchief or pinch the edges of the wound with your fingers. As soon as possible, apply a pressure bandage.

Burns

8-41. Most burns occur when an animal comes into contact with hot water, grease, tar, or other scalding liquids. MWDs may also get electrical burns by chewing electrical wires. If an MWD is trapped in a burning building, it may suffer from smoke inhalation in addition to surface burns. The handler can prevent all of these situations through positive control of the MWD. If a burn does occur, applying cold-water soaks or ice packs to the burn for approximately 20 minutes may treat minor burns. The cold helps to reduce the pain. The hair around the burn should be clipped away and the burn should be washed gently with a surgical soap. Blot it dry with sterile or clean soft gauze or cloth. Protect the burned area from rubbing by applying antibiotic ointment and a loose fitting gauze dressing.

Shock

8-42. An animal may go into shock after injuries to internal organs, excessive bleeding, or trauma. Shock can be recognized by the following symptoms:

- A glassy look to the eyes.
- A rapid or weak pulse or rapid shallow breathing.
- A dropping body temperature. The lips and feet may feel cold.
- A paleness of the membranes of the mouth and eyes.
- Slow capillary refill time. To determine capillary refill time, press firmly against the MWD's gums until they turn white. Release the pressure and count the number of seconds until the gums return to their normal color. If it is more than two to three seconds, the MWD may be going into shock. Failure to return to the red-pink color at all indicates that the MWD may be in serious trouble and needs immediate assistance.

8-43. If a handler suspects that an MWD has internal injuries or is going into shock, request help from a VCO immediately. Keep the MWD warm and quiet, and lower his head to prevent possible brain damage. If it is necessary to move the MWD, use a litter. Administration of intravenous fluids by appropriately trained personnel will help to prevent permanent injury from shock.

Artificial Respiration

8-44. There are conditions that may cause an MWD to stop breathing. When this happens, do not panic. Follow these procedures:

- Open the MWD's mouth and check for any obstructions. Extend the MWD's tongue and examine his throat.
- Clear the MWD's mouth of any obstructions then, close his mouth, and hold it gently closed.
- Inhale, and then cover the MWD's nose and mouth with your mouth.
- Exhale gently. Do not blow hard. Carefully force air into the MWD's lungs and watch for his chest to expand. Repeat every five to six seconds or at a rate of 10 to 12 breathes per minute.

Snakebites

8-45. The bite of a poisonous snake can cause serious illness and death if not treated immediately. Keep the MWD quiet and calm, and request veterinary assistance or move the MWD as quickly as possible to a treatment facility. Panic or exertion causes snake venom to move more rapidly through the bloodstream. If possible, kill the snake so it can be shown to the VCO. The handler should be careful not to be bitten by the snake or come into contact with the snake venom. Bites may occur on the face or neck of the MWD. When this happens, remove the choke chain and loosen or remove the collar and muzzle. Swelling occurs rapidly after a snakebite, and this equipment may restrict breathing. Position the MWD's head extended from his body to allow him to get maximum airflow. An ice pack applied to the bite area helps to slow the flow of blood and helps to keep the venom from spreading.

CAUTION

A tourniquet should not be applied. Constricting the blood flow could increase tissue death.

Foreign Objects in the Mouth

8-46. An MWD may get a stick or other foreign object stuck in his mouth or throat. The most common foreign object that becomes stuck is a Kong® or ball reward that is too small for the MWD. The MWD may cough, gag, have difficulty swallowing or breathing, paw at the mouth, and drool. If the MWD is having difficulty breathing, cautiously and gently open his mouth. Look for any abnormal object in the throat, under the tongue, between the teeth, in the gums, or in the roof of the mouth. If the MWD is having trouble breathing, removing the object is an emergency. The ball, Kong, or other object can be removed using pliers or clamps. If the object is not causing breathing problems, it is best to leave it in place and get veterinary assistance.

Poisonous Substances

8-47. There are many toxic chemicals an MWD may come in contact with and/or ingest. These include insecticides, herbicides, rodenticides, and antifreeze. The symptoms of poisoning vary. Unless the handler is certain the MWD has eaten a poison, do not treat for poison. Occasionally, MWDs accidentally swallow narcotics or explosives training aids. Exposure to poisons and accidental ingestion of training aides should be prevented by positive control of the MWD by the handler. If the handler knows that the MWD has eaten a poison, narcotic, or explosive, take the following actions:

- Request veterinary assistance immediately.
- Determine the type and quantity of poison, chemical, narcotic, or explosive that has been swallowed. If any part of the substance or container is available, keep it for the VCO to examine.
- If the Army VCO is not available, immediately make contact with a civilian veterinary to determine if vomiting is appropriate. Causing an animal to vomit is not recommended in all types of poisoning as it may cause more harm to the animal. When vomiting is appropriate, the animal may be induced to vomit by giving him one-quarter to one-half of a cup of hydrogen peroxide.
- Keep the MWD quiet and warm until the VCO arrives.

8-48. For future reference, keep a copy of the material safety data sheet for any training aide that may be ingested available. MWD first aid should be according to the material safety data sheet for people.

Overheating

8-49. Overheating results when an MWD is unable to eliminate body heat rapidly enough. This condition requires immediate action by the handler to save the MWD's life. During hot, humid weather, an MWD may easily become overheated during training, during operations, or while being transported. A body temperature of 105° Fahrenheit or more, combined with poor response to commands, weakness, unsteady movement, vomiting, difficult or labored breathing, convulsions, and collapse are all symptoms of overheating. It is helpful to take the MWD's temperature in order to monitor the severity of overheating and progress in cooling, but temperature alone does not determine the need for or method of treatment.

8-50. For mild to moderate heat injury, the MWD's temperature is generally lower than 104° Fahrenheit. The MWD may have mild symptoms, to include uncontrolled panting, a rapid but somewhat weak pulse, mild to moderate difficulty breathing (but not cyanotic [blue]), and exercise intolerance or loss of stamina. This is equivalent to heat exhaustion in people.

8-51. When the symptoms of overheating occur, loosen the leash, collar, and muzzle as much as possible without loosing control of the MWD. This allows for more efficient panting and cooling. Follow these guidelines at the onset of symptoms:

- Carry the MWD to the nearest shade and fan him if no wind is present.
- Try to quickly lower the MWD's body temperature by running and sponging cool water over the MWD's head, body, and legs. If a body of water is available, allow the MWD to stand in the water but make sure the MWD's head remains above water. Running water works better than standing water.
- Place the MWD in an air-conditioned kennel, vehicle, or shelter.
- Apply ice packs to the MWD's groin, armpits, and head to increase speed of cooling.
- Dampen, but don't soak, the MWD's feet, ears, and abdomen with rubbing alcohol. Allow it to evaporate and then reapply more.
- Do not let the MWD over drink. He should not consume more than one cup of water until he is calm and his temperature returns to normal.
- Monitor the MWD's temperature every 5-10 minutes, and stop cooling efforts when it reaches 100-103° Fahrenheit. Keep monitoring his temperature for an additional 60 minutes.

8-52. If the MWD's temperature remains normal (about 100-103° Fahrenheit), treatment is complete; however the MWD must be examined by a VCO and a heat injury must be reported in his VTR. If his temperature climbs again, restart cooling efforts.

8-53. If the MWD's temperature drops below 100° Fahrenheit, begin warming by drying and wrapping him with sheets and blankets. Remove the heat when his temperature goes above 101° Fahrenheit.

8-54. For severe heat injuries, the MWD's temperature is generally over 104° Fahrenheit combined with poor response to commands, weakness, unsteady movement, vomiting, difficult or labored breathing, uncontrolled panting, convulsions, collapse, a dull or depressed attitude and behavior, a rapid but somewhat weak pulse, exercise intolerance or loss of stamina, and cyanosis (blueness) or dark red mucous membranes. This is equivalent to heatstroke in people. If this occurs take the following actions:

- Initiate all cooling and monitoring efforts as noted in *paragraph 8-51*.
- Administer 1 liter of intravenous fluids to the MWD as quickly as possible if appropriately trained personnel are available.
- Continue cooling and monitoring until a veterinarian can evaluate the MWD.

BLOAT (GASTRIC DILATION-VOLVULUS)

8-55. Bloat is an acute stomach enlargement and twisting that may be due to gas, food, or water. It may occur if the MWD is fed immediately before or after hard exercise or when he is returned to his kennel after work or exercise and allowed to drink too much water.

8-56. An enlargement may be seen just behind the ribs, primarily on the left side. The MWD may also be restless and show signs of pain in the abdominal region. He may attempt to vomit or have a bowel movement. His breathing may be difficult or labored due to pressure from the enlarged stomach.

8-57. Handlers should notify the VCO and stop all watering and feeding. Walking may enable the MWD to relieve himself through bowel movements or passing gas. Most cases require extensive treatment by the VCO.

8-58. To prevent bloat, MWDs should not be fed within a two-hour period before or after hard work or vigorous exercise. It is also safer to provide multiple small meals each day, feeding twice daily, rather than one large meal. In hot weather, give small amounts of water during training or working MWDs to prevent excessive thirst. For the first hour after working or training, only three inches of water in the bucket should be available. After this cooling-off period, more water may be given.

First Aid Kit

8-59. The responsible VCO determines the contents of MWD first aid kits. The VCO should consider the skill of the handlers, access to veterinary support, and emergency situations that may occur. The DODMWDVS does not have a standard first aid kit supply list because needs vary. The local VCO and KM are responsible for developing and putting together an appropriate kit and training the handlers how to use it. The kit should contain supplies for cleaning and dressing minor wounds, cleaning and protecting an eye injury, bandaging or splinting minor breaks, and removing mild superficial foreign bodies (splinters or glass). It should also contain peroxide, antibacterial solutions, and a small supply of any prescription medications needed for individual MWDs.

ENVIRONMENTAL EFFECTS ON THE MILITARY WORKING DOG

8-60. Just as the environment affects soldiers performing their duties, the MWD is affected by his surroundings. It is important for planners and leaders to become aware of the many different conditions that can affect the performance of MWDs. Not all MWDs are affected in the same manner. Acclimation times will vary between conditions and MWDs. Planning considerations should always include the VCO when the MWD will be used or trained in environments that he is not accustomed to.

Wind

8-61. Always consider the direction and velocity of the wind when using an MWD for any detection operations. This consideration holds true even indoors. There may be a slight breeze that will move a scent away from the location of its source. Fans, air conditioning units, and objects in motion can create winds that will affect the performance of a detector dog.

8-62. In addition to these operational concerns, wind may create medical issues (such as blowing sand) that are negative. Wind also creates increased cooling, which is positive in hot weather but negative in cold weather.

Inclement Weather

8-63. Precipitation can degrade the MWD's performance of detection operations; however, do not assume that an MWD will not be able to perform satisfactorily. The amount of moisture accumulated and the amount of time elapsed since the scent was left are variables. MWDs have been proven to be effective even in inclement weather. MWDs that become wet and are returned to their kennels or crates while wet are at increased risk of skin injury and infection.

Heat

8-64. When operating in locations of extreme heat, the MWD should be allowed sufficient time to acclimate. Consult with the VCO at the location to determine how long an MWD should be allowed to acclimate. Care should be given to ensure that the MWD does not sustain injuries to the pads of his paws due to excessive walking on concrete or pavement in extreme heat conditions. Always ensure that the MWD has a fresh and constant source of drinking water.

Cold

8-65. Extreme cold temperatures should not deter MWD operations. Special care should be taken to enable an MWD to perform in cold locations. The accumulation of moisture on the MWD's paw pads can freeze and injure him. Using paw pad protectors to aid in the prevention of injury due to cold weather may be helpful, but the MWD must be trained to wear them. The MWD should be allowed to rest and warm up on the same cycle as the handler.

TERRAIN

8-66. The safety of the handler and MWD must be given the highest consideration when operating in hazardous terrain. Care should be taken when employment will take place in rocky or mountainous areas. Unfamiliar locations can pose threats in the form of caves, holes, loose footing, and other hazards. Whenever possible, conduct a reconnaissance of the area before employing an MWD.

DISTRACTIONS

8-67. There are numerous distractions that can affect MWD operations. These are conditions that can cause the MWD to focus his attention away from the desired task at hand. Planning and consideration should be given to eliminate or reduce distractions that hinder operations. Distractions are natural and man-made. Some examples of distracters are:

- Other dogs (both MWD and indigenous).
- Other animals.
- Human presence and activities.
- Noise.
- Smells.

Appendix A

Metric Conversion Table

This appendix complies with current Army directives, which state that the metric system will be incorporated into all new publications. *Table A-1* is a conversion chart.

Table A-1. Metric Conversion Chart

US Units	*Multiplied By*	*Equals Metric Units*
	Length	
Feet	0.30480	Meters
Inches	2.54000	Centimeters
Inches	0.02540	Meters
Inches	25.40010	Millimeters
Miles (nautical)	1.85320	Kilometers
Miles (statute)	1.60930	Kilometers
	Area	
Square inches	6.45160	Square centimeters
Square feet	0.09290	Square meters
	Volume	
Cubic feet	0.02830	Cubic meters
Fluid ounces	29.57300	Milliliters
	Temperature	
Degrees Fahrenheit	Subtract 32, multiply by 5/9	Degrees Celsius
	Weight	
Ounces	28.34900	Grams
Pounds	0.45359	Kilograms
Metric Units	***Multiplied By***	***Equals US Units***
	Length	
Centimeters	0.39370	Inches
Kilometers	0.53960	Miles (nautical)
Kilometers	0.62137	Miles (statute)
Meters	3.28080	Feet
Meters	39.37000	Inches
Millimeters	0.03937	Inches
	Area	
Square centimeters	0.15500	Square inches
Square meters	10.76400	Square feet
	Volume	
Cubic meters	35.31440	Cubic feet
Milliliters	0.03380	Fluid ounces

Table A-1. Metric Conversion Chart (Continued)

Metric Units	*Multiplied By*	*Equals US Units*
	Temperature	
Degrees Celsius	Multiply by 9/5, add 32	Degrees Fahrenheit
	Weight	
Grams	0.03527	Ounces
Kilograms	2.20460	Pounds

Appendix B

Kennel Construction

This appendix is intended to be a starting point in the design process with the identification of requirements and spaces. It should serve as a supplement to the installation design guide and not as a replacement. This appendix provides commanders and planners with guidelines and specifications to consider for the construction and equipment of kennel facilities. The floor plan (*Figure B-1*) is an example of how an MWD facility works, how areas relate to each other, and what is required for each space. It is not intended to serve as a rigid model for duplication.

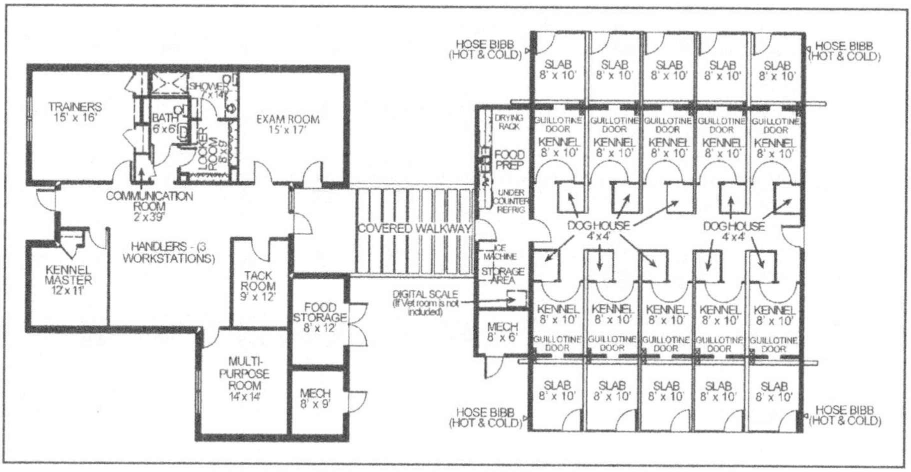

Figure B-1. Example of an MWD Kennel Floor Plan

TEMPORARY KENNELS

B-1. There are two major types of kennels, temporary (*Figure B-2*, page B-2) and permanent. According to *AR 190-12*, temporary kennels are to be used during deployments for no longer than one year, to allow sufficient time for permanent kennels to be built. Temporary kennels may be constructed out of any available resources. The design is not standard and varies from site to site depending on the location, mission, and duration of operations. Using a concept similar to constructing fighting positions, the temporary kennels will be improved continuously as needed or until a permanent facility is completed.

B-2. Using tents and shipping crates may be a hasty method of setting up temporary kennels. Although this method is not optimal for sustained operations, it may be useful until other accommodations can be coordinated. Erect the tents at a location that allows the MWDs a safe place to rest away from other troop activities after operations. Whenever possible, use terrain or buildings to place a barrier between troop activity and the temporary kennel. Never place the temporary kennel site near motor pools, dining facilities, or high noise areas.

Figure B-2. Temporary Kennel

B-3. In some areas of operation, it may be possible to obtain the use of a building or other structures for temporary kennels. In these cases, leaders must ensure that the building is well ventilated, safe, and structurally sound. Ensure that there are no hazards (such as chemicals, bare wires, holes, or debris) that can cause injury to the MWDs or handlers.

B-4. Engineers are often used to assist in the improvement of temporary kennels. In addition, fences should be erected and dog runs created.

> ***Note*:** From the start of operations, sanitation measures must be employed in all temporary kennel configurations.

PERMANENT KENNELS

B-5. The following information and figures give specific details for the construction of a permanent kennel facility and adjacent areas. These guidelines may be altered to allow for special needs or uses unique to each location.

EXTERIOR AREAS

B-6. An MWD kennel complex (*Figure B-3*) consists of kennels, a support building, an obedience course, an exercise area, an MWD break area, exterior storage, parking areas, drives, and walks. Often, as a result of the location, utility services are limited at the site. Construction should provide the following utilities:

- Water (potable with backflow prevention).
- A sanitary sewer.
- Electric.
- Telephone.
- Network and communication service to the site.

Other systems should be incorporated according to local procedures.

B-7. The entire complex should be enclosed, at the minimum, with a heavy-duty, 8-foot-high chain-link fence with three strands of straight wire (no barbed) at the top to prevent an MWD from climbing or jumping out. A 10-foot (minimum) vehicle gate that can be padlocked is to be installed to allow for food deliveries to the kennel and other access requirements. A personnel entry gate that is visible from the KM's office is also required. The personnel gate should have a cipher lock and a notification buzzer that sounds in the KM's office, the obedience course, and the kennel. The buzzer in the kennel should have the capability to be turned on when personnel are in the area and turned off to allow the MWDs to rest when personnel are absent. All gates should be self-closing and self-latching.

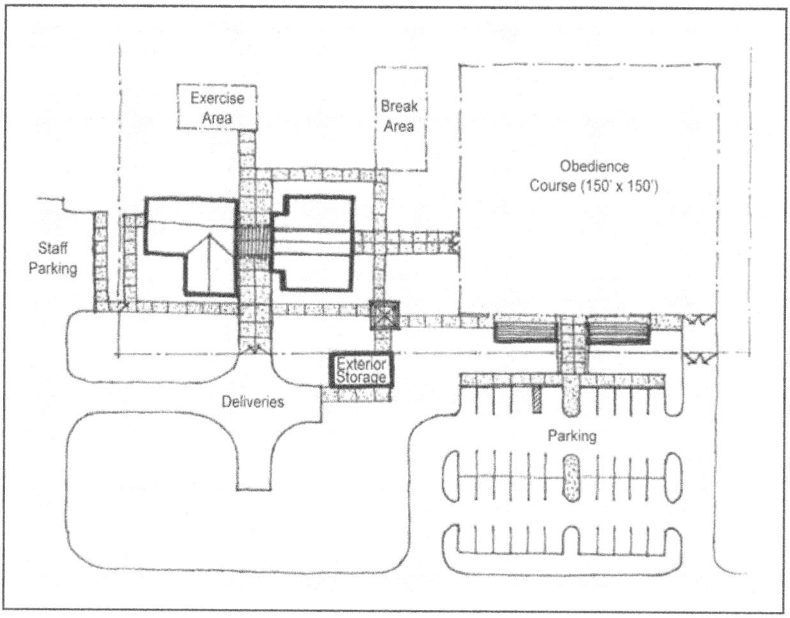

Figure B-3. MWD Kennel Complex

B-8. Security lighting for the area should be according to local policy. At a minimum, the lighting should be on from dusk to dawn and placed on all corners of the fencing area. Security lights should be placed in areas that the other lights do not cover.

B-9. Warning signs will be posted on the exterior fencing and buildings of the MWD kennel and exercise area. Signs should contain the following words: "DANGER—OFF LIMITS–MILITARY DOG AREA." Personnel approaching the kennel area should be able to see and read the warning signs under normal daylight conditions from a distance of 50 meters.

B-10. Locate a trash dumpster for the complex outside of the perimeter fence. Screen the dumpster and provide low-maintenance landscaping for the complex according to local policy.

Exterior Storage, Parking Areas, Drives, and Walks

B-11. An enclosed, 400-square-foot exterior storage building is required to store maintenance equipment, portable kennels, and obstacles. Interior/exterior lighting, power, and a water faucet should be included.

B-12. Staff parking should be a paved parking lot large enough for eight POVs, three GOVs, two visitor vehicles, and one handicap parking space. Pave access drives to accommodate parking, exterior storage, food deliveries, emergency vehicles, and the transportation of MWDs.

B-13. Provide paved walks to all areas. When designing, give careful consideration to creating a one-way MWD traffic system. To prevent confrontations between the MWDs, avoid situations that allow MWDs to meet head on.

Obedience Course (150 Feet by 150 Feet)

B-14. The obedience course plays an important role in maintaining the MWD's agility and stamina as well as in reinforcing obedience and proficiency training. The course should be grassed and free of hazards (trees, large rocks, holes, and burrs) that may be harmful to MWDs and handlers. The site should be graded for drainage but minimally sloped to provide a level field for training. The area should be enclosed with an 8-foot chain-link fence. Gates should be self-closing and self-latching with a minimum width of 5 feet to allow for lawn maintenance equipment. The area should be well lighted to eliminate shadows. Light poles should be located on the exterior of the fence. On and off switches should be manual. Weatherproof outlets should be provided at each light to provide power at the site. Speaker units connected to the chosen public address system should also be provided. Water should be at the course site with water faucets installed to provide water for MWDs and lawn watering. An irrigation system is desirable for long-term lawn maintenance.

B-15. Obstacles for the course (*Figure B-4*) include barrels, tunnels, steps, jumps, framed windows, and dog walks. Each obstacle is 15 to 20 feet from the previous obstacle; the course is run in a sequence. Obstacles should be constructed to the following specifications:

- Barrel 1 is 35 inches long with a 23-inch opening.
- Barrel 2 is 70 inches long with a 23-inch opening.
- Barrel 3 is 105 inches long with a 23-inch opening.
- The tunnel is 146 inches long with a 19-inch opening.
- The steps are constructed out of wood and are 94 inches high with 5 steps on each side. Each step is 43 inches wide by 16 inches high by 24 inches long. The top platform is 48 inches long by 43 inches wide.
- Jumps 1, 2, and 3 are constructed of wood and are 36 inches high at the maximum raised level. There are 6 removable boards that are 51 inches long by 6 inches high and 1 inch thick.
- The framed window is constructed of wood and is 8 feet tall. The maximum height for the MWD to negotiate is 36 inches. The window opening is 48 inches wide. Removable boards are 5 1/7 inches wide by 51 inches long and 1 inch thick.
- The A-frame is constructed of wood. Its two sides are 8 feet long by 4 feet wide by 1 inch thick with 1-inch by 2-inch boards every 6 inches to secure footing. It has a maximum raised height of 6 feet with a base spread of 10 feet.
- The dog walk is constructed of wood and is 26 inches high by 224 inches long and 11 inches wide with 1-inch by 2-inch boards every 4 inches to secure footing. Each end has a 45° incline ramp that is 11 inches wide by 49 inches long (included in the 224 overall length measurement).

Note: See *DA Pam 190-12* for more complete construction guidelines including the bracing necessary for the various obstacles.

Military Working Dog Break Area (10 Feet by 20 Feet)

B-16. An MWD break area should be located near the kennels. A break area allows the handler to release the MWD immediately after exiting the kennels or before entering the kennels so that the MWD can relieve himself. The area should be enclosed with an 8-foot chain-link fence. Gates should be self-closing and self-latching and a minimum of 5 feet wide to allow for lawn maintenance equipment. The ground can be a sandy area for cleaning ease. A water faucet is required to maintain the area.

Exercise Area (20 Feet by 40 Feet)

B-17. The exercise area is a space where the MWD can be released without the handler being present. The exercise area should not be a shared area with the obedience course, as it would conflict with training objectives. The exercise area should be visually separated from the obedience course to prevent MWD from being distracted during training. The exercise area should be grassed, hazard free, and graded slightly for drainage purposes.

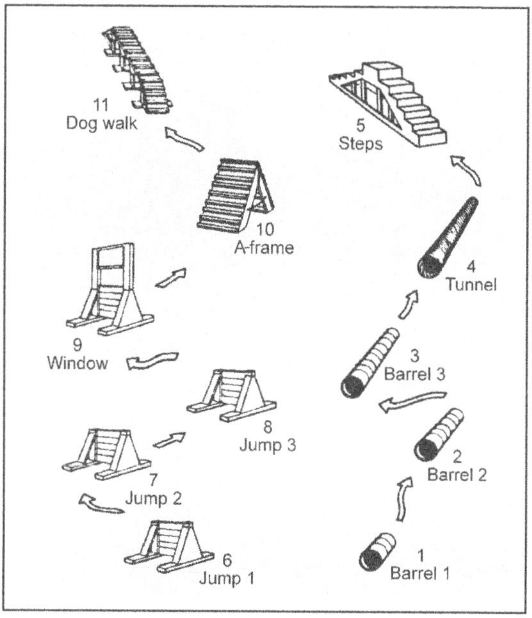

Figure B-4. Obstacle Course Layout

INTERIOR AREAS

B-18. Interior areas have heating, ventilation, and exhaust systems; general-purpose lighting; and power outlets in all spaces. Provide phone, fax, and local area network (LAN) connections to support each area and hot and cold water to all latrines, sinks, and showers. Construct the facility to meet all fire protection and life safety feature requirements.

Administration Areas

B-19. Since MWDs are likely to visit the administration area (*Figure B-5*, page B-6) at some point, finishes should be of durable material, require minimal maintenance, and be easily repaired or replaced in case of damage. (For example, flooring should be covered by carpet.)

Kennel Master's Office (132 Square Feet)

B-20. The KM's office serves as the nerve center for the complex. Ideally, the office should be located at the front of the administrative area with exterior windows that view the entry gate to the complex. It includes a lockable closet for supply storage. Furnishings include a desk, a chair, two visitor chairs, a four-drawer file cabinet, and a bookcase. Equipment includes a computer, a printer, a fax/copier, and a public address system or central intercom console.

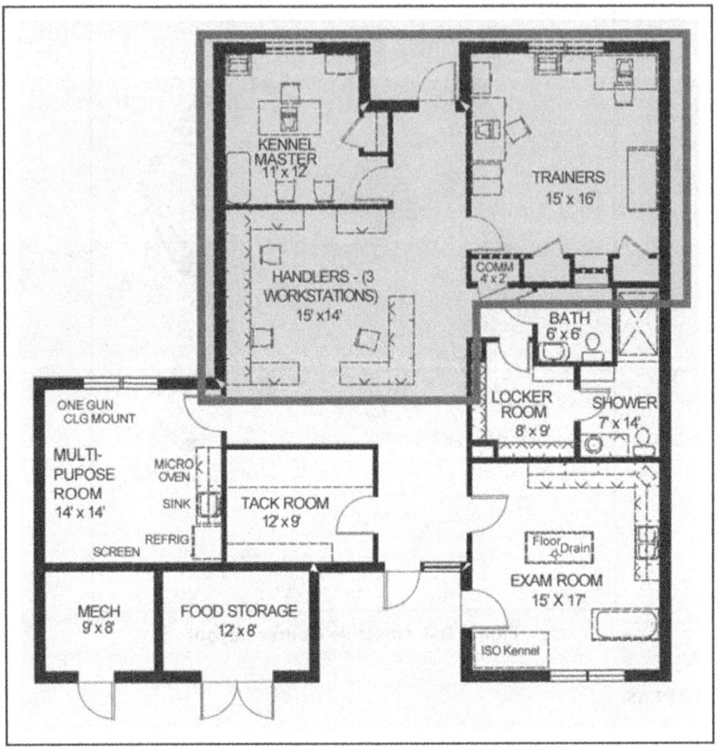

Figure B-5. Kennel Administrative Areas

Trainer's and Plans Noncommissioned Officer's Office (240 Square Feet)

B-21. The trainer office should provide workspace for two people. Lockable closets are required to store training aides for each trainer. Furnishings include a desk, a chair, and a four-drawer file cabinet for each trainer and a common bookcase. Equipment includes a computer, a printer, a copier, and a 4- by 8-foot dry erase board.

Handlers' Area (210 Square Feet)

B-22. The area for the handlers is an open area for general office functions. It includes—

- Three common workstations, depending on the size of the kennels, for handlers to update records and complete daily reports.
- Space for document storage.
- Furnishings (three workstations, file cabinets [one four-drawer vertical file per workstation or the equivalent], and a distribution counter.
- Equipment (computers and a printer).

As a general rule, one workstation should be available for every three handlers assigned to an MWD section.

Controlled-Substance Storage Room

B-23. This room is inside the administrative area and allows sufficient room for the storage safe in which controlled-substance training aids are stored. The doors are secured using approved locking devices. Lightweight doors are replaced with solid metal or wood doors or covered with a 9- to 12-gauge security screen or 16-gauge sheet steel. They are fastened with smooth-headed bolts and nuts and peened in place. All windows providing access to a storage room that is not staffed 24 hours a day are protected by a 9- to 12-gauge security screen or 3/8-inch or larger diameter steel bars spaced no more than 6 inches apart. The frames holding the screen or bars must be fastened to the window frame with smooth-headed bolts.

Special-Use Areas

B-24. Special-use areas (*Figure B-6,* page B-8) are the transition spaces between the administration area and the kennels. They are important in the day-to-day operations of the MWD facilities. They are not normally occupied on a daily basis. As such, these areas provide a secondary function of serving as a noise buffer between the kennels and the administrative areas. The special-use areas provide a workspace for the VCO and accommodate the unique storage requirements for the kennels.

Veterinary Treatment Room

B-25. A veterinary treatment room is used to perform health care and first aid for MWDs since many locations do not have veterinary treatment areas in close proximity to the kennels. In order to maintain sanitary conditions, treatment rooms require a higher degree of cleanliness and durability than the rest of the support building. Basic finish requirements are—

- Seamless floors with an integral base.
- Washable walls.
- Painted gypsum board ceilings.
- Cabinets faced with plastic laminate.
- Solid-surface countertops.

B-26. Interior partitions should provide a sound transmission class rating of 50 to 55. Floor drains are centered in the room to aid in cleaning, and a dedicated floor drain is located under the isolation kennel. Wall and base cabinets with drawers are installed, including stainless steel dual washbasins with hot and cold water. One section of wall cabinets should be lockable so that prescription medicines for MWDs may be stored in the room. In addition to the normal electrical outlets required by code, a 220-volt outlet is required to support other portable equipment that may be used. Equipment for the room includes the following:

- **An MWD isolation kennel.** The entire structure should be made of uniform, nonglare type 304 stainless steel. The frame, cross members, and vertical spokes should be welded together at every intersection. Latches and hinges are made of 12-gauge stainless steel. The floor of the pen is a raised, removable grate. The floor grate is a heavy-gauge stainless steel, plastic-coated section installed 1 3/4 inches off the floor. A dedicated floor drain is required for this kennel.
- **Standard table tub**. The tub is made of type 304 stainless steel with a minimum depth of 15 inches. The tub slopes to the drain end. The design includes a removable, bathing rack made of plastic-coated stainless steel that fits into the bottom of the tub to give MWDs better footing during treatments. Faucets include a flexible, stainless steel hose that will extend the length of the tub and spray powerfully to reach all areas of the tub.
- **Exam room light.** A ceiling-mounted exam room light with an arm extension and swivel capacity of 360° is recommended.
- **Walk-on platform scale.** The scale is a type 304 stainless steel platform with a wall- or post-mounted light-emitting diode (LED) display. The scale is able to perform calibration and weigh animals to the nearest 0.1 of a pound or kilogram.
- **Stationary exam table.** The table top is type 304 stainless steel die-formed from one piece of 20-gauge steel. It should have raised edges to prevent fluid runoff and be permanently attached

to a heavy base. There should be a minimum 35-inch work height from the floor to the top of the table. The heavy base should have adjustable leveling screws.

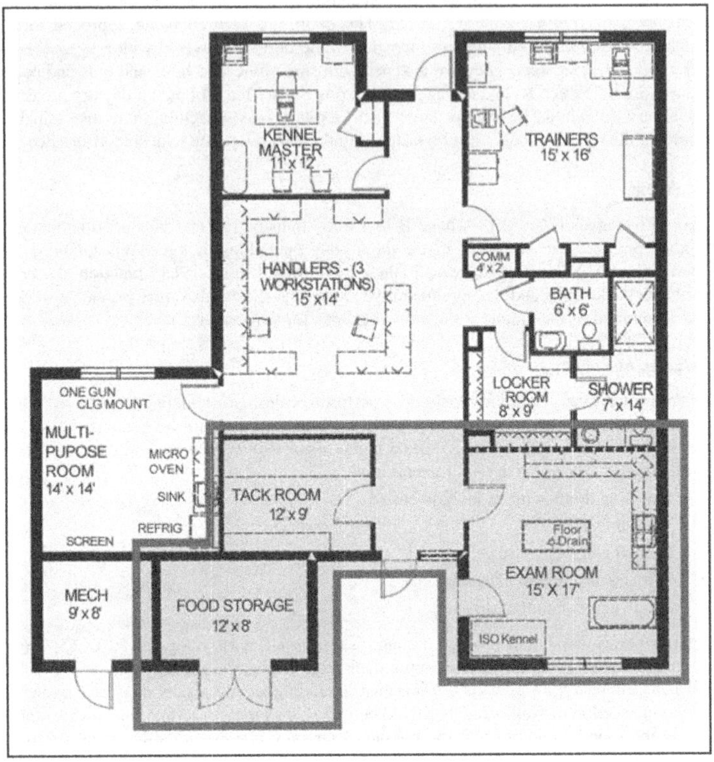

Figure B-6. Kennel Special-Use Areas

Tack Room (108 Square Feet)

B-27. The tack room is for the storage of extra MWD equipment, such as bite suits and portable kennels. The tack room is located near the entrance to the kennels but not the main building entrance. Requirements include 12 linear feet of shelves that are 18-inches wide and 36 inches deep and allow for 36 inches above each shelf; a 24-inch space with hooks to hang items on; and 6 linear feet of shelves that are 18 inches wide and 12 inches deep and allow for 18 inches above each shelf.

Food Storage Room (96 Square Feet)

B-28. The food storage room is normally for the bulk storage of a 30-day supply of food for the MWDs. The room is adjacent to the kennels with an exterior entry. The entry is a pair of 36-inch doors to allow for palletized food delivery and storage. Doors require weather stripping and thresholds to aid in insect and rodent control. Temperature and humidity control equivalent to the office area is required to control food spoilage. A minimum separation of 2 inches is required between the walls and the stored food to allow for air circulation. Shelving should be installed to meet local delivery operations.

SUPPORT AREAS

B-29. The support areas (*Figure B-7*) provide space for the common-use areas of the building. Areas for the multipurpose room, latrines, communications, and mechanical necessities form the core of the support areas. Support areas should be centrally located and easily accessible for building occupants or maintenance personnel in the case of the mechanical room.

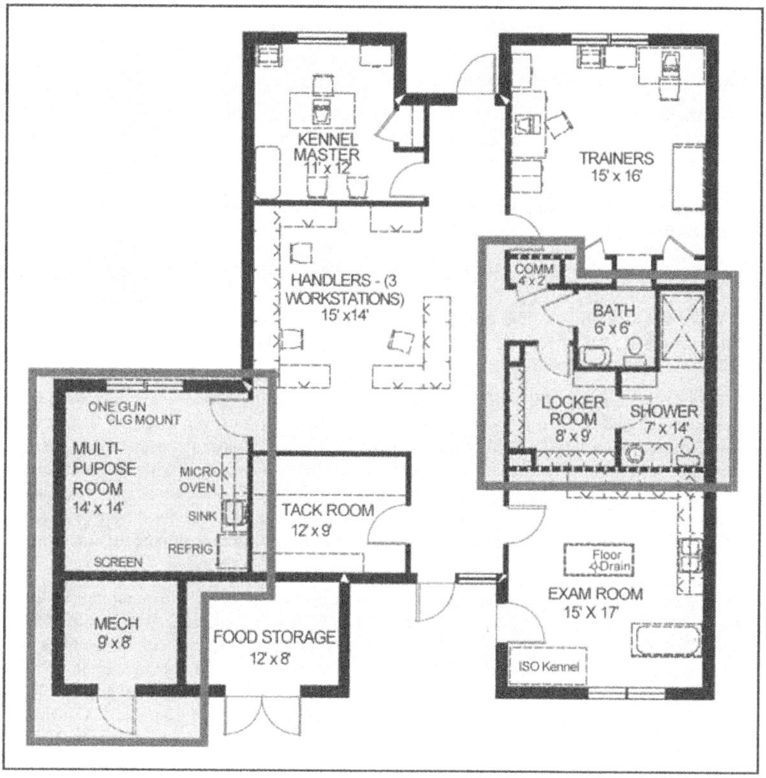

Figure B-7. Kennel Support Areas

Latrine, Shower, and Lockers (206 Square Feet)

B-30. A unisex area is divided into three rooms for the latrine, shower, and lockers. One room is a handicapped accessible latrine with a water closet and lavatory for general use. The locker room includes 10 full-height metal lockers and a bench for MWD handlers' use. The shower room includes a shower, bench, water closet, and lavatory. Do not neglect to include other accessories such as mirrors, soap dispensers, paper towel dispensers, disposal units, grab bars, and hooks.

Heating, Ventilation, Air Conditioning, Mechanical, Electrical, and Communications Systems

B-31. Heating, ventilation, and air conditioning (HVAC) is a process to add hot or cold air in a facility. It also introduces fresh air in to the facility to maintain indoor air quality and eliminate sick building syndrome. Design HVAC systems to optimize energy use and meet the design criteria for the area in which you are located. Place the hot water heater in the mechanical room and with a capacity that will meet the demands of the latrine and shower; multipurpose room; and veterinary treatment room. Locate electrical panels in this area as well. Although a recommended size is shown on the plans, the room should be big enough to support the equipment required for the location.

B-32. While the number of instruments and computers is not large, the communications closet is vital to the success of the organization. Provide a minimum of one prewired telephone and LAN outlet for each workstation, one in each common area, and one in the kennel food preparation area. Include telephone and LAN outlets for fax machines and printers. Conceal conduit in the walls.

Multipurpose Room (196 Square Feet)

B-33. The multipurpose room serves as a combination conference room, classroom, and break room. Food preparation requirements are a light, a countertop, and base and upper wall cabinets for storage. Lighting should have a variable-level control for use as a conference room and/or classroom. Furnishings include either a conference table for fifteen or three five-person tables; chairs; a bulletin board; and a 4- by 8-foot dry erase board. Equipment includes a video projection system, a projector screen, a sink, a microwave oven, and a refrigerator.

Kennel Area

B-34. The kennel area provides the daily living environment for the MWDs, giving each MWD a private place for eating and resting. Kennel areas should be built to accommodate large-breed dogs and designed as modular structures to allow for the future expansion of runs.

B-35. There are three types of kennel areas; indoor kennels, outdoor kennels, and indoor/outdoor kennels. The preferred standard is the combination indoor/outdoor kennel; however, there are factors that must be taken into consideration when selecting the kennel type. Cost is always a consideration, but climate also influences the selection. In cold weather climates, the benefit-to-cost ratio will drive the selection to an indoor kennel while the opposite is true for hot weather climates. The more temperate the climate the more suited the indoor/outdoor kennel is for use. Each type of kennel includes the same space requirements. A kennel includes kennel runs with doghouses, a food preparation area, a storage area, and a mechanical room.

B-36. The kennel should be separated from the administrative building by a minimum of 20 feet. This provides separation between different functions and aids in noise control. As a result of the constant interaction between the two facilities, a covered walkway is required to connect the facilities and provide protection from the elements. Care should be taken in the layout of the kennel in relation to the administrative building; a one-way system should be maintained in the movement of MWDs. The objective is to avoid situations where MWDs meet face to face and to prevent confrontations. Consider installing mirrors at corners, intersections, and blind spots to alert handlers of potential wrong-way traffic.

B-37. All entries and exits from the kennel area are required to be self-closing and self-latching. Doors exiting the kennel runs should open inward to aid in preventing MWD escape. Proper ventilation is important in the kennels to prevent the spread of diseases and to control odors. It is recommended that the ventilation standard be to cycle out the room air 10 to 15 times per hour. Kennel temperature should range from 45° to 85° Fahrenheit with humidity in the range of 40 percent to 70 percent. Temperature should be maintained within 10° of the exterior temperature. MWDs work more effectively and are more alert when the kennel temperature is close to the temperature of their working environment.

B-38. Potable water is required at the kennels. Hot and cold water lines are installed at ceiling level down the length of the 5-foot wide center corridor connecting to hose reels at each end of the corridor. A high-pressure washer is required to assist in sanitizing and cleaning kennel runs.

B-39. Since the kennel is basically a wet environment, all electrical receptacles should be provided with ground fault circuit interrupters and all-weather covers. Receptacles mounted at 48 inches above the finished floor are required at each end of the central corridor.

B-40. Each kennel run should have a waterproof light fixture that has an individual switch at the corridor entrance. Other areas require general-purpose lighting according to industry standards. All drain lines in the kennel should be a minimum of 6 inches in diameter and be designed to sustain flow velocities that will maintain a self-cleansing action. Due to the nature of the waste, cleanouts should be numerous and easily accessible. Floor drains should be included in the central corridor to aid in the cleaning of kennels.

Indoor/Outdoor Kennel

B-41. A combination indoor/outdoor kennel (*Figure B-8*, page B-12) simply implies that there is an interior run and an exterior run linked with each MWD living area. A guillotine type door will connect the interior run and the exterior run. Other door types may be used, but the idea is to permit isolation of the MWD during cleaning operations. Ideally, operation of the door is from the central corridor and from the outside of the exterior run. At 6 feet above the finished floor, use translucent wall panels in the wall separating the runs to provide natural light inside the kennel.

B-42. The roof should extend over the exterior runs and include gutters and/or downspouts to prevent excessive water in the runs. Exterior framing should be enclosed with a soffit to prevent birds from roosting. Exterior water faucets (with freeze protection where required) with both hot and cold water are required at each corner of the building. These are necessary for cleaning exterior runs. Floor drains (*Figure B-9*, page B-12) should be located in the corner of each run adjacent to the common wall separating exterior and interior runs. Drains should tie-in to a common waste line. Drain covers should be flush with the floor. Kennel floors should slope (at least 1/4- inch per foot) toward the drains to allow for quick water drainage and drying.

Indoor or Outdoor Kennel

B-43. The plan for the indoor or outdoor kennel (*Figure B-10*, page B-13) is the same for either style. The obvious difference is the wall that encloses the kennels. For indoor kennels, the exterior walls along the runs should have windows beginning at 6 feet above the finished floor to allow for natural light. Windows should be operable and hinged at the sill to tilt inward to prevent the escape of MWDs. For outdoor kennels, the exterior wall is simply an 8-foot chain-link fence. Water faucets with hot and cold water are required at each corner of the building. These are required to clean the gutter and drain system (*Figure B-11*, page B-13) for the kennel. A gutter and drainage system is provided at the back of the kennel run for either type of kennel for sluicing wastewaters from cleaning operations. Kennel run floors should slope to the gutter and the gutter should slope to the drain. Gutters should be sufficient to allow for easy cleaning. A minimum of 3 feet should be allowed between the end of the kennel and the wall or fence.

Kennel Run

B-44. Runs should be arranged so that openings are staggered in order to avoid MWDs facing each other across the corridor. Full-height partitions are required between runs for all kennel types. The first 6 feet of wall above the floor should be sealed concrete. For interior runs, steel clad acoustical panels should be considered from the top of the concrete partition to the ceiling. Steel clad acoustical ceiling panels should also be considered above the kennels.

B-45. Heavy-gauge chain-link fence is used for exterior runs and is also an option for interior runs. Because of the noise generated in the enclosed area of an interior run, it is important that noise-reducing materials be introduced to prevent stress to the MWD and hearing loss to personnel.

B-46. As a minimum, end walls should be full-height, galvanized chain-link fencing. Take care to prevent sharp edges from bolts, hinges, and tie wires on the interior of the runs where MWDs may be cut or harmed. Self-latching entry gates should swing 180° and be able to cover the opening of the MWD house. This allows the MWD to be penned in the house during run-cleaning operations. A stainless steel water bucket is required for each kennel, including a holder that prevents the MWD from overturning the bucket.

Figure B-8. Combination Indoor/Outdoor Kennel

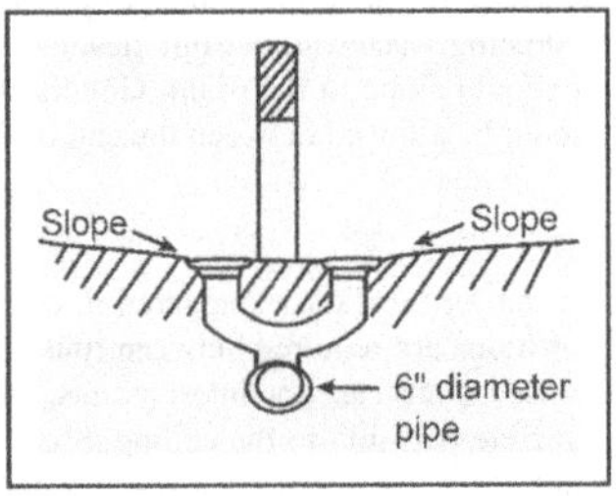

Figure B-9. Example of a Combination Indoor/Outdoor Kennel Floor Drain

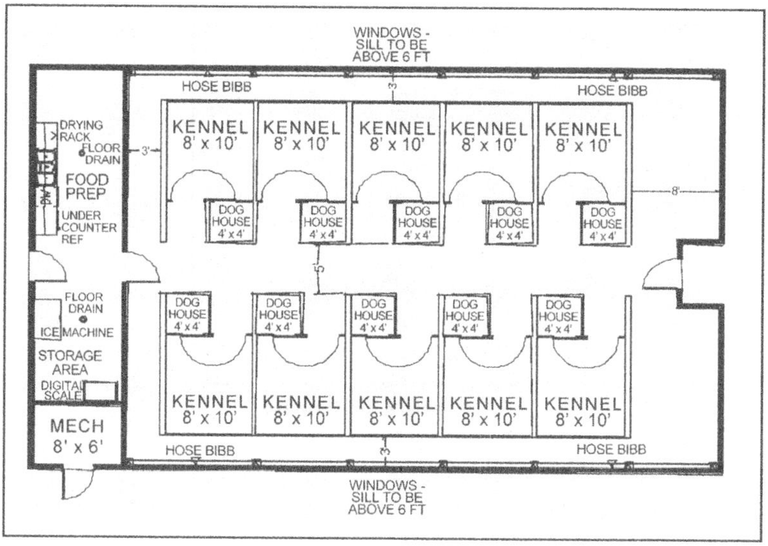

Figure B-10. Indoor or Outdoor Kennel

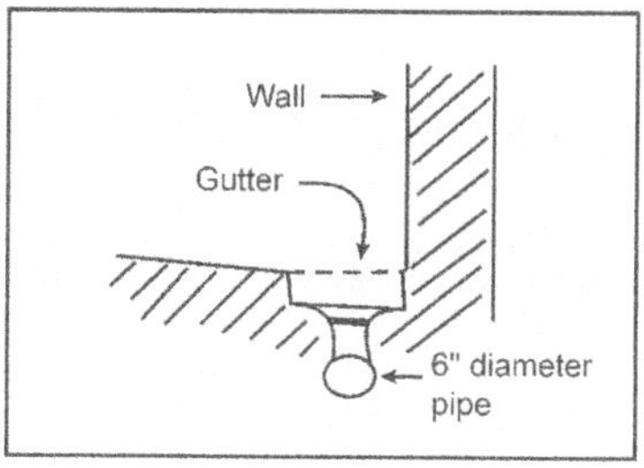

Figure B-11. Example of an Indoor or Outdoor Kennel Floor Drain

B-47. Flooring should be concrete and slope toward floor drains and/or gutters to provide rapid runoff of water and drying. Since runs are subject to frequent wash downs to clean and prevent disease, a smooth epoxy-type finish is needed. Temperature-controlled floors should be emplaced for locations that have inclement weather.

B-48. All materials in areas accessible to MWDs should be resistant to damage by scratching, biting, and chewing. Materials should be durable and easy to sanitize. Avoid using material such as angle iron that rusts over a period of time and generates sharp edges that pose a danger to MWDs.

B-49. MWD houses are 4 feet by 4 feet by 4 feet sealed, concrete boxes with a metal top. Tops should hinge toward the end wall in order to be raised to clean the house or treat an MWD. Once lifted, a holding device should be available to keep the top open. Wall opening dimensions are shown in *Figure B-12*, page B-14. The house should be a minimum of 8 inches and a maximum of 12 inches above the run floor. The house floor should have positive drainage toward the wall opening.

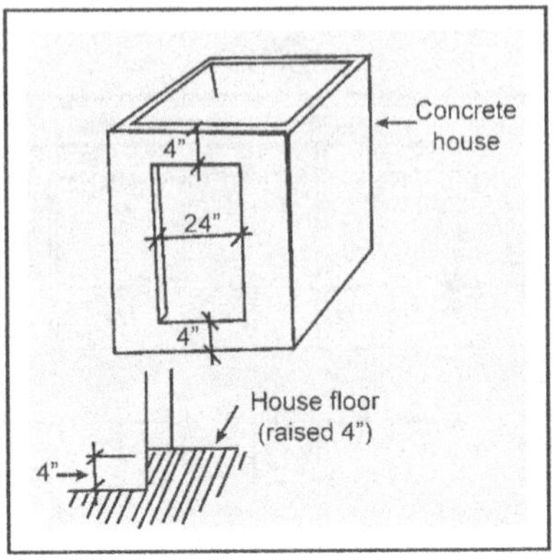

Figure B-12. MWD House

Storage and Mechanical Area

B-50. Additional storage for the kennel is provided. A commercial quality ice machine is required to aid in cooling down MWDs. Include a floor drain for the area. If a veterinary treatment room is not included, a digital scale should be located in this area. Design mechanical systems to optimize energy use and meet the design criteria for the area where the kennel is located. Include the hot water heater in the mechanical room and with a capacity to meet the demands of the kennel. Allow for the location of electrical panels in this area as well.

Food Preparation Room (112 Square Feet)

B-51. The food preparation area is where feedings are prepared and feeding pans/bowls are stored and cleaned. Food preparation for MWDs requires a large countertop with a deep, three-compartment stainless steel sink. Upper wall cabinets are required for general storage. An under the counter refrigerator, a dishwasher, and space to store food for daily use are necessary for operation. A 2-foot space at the end of the counter should accommodate a storing and drying rack for pans and bowls. Lighting and receptacles and a floor drain should be installed per the code. A telephone, bulletin board, feeding chart, and large garbage can should also be provided. Optional equipment includes a garbage disposal unit and a stackable washer and dryer.

Appendix C
Inspection Memorandum

This appendix provides a sample memorandum (see *Table C-1*, page C-2) for documenting the results of inspections. Commanders and program managers will find this useful in preparing for inspections. KMs, plans NCOs, and handlers will find this list helpful in conducting daily duties and adhering to requirements for inspections. Maintain completed memorandum records on file to assist in future inspections.

Table C-1. Sample Inspection Memorandum

DEPARTMENT OF THE ARMY
UNIT
ADDRESS OF UNIT

OFFICE SYMBOL 15 May 2004

MEMORANDUM FOR RECORD

SUBJECT: MACOM Technical Assistance Visit or Provost Marshal Inspection

1. SECTION I. GENERAL INFORMATION.

UNIT: 529th Military Police Company LOCATION: Kaiserslautern
DATE: 3-14-05 TIME: 0900

INSPECTOR (S): LTC Jeffery S. Captain ESCORT: SFC James Majors

TYPE OF VISIT: __X__ ANNOUNCED______UNANNOUNCED

TYPE	AUTHORIZED	ON HAND
Patrol		
Patrol Explosive	6	4
Patrol Narcotic	3	2
Explosive Detector	0	0
Narcotic Detector	0	0

2. SECTION II. PERSONNEL.	**YES**	**N/A**	**NO**
a) Are all authorized MWDs on hand (unit TDA/MTOE document)?	___	___	___
b) Are all MWD handlers certified by the DOD Dog Center?	___	___	___
c) Are ASI trained handlers being used as specified in AR 614-200?	___	___	___
d) Is the assignment policy of one dog and one handler being followed?	___	___	___
e) Are all MWDs assigned to a handler?	___	___	___
f) Has an excess of dogs been reported to the MACOM?	___	___	___
g) Does the KM meet the minimum criteria to be appointed as a KM?	___	___	___
h) Are handlers being assigned duties that take away from their handling responsibilities (excluding PT, staff duty, field duty, and unit training)?	___	___	___
3. RESPONSIBILITIES OF THE PROVOST MARSHAL AND COMMANDER.			
a) Is there an updated SOP signed by the PM explaining policies, procedures, and responsibilities for the direction, management, and control of the kennels?	___	___	___
b) Has the PM initiated an MWD program based on the evaluation of the installation support, customs, HRP support, and combat support missions?	___	___	___

Table C-1. Sample Inspection Memorandum (Continued)

OFFICE SYMBOL
SUBJECT: MACOM Technical Assistance Visit or Provost Marshal Inspection

		YES	N/A	NO
c)	Has the PM issued written guidelines governing the release of an MWD to apprehend a suspect?	____	____	____
d)	Are there specific criteria in contingency plans governing the use of dog teams during civil disturbance situations?	____	____	____
e)	Does the PM provide guidance on procedures for explosive detector dog support to civil authorities?	____	____	____
f)	Is the proficiency of each MWD team evaluated periodically by the PM?	____	____	____
g)	Has the PM and/or unit commander checked one or more of these areas monthly: kennels, kennel support building, training and exercise areas, equipment, handlers, dogs, training, utilization and team proficiency?	____	____	____
h)	Has the PM and/or unit commander inspected all areas of the MWD kennels quarterly, and is a written record on file?	____	____	____
i)	Are MWD requisitions submitted according to *AR 190-12* and *AR 700-81*?	____	____	____
j)	Are MWDs placed on the unit property book by NSN, name, and tattoo number?	____	____	____
k)	Are MWDs assigned to the KM on a *DA Form 3161* (*Request for Issue or Turn-In)* hand receipt (according to *DA Pam 710-2-1*)?	____	____	____
4.	**VCO RESPONSIBILITIES.**			
a)	Are the dogs examined by the VCO every six months (according to *AR 40-905*)?	____	____	____
b)	Does the VCO inspect the kennel facility and area for proper sanitary standards on a quarterly basis?	____	____	____
c)	Is the issued dog food and diet checked periodically?	____	____	____
d)	Is there a program in effect for insect and rodent inspection?	____	____	____
e)	Does the VCO concur with the kennel standards/facilities?	____	____	____
	If not what course of action has been taken? ______________________________			
f)	Does the VCO provide regular training to handlers on MWD health care, feeding, and first aid?	____	____	____
g)	Does the VCO provide a list to the KM concerning the deployable status of all MWD?	____	____	____
h)	Does the VCO issue health certificates for dogs pending shipment?	____	____	____

Table C-1. Sample Inspection Memorandum (Continued)

OFFICE SYMBOL SUBJECT: MACOM Technical Assistance Visit or Provost Marshal Inspection	**YES**	**N/A**	**NO**
5. WARNING SIGNS.			
a) Are there warning signs with the following words posted at each instillation entrance, "**CAUTION—THIS AREA PATROLLED BY MILITARY WORKING DOGS**?"	____	____	____
b) Are kennels located away from disturbing influences and adequately posted with "**DANGER—OFF LIMITS—MILITARY DOG AREA**?"	____	____	____
c) Are the shipping crates stenciled with "**CAUTION—MILITARY WORKING DOG—DO NOT TAMPER WITH ANIMAL**?"	____	____	____
d) Are shipping crates stenciled with "**LIVE ANIMAL**?"	____	____	____
e) Are removable or permanent markings placed on the sides and rear of vehicles with the following phrase "**CAUTION—MILITARY WORKING DOGS**?"	____	____	____
6. KENNEL AREA.			
a) Has a risk analysis been conducted on the kennels?	____	____	____
b) Is the kennel manned both day and night?	____	____	____
c) Does the kennel have an IDS/CCTV system if not manned both day and night?	____	____	____
d) Are approved locks and padlocks used to secure government equipment?	____	____	____
e) Have primary and alternate key custodians been appointed, in writing, by the PM or commander?	____	____	____
f) Are personnel authorized to issue and receive keys listed on an access roster?	____	____	____
g) Are all keys controlled and accounted for on a *DA Form 5513-R* (*Key Control Register and Inventory)?*	____	____	____
h) Is the *DA Form 5513-R* secured in an approved, lockable container?	____	____	____
i) Is the key depository box permanently affixed to a wall?	____	____	____
j) Are keys inventoried semiannually, and is the last report on file?	____	____	____
k) Are all keys inscribed with a serial number?	____	____	____
l) Are kennels used to house or care for other animals?	____	____	____

Table C-1. Sample Inspection Memorandum (Continued)

OFFICE SYMBOL SUBJECT: MACOM Technical Assistance Visit or Provost Marshal Inspection	**YES**	**N/A**	**NO**
m) Are adequate training areas adjacent to the kennel areas?	____	____	____
1) Is the training area fence 8 feet high?	____	____	____
2) Does the overhanging fence face inward?	____	____	____
3) Is the fence staked to the ground?	____	____	____
n) Does the kennel area have an adequate water supply?	____	____	____
o) Are the kennels of a standard design?	____	____	____
7. EQUIPMENT.			
a) Is all authorized equipment on hand?	____	____	____
b) Are Air Force shipping crates returned within 10 days after receiving a new or replacement MWD?	____	____	____
c) Is there one shipping crate available for each dog authorized?	____	____	____
d) Is leather equipment properly maintained and cleaned?	____	____	____
e) Is all metal equipment free of rust?	____	____	____
f) Is the web equipment cleaned with proper cleaning material to prevent deterioration?	____	____	____
g) Are there adequate medical supplies on hand for first aid treatment and training?	____	____	____
8. USE OF MWDs.			
a) Are the dog teams being employed a minimum of 24 hours per week?	____	____	____
b) Are PDs employed in all areas of the installation during both day and night?	____	____	____
9. NARCOTICS.			
a) Is the MWD section registered with the Drug Enforcement Administration?	____	____	____
b) Is a *DEA 223 (Controlled Substance Registration Certificate)* on file and current?	____	____	____
c) Have all personnel been designated, in writing, by the PM to handle training aids?	____	____	____
d) Has primary and alternate narcotic training aid custodians been appointed, in writing, by the PM?	____	____	____

Table C-1. Sample Inspection Memorandum (Continued)

OFFICE SYMBOL SUBJECT: MACOM Technical Assistance Visit or Provost Marshal Inspection	**YES**	**N/A**	**NO**
e) Have primary and alternate narcotic custodians been cleared by a favorable Crime Records Center name check?	____	____	____
f) Is the Crime Records Center name check maintained on file?	____	____	____
g) Have all personnel authorized to handle or use training aids been briefed at least annually?	____	____	____
h) Is a written record of the briefing being maintained on file?	____	____	____
i) Are the monthly/quarterly inventories being conducted by a disinterested E7 or above?	____	____	____
j) Are the monthly/quarterly inventories being maintained with the controlled substances?	____	____	____
k) Are the monthly/quarterly inventories being maintained for at least two years?	____	____	____
l) Are joint inventories being conducted on the change of primary or alternate custodians and is the inventory maintained on file?	____	____	____
m) Is a daily issue/return green logbook on hand?	____	____	____
n) Is the property book properly maintained to account for all authorized explosive training aids?	____	____	____
o) Is an *SF 702 (Security Container Check Sheet)* posted on the safe?	____	____	____
p) Are *SFs 702* maintained for 30 days?	____	____	____
q) Is a controlled-substance accountability folder on hand?	____	____	____
r) Is the controlled-substance accountability folder properly maintained?	____	____	____
10. EXPLOSIVES.			
a) Have primary and alternate explosive custodians been appointed, in writing, by the PM?	____	____	____
b) Have primary and alternate custodians been cleared by a favorable Crime Records Center name check?	____	____	____
c) Are records of the Crime Records Center checks on file?	____	____	____
d) Are all personnel authorized to receive and handle explosive training aids designated, in writing, by PM?	____	____	____
e) Is annual explosive safety training being conducted by EOD personnel or other qualified personnel?	____	____	____

Table C-1. Sample Inspection Memorandum (Continued)

OFFICE SYMBOL SUBJECT: MACOM Technical Assistance Visit or Provost Marshal Inspection		YES	N/A	NO
f)	Is a written record of the safety training being maintained on file for a minimum of two years?	____	____	____
g)	Is there an SOP that outlines safety procedures in training areas, transportation of explosives, severe weather conditions, and accidents?	____	____	____
h)	Are the monthly/quarterly inventories being conducted by a disinterested E7 or above?	____	____	____
i)	Are the monthly/quarterly inventories maintained on file for at least two years?	____	____	____
j)	Are all required explosive training aids available to ensure that proficiency standards are met?	____	____	____
k)	Are chlorates being replaced quarterly?	____	____	____
l)	Is each type of explosive being stored in a separate vapor proof container?	____	____	____
m)	Is each vapor proof container properly marked?	____	____	____
n)	Has the vehicle that is used to transport explosive training aids been approved by safety personnel?	____	____	____
o)	Does the installation have an explosive detector training plan that has been approved by the installation commander which lists buildings and facilities authorized for training the PEDD teams?	____	____	____
p)	Is a daily issue/return green logbook on hand?	____	____	____
q)	Is the logbook properly maintained to account for all authorized explosive training aids?	____	____	____
r)	Is a *DA Form 7281-R (Command Oriented Arms, Ammunition, & Explosives Security Screening and Evaluation Record)* on hand for each handler that is authorized to handle or receive explosive training aids?	____	____	____
11.	**TRAINING.**			
a)	Does the KM (or handlers) give training to military police on MWD team functions and operational considerations?	____	____	____
b)	Are the MWDs receiving a minimum of 4 hours of patrol and 4 hours of narcotic/explosives proficiency training each week?	____	____	____
c)	Have all certified narcotic/explosive detector teams received quarterly validation testing by the KM?	____	____	____
d)	Is the minimum proficiency standard for assigned patrol/narcotic detector dogs 90 percent and patrol/explosive detector dogs 95 percent?	____	____	____
e)	Was the dog decertified when his proficiency standard was less than 90 percent and less than 95 percent for more than three consecutive months?	____	____	____

Table C-1. Sample Inspection Memorandum (Continued)

OFFICE SYMBOL
SUBJECT: MACOM Technical Assistance Visit or Provost Marshal Inspection

	YES	N/A	NO
12. RECORDS.			
a) Is the handler's information being recorded on the back of *DD Form 1834*?	____	____	____
b) Have all *DA Forms 2807-R* been completed properly?	____	____	____
c) Have all deficiencies for the MWDs been annotated and explained on DA Forms 2807-R?	____	____	____
d) Is a *DA 3992-R* being maintained to record the MWD's reliability?	____	____	____
e) Does each MWD team have a probable cause folder?	____	____	____

13. THE RESULTS OF THIS INSPECTION ARE—

____EXCELLENT

____SATISFACTORY

____UNSATISFACTORY

14. REMARKS. All areas noted with a deficiency will be annotated in the remarks section. A copy of this technical assistance visit will be provided to OPMG, the MACOM program manager, and the KM. All records of technical assistance visits will be maintained on file at the installation for a minimum of two years after the inspection date with the corrective action memorandum.

15. POINT OF CONTACT. The point of contact for this action is the undersigned at COMM: 555-555-1511 or e-mail address.

(Original Signed)
JEFFREY S. CAPTAIN
LTC, USA
MACOM PM

Appendix D
Training Records

This appendix provides samples and details of how to properly use MWD training records. Records must be maintained for the life of the MWD. At least one year of historical documents must be maintained on file at each kennel for every MWD. Records can be discarded two years after the death, adoption, or euthanasia of the MWD. Nearly all MWDs have a dual purpose (PD and PEDD or PNDD); therefore, every MWD should have a PD and a PEDD/PNDD form on file. Kennels that have internet access can obtain records via the Military Working Dog Management System (MWDMS).

DEPARTMENT OF THE ARMY *FORM 2807-R*

D-1. *DA Form 2807-R (Military Working Dog Training and Utilization Record)* (*Figure D-1*, page D-3) is a standard form used by all MWD personnel. It is used to record the training and utilization of PDs. Use the following instructions when preparing a *DA Form 2807-R*:

ADMINISTRATIVE DATA

D-2. Enter the following information in the appropriate blocks:

- The date.
- The MWD's name and type, tattoo number, and age.
- The handler's name and grade.
- The organization and the location.

TRAINING

D-3. The training section of the form provides complete information on all of the training tasks that must be accomplished for a PD to maintain proficiency. It is not necessary to train on all of the controlled-aggression tasks during the same day; however, all of the controlled-aggression tasks must be trained every week.

D-4. Entries for the amount of time spent training on specific tasks during a particular day of the month are made in minutes. The monthly total for each specific task will be in hours and will be automatically entered in the far right-hand column titled "Total Hours." The handler's or trainer's evaluation of the MWD's daily rating is entered as S or U. Remarks are mandatory for any area the MWD fails to perform to standard. These remarks will be annotated on the reverse side of the form using the following format:

- Deficiency (DF). The MWD's shortcoming.
- Corrective action (CA). The handler's response to the MWD's deficiency.
- Corrective response (CR). The MWD's response to the CA.

D-5. Enter the appropriate information on—

- **Line 1.** Record the number of minutes per day that the MWD is trained on obedience on leash.
- **Line 2.** Record the number of minutes per day that the MWD is trained on obedience off leash.
- **Line 3.** Record the number of minutes per day that the MWD is trained on the obedience course.

- **Line 4.** Record the daily rating of the MWD on controlled aggression (either S or U). Write an explanation of any U rating on the reverse side of the form. In 4a to 4e, record the number of minutes per day that the MWD trained on each controlled-aggression task.
- **Line 5.** Record the number of minutes per day that the MWD trained on building searches on line 5.
- **Line 6.** Record the number of minutes per day that the MWD trained on gunfire with the handler firing the weapon. The number of rounds will be annotated on the reverse side of the form.
- **Line 7.** Record the number of minutes per day that the MWD is trained on gunfire with a third party firing a weapon. The number of rounds will be annotated on the reverse side of the form.
- **Line 8.** Record the number of minutes per day that the MWD is trained on scouting/patrolling. Record the distance (in feet) for scent, sight, and sound detection on lines 8a, b, or c.
- **Line 9.** Record the number of minutes per day that the MWD is trained on vehicle patrol.
- **Line 10.** Record the number of minutes per day that the MWD is trained on tracking.
- **Line 11.** Record the daily training rating (S or U). Explain deficiencies on the reveres side of the form. Any failure of a critical task requires a U daily training rating. See *DA Pam 190-12* for critical tasks.

UTILIZATION

D-6. The utilization section of the form provides a daily record of the time spent performing military police duties in the three general categories of combat support operations, patrol–law enforcement, and patrol–security.

D-7. The combat support operations category may be used to record time spent performing military police missions (actual or training) in support of combat units. Such operations include field training exercises, CP exercises, mobilization exercises, and other activities related to the tactical and strategic missions of the Army, provided that these missions are accomplished with a handler and an MWD together as a certified team.

D-8. The patrol–law enforcement category may be used to record time spent performing law enforcement or force protection type missions. These missions include but are not limited to—

- Random vehicle searches.
- Law enforcement patrols.
- Other duties performed while working in a law enforcement type mission.

D-9. The patrol–security category may be used to record time spent performing security type missions such as—

- Dignitary security.
- Critical asset patrolling in a peacetime environment.
- USSS, Department of State, or other explosive type searches.

D-10. The daily rating of the MWD's performance on military police duties is recorded as S or U. All U performance annotations will have an explanation for the DF and have the CA taken recorded on the reverse side of the form and continuation sheets, if necessary. The total monthly utilization hours are recorded in the last column, titled "Total Hours."

D-11. When training is conducted during the actual use and employment of an MWD, the time used for training may also be reported in the training section. This training is not an automatic entry every time the team has utilization hours. Time annotated as training during utilization hours will be actual time training on patrol or detector tasks. It is unethical to count all or even half of an MWD team's utilization hours as both utilization and training time for every shift, and unrealistic numbers will not be accepted. When time is recorded as utilization and training, a notation is made on the reverse side of the form explaining the double time entry and the type of training conducted. This will make it possible to differentiate between training time, training time while on duty, and duty (utilization) time. Although additional training during utilization hours is acceptable, it will not be accepted as the primary amount of mandatory training hours each week.

MILITARY WORKING DOG TRAINING AND UTILIZATION RECORD

For use of this form, see AR 190-12; proponent agency is ODCSOPS

NAME/TYPE OF DOG	TATTOO/NUMBER	AGE	NAME OF HANDLER	GRADE	ORGANIZATION AND LOCATION	MONTH AND YEAR
NATZ/PEDD	E309	2	BOOKSTEIN, DAVID B.	SGT	HHC 98TH ASG, BAMBERG, GE	SEP 2003

DAILY RATINGS: S - SATISFACTORY, U - UNSATISFACTORY (Explain deficiency and corrective action on reverse)

(Use reverse side of this form for any remarks or notes)

TRAINING	1	2	3	4	5	6	7	8	9	10	11	12	13	14	15	16	17	18	19	20	21	22	23	24	25	26	27	28	29	30	31	TOTAL HOURS
1. ON LEASH OBEDIENCE	15	15	15	15	30		30	40	10	15			20	15		15	20	30	15		15	15	15	15					15	15		6.50
2. OFF LEASH OBEDIENCE	20	20	15	15	40		30	40	10	15			20	30		15	30	30	15		15	20	20	20					15	15		7.50
3. OBEDIENCE COURSE					20		20			15																						0.92
4. CONTROLLED AGGRESSION (S OR U)		S	S					S	S				S	S																		
a. FALSE RUN		5	5					15	5				5	5																		0.67
b. ATTACK		10	10					20	10				10	10																		1.17
c. SEARCH AND ATTACK		5	5					15	10				10	10																		0.92
d. STAND OFF		10	15					20	5				10	10																		1.17
e. ESCORT		5	5					15	5				5	5																		0.67
5. BUILDING SEARCH		15						15																20								0.83
6. GUNFIRE - HANDLER								10																								0.17
7. GUNFIRE - DECOY/AGITATOR					10														30													0.67
8. SCOUTING/PATROLLING (TIME)					15																			15					15			0.75
a. SCENT DETECTION (DISTANCE)					40																			30					30			
b. SIGHT DETECTION (DISTANCE)																																
c. SOUND DETECTION (DISTANCE)																																
9. VEHICLE PATROL																																0.00
10. TRACKING																																0.00
11. DAILY TRAINING RATING (S OR U)	S	S	S	S	S		S	S	S	S						S	S	S	S		S	S	S	S					S	S		21.92

UTILIZATION (HOURS PER DAY)	1	2	3	4	5	6	7	8	9	10	11	12	13	14	15	16	17	18	19	20	21	22	23	24	25	26	27	28	29	30	31	TOTAL HOURS
1. COMBAT SUPPORT OPERATIONS																																0
2. PATROL - LAW ENFORCEMENT	8	8	8	8	8			8	8	8			8	8	8	8				8	8	8	8						8	8		144
3. PATROL - SECURITY																																0
4. DAILY UTILIZATION RATING (S OR U)	S	S	S	S	S			S	S	S			S	S	S	S				S	S	S	S						S	S		144.0

DAILY FEEDING (QUANTITY OF FOOD (BY WEIGHT))	1	2	3	4	5	6	7	8	9	10	11	12	13	14	15	16	17	18	19	20	21	22	23	24	25	26	27	28	29	30	31	WT OF DOG: DATE	LBS
2 CU S/D	A	A	A	A	A	A	A	A	A	A	A	A	A	A	A	A	A	A	A	A	A	A	A	A	A	A	A	A	A	A	A	1ST	75
2 CU S/D	A	A	A	A	A	A	A	A	A	A	A	A	A	A	A	A	A	A	A	A	A	A	A	A	A	A	A	A	A	A	A	15TH	73.5

DA FORM 2807-R, Oct 84 EDITION DEC 72 IS OBSOLETE, AND REPLACES DA FORMS 2810-R AND 2815-R, DEC 72

Sample

Figure D-1. Sample DA Form 2807-R

DA Form 2807-R
Continuation Sheet

DATE	REMARKS
2-Sep-03	**DF1:** MWD WAS DISTRACTED BY OTHER HANDLER; WOULD NOT FOCUS ON COMMANDS.
	CA1: CONDUCTED PROPER AVOIDANCE TRAINING TECHNIQUE.
	CR1: AFTER MWD DID NOT RESPOND TO INITIAL AVOIDANCE TRAINING TECHNIQUE, A PINCH COLLAR WAS USED, RESULTING IN REGAINED FOCUS.
	DF2: MWD BIT SLEEVE ON STAND OFF.
	CA2: CONDUCTED PROPER AVOIDANCE TRAINING TECHNIQUE.
	CR2: MWD PERFORMED STAND OFF CORRECTLY.
5-Sep-03	GUNFIRE TRAINING: 5 ROUNDS FIRED BY THIRD PARTY.
	DF1: MWD REPEATEDLY BROKE POSITION (SIT TO DOWN AND DOWN TO SIT).
	CA1: CONDUCTED PROPER AVOIDANCE TRAINING TECHNIQUE.
	CR1: MWD REMAINED IN POSITION.
	DF2: MWD WAS HESITANT TO OUT REWARD (BITE SLEEVE).
	CA2: CONDUCTED PROPER AVOIDANCE TRAINING TECHNIQUE.
	CR: MWD OUTED REWARD ON COMMAND.
7-Sep-03	MWD HAS BEEN BREAKING POSITION AFTER SHORT TIME DURING OFF/L OB WHEN HANDLER IS POSITIONED AWAY FROM HIM. TODAY MWD REMAINED IN THE SIT FOR 1 MIN 50 SEC, BROKE SIT, WAS VERBALLY CORRECTED AND REMAINED IN SIT FOR AN ADDITIONAL 25 SEC.
	DF: MWD BROKE FROM DOWN SEVERAL TIMES DURING PARADE REST.
	CA: CONDUCTED PROPER AVOIDANCE TRAINING TECHNIQUE.
8-Sep-03	GUNFIRE TRAINING: 10 ROUNDS FIRED BY HANDLER.
	CR: AFTER MWD DID NOT RESPOND TO AVOIDANCE TRAINING, ESCAPE TRAINING WAS CONDUCTED. EVENTUALLY MWD REMAINED IN DOWN. MWD IS SO EXCITABLE THAT HE HAS A TENDENCY TO CUE OFF HANDLER'S MOVEMENTS TO BREAK FROM POSITION. HANDLER IS CONCENTRATING ON TEACHING MWD NOT TO BE SO DISTRACTED BY OUTSIDE STIMULUS.
	DF1: MWD WAS GETTING DISTRACTED AND NOT HEELING IN THE PROPER POSITION.
	CA1: CONDUCTED PROPER AVOIDANCE TRAINING TECHNIQUE (HEEL-A-WAYS).
	CR1: MWD HEELS CORRECTLY.
	DF2: MWD WAS BITING SLEEVE DURING STAND OFF.
11-Sep-03	**CA2:** CONDUCTED PROPER AVOIDANCE TRAINING TECHNIQUE.
	CR2: MWD PERFORMED STAND OFF CORRECTLY.
	NO TRAINING DUE TO MWD GOING TO VET. MWD HAS BEEN DIAGNOSED WITH A POSSIBLE PULLED/STRAINED MUSCLE IN HIS RIGHT LEG, WHICH HE IS FAVORING AS HE WALKS. X-RAYS SHOW NO BREAKS OR FRACTURES. MWD CANNOT DO ANYTHING STRENUOUS FOR THE NEXT WEEK (NO RUNNING OR JUMPING UNTIL 18 SEP) INCLUDING BITE WORK OR OB COURSE. MWD CAN DO BASIC OB STARTING TUES, SEP 16.
19-Sep-03	GUNFIRE TRAINING: 15 ROUNDS FIRED BY DECOY.

Figure D-1. Sample DA Form 2807-R (Continued)

D-12. Record the appropriate utilization information on—

- **Line 1.** Record the hours per day that the MWD is used for combat support operations.
- **Line 2.** Record the hours per day that the MWD is used for patrol–law enforcement.
- **Line 3.** Record the hours per day that the MWD is used for patrol–security.
- **Line 4.** Record the daily utilization rating (S or U). Explain any U on the reverse side of the form.

Total Hours

D-13. Record the total hours of training and use in the last column on the right side of the form. The time should be recorded in hours and minutes. The MWDMS makes this entry automatically.

DAILY FEEDING

D-14. Record the quantity of food that the MWD eats every day in this section. If the MWD is fed once daily, the type and amount of food will be noted in the first block to the left. Under quantity of food, the amount eaten for each feeding will be annotated as follows:

- "A" if all food was eaten.
- "3/4" if three-quarters was eaten.
- "1/2" if half was eaten.
- "1/4" if one-quarter was eaten.
- "0" if none of the food was eaten.

D-15. If the MWD is fed twice daily, the second block is used to record the second daily feeding (which could be a different type of food). If the MWD is fed once daily, the second block may be used to record a change in the diet of the MWD, directed by the veterinarian. Block 2 may also be used to indicate the type and amount of food used daily for MWDs on a food reward system. Monitoring MWD food consumption is a critical portion of the required daily health checks.

D-16. The last column, titled "Wt of Dog," is used to record a semimonthly weight for the MWD. The weight checks occur on or around the first and the fifteenth day of each month. There is sufficient room under the headings "Date" and "Lbs" to record both the date when the MWD is weighed and the weight in pounds. Any drastic changes in food consumption or weight gain or loss are reported to the veterinarian.

D-17. Enter the following information in the appropriate blocks:

- Enter the VCO's prescribed amount of food to be given daily directly under "Daily Feeding." Both blocks are used when an MWD is fed twice daily.
- Record the amount of food eaten per day, in the appropriate day's block under "Quantity of Food (by Weight)."
- Record the date and weight of the MWD in the respective blocks in the bottom right-hand corner.

REMARKS

D-18. The reverse side of *DA Form 2807-R* (*Figure D-1*) is used to annotate advanced techniques used during training and DFs with CAs and CRs listed. This section is very important for probable cause and shows the time line of events for training. Each day is listed with the action taken on that day. If no training is conducted due to company type training, the words "soldier development" are written on the line followed by the type of training conducted. "Soldier development" is only used when MWD handlers are conducting training in which the MWD is not needed or not taken. If no training is conducted due to leave, passes, or scheduled days off, no entry is needed; *DA Form 6 (Duty Roster)* can be used to verify these days.

DEPARTMENT OF THE ARMY *FORM 3992-R* (FIRST PAGE)

D-19. *DA Form 3992-R (Narcotics or Explosives Detector Dog Training and Utilization Record)* (*Figure D-2*, page D-9) is a standard form used by all MWD personnel. It is used to record the use and training of detector dogs. Use the following instructions when preparing a *DA Form 3992-R:*

ADMINISTRATIVE DATA

D-20. Enter the following information in the appropriate blocks:

- The date.
- The MWD's name and type, tattoo number, and age.
- The handler's name and grade.
- The organization and location.

TRAINING

D-21. The training section of the form is used for recording all training of the detector dog team in the particular detection skill. The detection areas of buildings, containers, and vehicles are listed. There is one blank space for detector training in other areas. Examples of other areas may include—

- Mail facilities.
- Open or wooded areas.
- Warehouses.
- Aircraft.
- Areas considered significant or unique to the command to which the MWD team is assigned.

D-22. Although some of the examples may be included under one of the three prelisted categories, the others should be listed separately. For areas covered under the three general headings, the handler or trainer annotates the specific location under the appropriate day in the remarks (for example, Smith High School, NCO Club, or Corps Auditorium).

D-23. The top line of each training category is divided diagonally into two blocks. The top of each block is used to record the number of training aids planted for that day's detector training. The bottom of each block is used for the number of training aids found by the proper response of the detector dog. If the MWD responds on the same training aid more than once in the same training scenario, it will not be counted as a find unless the number of plants is changed as well. At no time will there be more finds annotated than there are plants. Continually allowing the MWD to return to the last aid found creates false response problems in the MWD and should not be encouraged. It is the handler's responsibility to keep the MWD from returning to the scent cone of the last found training aid, whether the MWD is being worked on or off leash. Once the MWD has entered the scent cone, he must be allowed to work to the source and final respond. At no time will an MWD be yanked off an aid once he is working the scent cone.

D-24. The bottom line of each training category is used to record the search time. Search time is the amount of time devoted to detector training that day. Detector training is the time from the moment the handler enters a problem and commands the MWD to "Seek" until the MWD has completed the search of that training problem, excluding handler or MWD breaks. Add the daily time entries in each training category, and enter the results in the last column, titled "Total Hours."

D-25. Record the following information beside the appropriate training location (building, containers, vehicles, or other):

- **Plants/Finds.** Record the number of plants in the top half of the block and the number of finds in the lower half.
- **Search Time.** Record the amount of time devoted for detector training that day.
- **Training Total Hours.** Record the sum of hours performing detection missions in the far right

UTILIZATION

D-26. All operational detection missions are recorded in the utilization section. The detection missions are listed. Detection missions that are significantly different from these are listed separately to give an accurate record of the types of missions for which the detector dog team is used. There are two blocks under each day of the month for each utilization category. The top block is for recording the number of times the MWD detected a substance. The bottom block is for recording the total search or inspection time for the detection mission. The last column is the sum of the daily times. It gives a record of the total amount of time spent performing the detection missions for that month. The number of finds for the month can also be totaled and entered in the last column, above the time entry.

D-27. Record the following information beside the appropriate location (building, containers, vehicles, or other):

- **Finds.** Record the number of times the MWD finds the substance on a detection mission.
- **Search Time.** Record the total search time.

Detector Dog Search Data

D-28. All relevant information about the productivity and success of each detection mission is recorded using the "Detector Dog Search Data" section. The location may be a building number, a unit designation, map grid coordinates, and/or any other information that helps identify where the detection mission was performed.

D-29. The MPR number is the number assigned to the case by military police to account for the custody and disposition of the substances found. The substance is the identification of the found material by a common name, such as heroin, marijuana, dynamite, or detonating cord.

D-30. Quantity is the measure of the amount of the substance found. Weight, volume, overall dimensions, length, or any other appropriate measure may be used.

D-31. The "Remarks" section may be used for adding any other relevant information about the substance found. This would include a field or laboratory verification of the type of substance, an EOD evaluation of an explosive device or explosive substance as live or inert, or the presence of other hazardous material in or around the substance found (such as razor blades, trip wires, or poisons). Record any information that may be useful in preparing the team for future searches or that may be applicable or useful to other detector dog teams.

D-32. Record the following information in the appropriate boxes:

- **Time.** Record the time the detector dog performed the mission.
- **Date.** Record the date the detector dog performed the mission.
- **Location.** Record the exact location of the detector mission (such as front gate, Fort Sill, Oklahoma; National Airport, Gabon, Africa; or building 442, Fort Bragg, North Carolina).
- **Military Police Report (MPR) Number.** Record the MPR number in this column, if applicable.
- **Substance.** Record the common name of the substance found, if applicable.
- **Quantity.** Record the quantity of the substance found, if applicable.
- **Remarks.** Use this column to record any other relevant information about the substance found (such as a positive response, no response, or a lab verification).

Detector Dog Proficiency (Second Page)

D-33. A correct response occurs when the MWD detects the substance and responds with the proper final response, and the training aid or substance is found in the general location where the MWD has responded. A not-at-source response is a training tool used to help an MWD team move closer to the source location of a training aid plant. The distance away from the source in order for a not-at-source response to be given is six to ten feet.

D-34. A missed response is when the MWD fails to detect and respond to the presence of a training aid with a final response. A handler miss or error is when—

- The handler fails to search or clear an area with the MWD and a training aid plant is missed.
- The handler fails to read an MWD's change of behavior and pulls the MWD away from the area where a training aid is planted and a training aid plant is missed.
- The handler presents an area but fails to ensure that the MWD is sniffing and a training aid plant is missed.

Handler misses or errors are considered missed explosives or narcotics and are counted against the find rate percentages of the MWD team.

D-35. False responses are noted in the second to last section of *DA Form 3992-R*. A false response is when the MWD responds as if he has detected a substance and the handler accepts that response, but no training aid or substance can be found within a reasonable distance of the MWD's final response.

D-36. The detector dog team's proficiency is computed monthly by adding the total number of correct responses ("a" for the month) and adding the total number of false/missed (MWD or handler) responses ("b" for the month). These two numbers are used in the following formula to obtain the detector dog team's proficiency rating:

$$\left(\frac{a}{a+b}\right)100 = proficiency$$

D-37. For example, during the month the detector dog team made 93 correct responses on training aids, the MWD team also had 4 false responses, 2 MWD and 1 handler missed responses for a total of 7 false/missed responses. Applying these two numbers (a=93 and b=7) to the formula, the following is obtained:

$$\left(\frac{93}{93+7}\right)100 = 93\%$$

D-38. The computation above shows that the MWD team is working at a 93 percent proficiency rate. This rate is above the minimum for a PNDD team (90 percent), but it is below the minimum standard for a PEDD team (95 percent). A PNDD handler should continue training to maintain and possibly increase the team's proficiency. A PEDD handler needs to identify the causes of the team's substandard performance and immediately begin corrective training to bring the team up to and over the minimum 95 percent detection proficiency standard.

> ***Note:*** Responses on actual substances during actual searches are not included in this computation. Search data cannot be included because it is impossible to determine the number of false or missed responses during an actual search.

D-39. Record the following information on—

- **Line a.** Record the total correct alerts. The MWDMS makes these entries automatically.
- **Line b.** Record the total false or missed alerts. The MWDMS makes these entries automatically.

NARCOTICS OR EXPLOSIVE DETECTOR DOG TRAINING AND UTILIZATION RECORD (THIRD AND FOURTH PAGES)

D-40. The first part of section of the Narcotics or Explosive Detector Dog Training and Utilization Record section is used to record the number of each training aid used in training on each day of the month. The entry for each training aid type is divided into three rows to record the day of the month, correct alerts, and missed alerts, respectively. Record the appropriate information in the following sections.

- **Administrative Data.** Record the MWD's name and tattoo number, the month, and the year in the appropriate blocks.
- **Aid Type.** Enter the specific aid type in the upper left-hand corner of each "Detector Dog Proficiency" section.
- **Total Correct Alerts.** Record the number of correct alerts the MWD made during training.
- **Missed Alerts.** Record the number of missed and false alerts the MWD made during training.
- **Proficiency.** Record the actual proficiency in this block. The MWDMS makes this entry automatically.

REMARKS

D-41. Remarks for *DA Form 3992-R* are recorded on a continuation sheet. Each DF (for example, a missed aid or false sit) is listed by the date it occurred. The CA and CR are also listed. If more than one DF is noted per day, number the DF, CA, and CR series to avoid confusion.

NARCOTICS OR EXPLOSIVES DETECTOR DOG TRAINING AND UTILIZATION RECORD

For use of this form see AR 190-12. The proponent is DCSPER.

NAME/TYPE OF DOG	TATTOO NUMBER	AGE	NAME OF HANDLER	GRADE	ORGANIZATION AND LOCATION	MONTH AND YEAR
NATZ/PEDD	E309	2	BOOKSTEIN, DAVID B.	SGT	HHC 98TH ASG, BAMBERG, GE	SEP 2003

TRAINING		1	2	3	4	5	6	7	8	9	10	11	12	13	14	15	16	17	18	19	20	21	22	23	24	25	26	27	28	29	30	31	TRAINING TOTAL HOURS 24.25
BUILDINGS	PLANTS / FINDS			7 / 7	8 / 5	4 / 4				4 / 4	6 / 6						11 / 10		5 / 4						4 / 4	8 / 8	6 / 6				7 / 7		70 / 65
	SEARCH TIME			120	80	60				60	60						130		70						60	120	120				120		16.67 HRS
CONTAINERS	PLANTS / FINDS									19 / 19	57 / 57						19 / 19		19 / 19														114 / 114
	SEARCH TIME									45	120						45		45														4.25 HRS
VEHICLES	PLANTS / FINDS					4 / 4													5 / 5														9 / 9
	SEARCH TIME					60													80														2.33 HRS
OTHER	PLANTS / FINDS									4 / 4																							4 / 4
	SEARCH TIME									60																							1.00 HRS

UTILIZATION		1	2	3	4	5	6	7	8	9	10	11	12	13	14	15	16	17	18	19	20	21	22	23	24	25	26	27	28	29	30	31	UTILIZATION TOTAL HOURS 0.00
BUILDINGS	FINDS																																0
	SEARCH TIME																																0.00 HRS
CONTAINERS	FINDS																																0
	SEARCH TIME																																0.00 HRS
VEHICLES	FINDS																																0
	SEARCH TIME																																0.00 HRS
OTHER	FINDS																																0
	SEARCH TIME																																0.00 HRS

DA FORM 3992-R, Oct 84 EDITION OF 1 DEC 72 IS OBSOLETE

Figure D-2. Sample DA Form 3992-R

DETECTOR DOG SEARCH DATA

NO.	TIME	DATE	LOCATION	MPR NUMBER	SUBSTANCE	QUANTITY	REMARKS
1							
2							
3							
4							
5							
6							
7							
8							
9							
10							
11							
12							
13							
14							
15							
16							
17							
18							
19							
20							
21							
22							
23							
24							
25							
26							
27							
28							
29							
30							
31							
32							
33							

DETECTOR DOG PROFICIENCY
(Alerts on training aids during training and actual searches)

DAY OF MONTH	1	2	3	4	5	6	7	8	9	10	11	12	13	14	15	16	17	18	19	20	21	22	23	24	25	26	27	28	29	30	31	PROFICIENCY = a / (a + b) X 100	
a. TOTAL CORRECT ALERTS	0	0	7	5	8	0	0	0	26	63	0	0	0	0	0	29	0	28	0	0	0	0	0	4	8	6	0	0	0	7	0	191	
b. TOTAL FALSE/MISSED ALERTS	0	0	0	3	0	0	0	0	1	0	0	0	0	0	0	3	0	1	0	0	0	0	0	0	0	0	0	0	0	0	0	199	96%

DA Form 3992-R, Oct 84

Figure D-2. Sample DA Form 3992-R (Continued)

NARCOTICS OR EXPLOSIVES DETECTOR DOG TRAINING AND UTILIZATION RECORD

NATZ/PEDD **TATTOO:** E309 **MONTH:** SEP **YEAR:** 2003

DETONATING CORD — DETECTOR DOG PROFICIENCY (ALERTS ON TRAINING AIDS DURING TRAINING AND ACTUAL SEARCHES)

DAY OF MONTH	1	2	3	4	5	6	7	8	9	10	11	12	13	14	15	16	17	18	19	20	21	22	23	24	25	26	27	28	29	30	31	PROFICIENCY A / A + B X 100
a. TOTAL CORRECT ALERTS			2	1						20						1									3							27
b. TOTAL MISSED ALERTS				1																												28 96%

C-4 PLASTIC EXPLOSIVE — DETECTOR DOG PROFICIENCY (ALERTS ON TRAINING AIDS DURING TRAINING AND ACTUAL SEARCHES)

DAY OF MONTH	1	2	3	4	5	6	7	8	9	10	11	12	13	14	15	16	17	18	19	20	21	22	23	24	25	26	27	28	29	30	31	PROFICIENCY A / A + B X 100
a. TOTAL CORRECT ALERTS				1					22	1						3		2								1						30
b. TOTAL MISSED ALERTS				1					1																							32 94%

POTASSIUM CHLORATE — DETECTOR DOG PROFICIENCY (ALERTS ON TRAINING AIDS DURING TRAINING AND ACTUAL SEARCHES)

DAY OF MONTH	1	2	3	4	5	6	7	8	9	10	11	12	13	14	15	16	17	18	19	20	21	22	23	24	25	26	27	28	29	30	31	PROFICIENCY A / A + B X 100
a. TOTAL CORRECT ALERTS									1							20		2						2	1							26
b. TOTAL MISSED ALERTS																1																27 96%

SODIUM CHLORATE — DETECTOR DOG PROFICIENCY (ALERTS ON TRAINING AIDS DURING TRAINING AND ACTUAL SEARCHES)

DAY OF MONTH	1	2	3	4	5	6	7	8	9	10	11	12	13	14	15	16	17	18	19	20	21	22	23	24	25	26	27	28	29	30	31	PROFICIENCY A / A + B X 100
a. TOTAL CORRECT ALERTS					1				2									21						1	1							26
b. TOTAL MISSED ALERTS																		1														27 96%

WATER GEL (TOVEX) — DETECTOR DOG PROFICIENCY (ALERTS ON TRAINING AIDS DURING TRAINING AND ACTUAL SEARCHES)

DAY OF MONTH	1	2	3	4	5	6	7	8	9	10	11	12	13	14	15	16	17	18	19	20	21	22	23	24	25	26	27	28	29	30	31	PROFICIENCY A / A + B X 100
a. TOTAL CORRECT ALERTS				1						21						1		1								2				2		28
b. TOTAL MISSED ALERTS				1																												29 97%

DA FORM 3992-R, OCT 84 (CONTINUATION SHEET)

Figure D-2. Sample DA Form 3992-R (Continued)

NARCOTICS OR EXPLOSIVES DETECTOR DOG TRAINING AND UTILIZATION RECORD

NATZ/PEDD TATTOO: E309 MONTH: SEP YEAR: 2003

SEMTEX	DETECTOR DOG PROFICIENCY (ALERTS ON TRAINING AIDS DURING TRAINING AND ACTUAL SEARCHES)																															PROFICIENCY A/(A + B) X 100
DAY OF MONTH	1	2	3	4	5	6	7	8	9	10	11	12	13	14	15	16	17	18	19	20	21	22	23	24	25	26	27	28	29	30	31	A + B
a. TOTAL CORRECT ALERTS			1		2				1							2		2							1					2		11
b. TOTAL MISSED ALERTS																																11 100%
AMMONIUM NITRATE	DETECTOR DOG PROFICIENCY (ALERTS ON TRAINING AIDS DURING TRAINING AND ACTUAL SEARCHES)																															PROFICIENCY A/(A + B) X 100
DAY OF MONTH	1	2	3	4	5	6	7	8	9	10	11	12	13	14	15	16	17	18	19	20	21	22	23	24	25	26	27	28	29	30	31	A + B
a. TOTAL CORRECT ALERTS																																0
b. TOTAL MISSED ALERTS																																0 #DIV/0!
NITROGLYCERIN DYNAMITE	DETECTOR DOG PROFICIENCY (ALERTS ON TRAINING AIDS DURING TRAINING AND ACTUAL SEARCHES)																															PROFICIENCY A/(A + B) X 100
DAY OF MONTH	1	2	3	4	5	6	7	8	9	10	11	12	13	14	15	16	17	18	19	20	21	22	23	24	25	26	27	28	29	30	31	A + B
a. TOTAL CORRECT ALERTS				2						21															1							24
b. TOTAL MISSED ALERTS																																24 100%
TNT (Trinitrotoluene)	DETECTOR DOG PROFICIENCY (ALERTS ON TRAINING AIDS DURING TRAINING AND ACTUAL SEARCHES)																															PROFICIENCY A/(A + B) X 100
DAY OF MONTH	1	2	3	4	5	6	7	8	9	10	11	12	13	14	15	16	17	18	19	20	21	22	23	24	25	26	27	28	29	30	31	A + B
a. TOTAL CORRECT ALERTS			3		3											2										1				2		11
b. TOTAL MISSED ALERTS																																11 100%
SMOKELESS POWDER	DETECTOR DOG PROFICIENCY (ALERTS ON TRAINING AIDS DURING TRAINING AND ACTUAL SEARCHES)																															PROFICIENCY A/(A + B) X 100
DAY OF MONTH	1	2	3	4	5	6	7	8	9	10	11	12	13	14	15	16	17	18	19	20	21	22	23	24	25	26	27	28	29	30	31	A + B
a. TOTAL CORRECT ALERTS			1		2																			1	1	2				1		8
b. TOTAL MISSED ALERTS																																8 100%
Totals from TNG Aids (Cont. Sheet)																																
TOTAL PLANTS	0	0	7	8	8	0	0	0	27	63	0	0	0	0	0	30	0	29	0	0	0	0	0	4	8	6	0	0	0	7	0	197
TOTAL FINDS	0	0	7	5	8	0	0	0	26	63	0	0	0	0	0	29	0	28	0	0	0	0	0	4	8	6	0	0	0	7	0	191 97%
a. TOTAL FALSE ALERTS																2																2 1%
Totals from TNG locations (3992 Front)																																
TOTAL PLANTS	0	0	7	8	8	0	0	0	27	63	0	0	0	0	0	30	0	29	0	0	0	0	0	4	8	6	0	0	0	7	0	197
TOTAL FINDS	0	0	7	5	8	0	0	0	27	63	0	0	0	0	0	29	0	28	0	0	0	0	0	4	8	6	0	0	0	7	0	192 97%

DA FORM 3992-R, OCT 84 (CONTINUATION SHEET)

Sample

Figure D-2. Sample DA Form 3992-R (Continued)

DA Form 3992-R
Continuation Sheet

DATE	REMARKS
3-Sep-03	TRAINING AT SCHWEINFURT BOWLING ALLEY. (TRAINING AIDS PLANTED FROM 6 FT H x 5 FT D)
4-Sep-03	TRAINING AT SCHWEINFURT FMO. (TRAINING AIDS PLANTED FROM 8 FT H x 3 FT D)
	DF1: MWD MISSED AID (1 STRAND DET CORD - 1 FT H x 1 FT D) DUE TO HANDLER ERROR.
	CA1: HANDLER TOLD TO WORK A REVERSE PATTERN UNTIL THE TEAM REACHED THE AID LOCATION.
	CR1: MWD FOUND AID AFTER HANDLER MADE PROPER PRESENTATION.
	DF2: MWD MISSED AID. (1/4 BLOCK C4 2 FT H x 1 FT D)
	CA2: MWD BROUGHT BACK FOR RE-PRESENTATION.
	CR2: MWD FOUND AID AFTER HANDLER MADE SECOND PRESENTATION.
	DF3: MWD MISSED AID. (1 STICK NITRO DYNAMITE- 7 FT H x 1 FT D)
	CA3: MWD BROUGHT BACK FOR RE-PRESENTATION.
	CR3: MWD FOUND AID AFTER HANDLER MADE HIGH PRESENTATION AND DRAGGED HAND UPWARD.
	DF4: MWD MISSED AID. (1 STICK WATER GEL - 5 FT H x 1 FT D)
	CA4: MWD BROUGHT BACK FOR RE-PRESENTATION.
	CR4: MWD FOUND AID AFTER HANDLER MADE SECOND PRESENTATION.
	MWD WILL BE PROTOCOLED ON ALL MISSED AIDS AT A LATER DATE DUE TO TIME CONSTRAINTS.
5-Sep-03	TRAINING AT 630TH MILITARY POLICE COMPANY MOTOR POOL. (TRAINING AIDS PLANTED FROM 3 FT H x 7 FT D)
9-Sep-03	TRAINING AT SCHWEINFURT OPEN AREA BY CONN CLUB & OUTDOOR REC. (TRAINING AIDS PLANTED FROM 7 FT H x 4 FT D)
	CA: (REFERENCE DF2, 4 SEP) MWD WAS PROTOCOLED ON 1/2 BLOCK C4.
	CR: COMPLETED 4 HV ASSISTED FOR 4 TRIALS AND 15 CONSECUTIVE UNASSISTED TRIALS.
10-Sep-03	TRAINING AT ANSBACH CMFO WAREHOUSES (TRAINING AIDS PLANTED FROM 4 FT H x 1 FT D).
	CA: (REFERENCE DF2-4, 4 SEP) MWD WAS PROTOCOLED ON 1 STRAND DC, 1 STICK ND, AND 1 STICK WG.
	CR: COMPLETED 4 HV ASSISTED FOR 4 TRIALS AND 15 CONSECUTIVE UNASSISTED TRIALS FOR EACH AID.
11-Sep-03	NO TRAINING DUE TO MWD GOING TO VET. MWD HAS BEEN DIAGNOSED WITH A POSSIBLE PULLED/STRAINED MUSCLE IN HIS RIGHT LEG WHICH HE IS FAVORING AS HE WALKS. X-RAYS SHOW NO BREAKS OR FRACTURES. MWD CANNOT DO ANYTHING STRENUOUS FOR THE NEXT WEEK, BUT IS CLEARED TO WORK DETECTION NEXT TUES (16 SEP).
16-Sep-03	TRAINING AT KITZIGEN TISA WAREHOUSE & LEIGHTON THEATER. (TRAINING AIDS PLANTED FROM 9 FT H x 6 FT D.)
	DF1: MWD MISSED AID. (PC - 1 FT H x 6 FT D)
	CA1: MWD BROUGHT BACK FOR REPRESENTATION. WILL CONDUCT PROTOCOL.
	CR1: MWD RESPONDED ON AID AFTER HANDLER MADE SECOND PRESENTATION. MWD WAS PROTOCOLED ON PC COMPLETING 4 HV ASSISTED FOR 4 TRIALS AND 15 CONSECUTIVE UNASSISTED TRIALS.
	DF2: MWD FALSE SAT IN THEATER LOBBY.
	CA2: MWD WAS ESCAPED FROM AREA AND CONTINUED TO WORK.
	CR2: MWD CONTINUED PATTERN WITHOUT FALSE SITTING.
	DF3: MWD FALSE SAT IN THEATER BACK WALL.
	CA3: MWD WAS ESCAPED FROM AREA AND CONTINUED TO WORK. MWD WILL BE WORKED ON BLANK PROBLEM.
	CR3: MWD CONTINUED PATTERN WITHOUT FALSE SITTING.
18-Sep-03	TRAINING AT SCHWEINFURT STRIP LOT & OUTDOOR REC. (TRAINING AIDS PLANTED FROM 8 FT H x 6 FT D)
	CA1: (REFERENCE DF2, 16 SEP) CONDUCTED BLANK PROBLEM.
	CR1: MWD DID NOT ATTEMPT TO FALSE SIT.
	DF2: MWD MISSED AID. (SC - 7 FT H x 3 FT D)
	CA2: MWD BROUGHT BACK FOR RE-PRESENTATION. MWD WAS PROTOCOLED ON SC.
	CR2: MWD FOUND AID AFTER HANDLER MADE SECOND PRESENTATION. COMPLETED 4 HV ASSISTED FOR 4 TRIALS AND 15 CONSECUTIVE UNASSISTED TRIALS.
24-Sep-03	TRAINING AT SCHWEINFURT FMO. (TRAINING AIDS PLANTED FROM 5 FT H x 5 FT D)
25-Sep-03	TRAINING AT THEATER. (TRAINING AIDS PLANTED FROM 2 FT H x 2 FT D)
26-Sep-03	TRAINING AT COMMISSARY. (TRAINING AIDS PLANTED FROM 9 FT H x 6 FT D)
30-Sep-03	TRAINING AT FURNITURE WAREHOUSE. (TRAINING AIDS PLANTED FROM 6 FT H x 7 FT D)

Figure D-2. Sample DA Form 3992-R (Continued)

Appendix E

Deployment Equipment List

Table E-1 is a list to be used as a general guide to aid leaders in the planning process prior to MWD deployments. Mission, enemy, terrain, troops, time available, and civilian considerations (METT-TC) should be considered when determining what equipment is necessary for mission success. Required and authorized equipment is determined by tables of organization and equipment (TOEs), modification tables of organization and equipment (MTOEs), and tables of distribution and allowances (TDAs).

Table E-1. Deployment Equipment List

	Worn or Carried	
M9 with LCE (LBV)	Identification card	Body armor
Canteen cup	Identification tag	Black gloves with liners
First aid pouch with dressing	Protective mask with a carrier	Military police brassard (subdued)
Kevlar® helmet		
	MWD Equipment	
Choke chain	Grooming brush/comb	Feed pan
Leather or nylon collar	Reward ball/Kong	Kennel crate
Stake-out chain	360-inch nylon leash	16-foot leash
Plastic or leather muzzle	Water bucket	Harness
	Alice Pack	
BDU (2 sets)	Towels (2)	Sleeping bag with a cover
Brown T-shirt (2)	Gloves, black, cold weather	Shower shoes
Undergarments (2)	BDU Cap	PT uniform (complete set)
Green and black socks (2)	Boot polish kit	MOPP gear in an NBC bag
Poncho with a liner	Flashlight	Boots
Shaving kit	Lip balm	Personal hygiene gear
Waterproof bag	Foot powder	
	A/B Bag	
Duffel bag with a padlock	Brown T-shirt (5)	White socks (3)
Cold weather boots	BDU (2)	Green and black socks (5)
Cold weather gear	Towels (3)	Personal hygiene gear
Waterproof bag	Overboots	Undergarments (5)
Military police brassard	Laundry bag (2)	
	Additional Equipment	
30-90 day supply MWD food	50-gallon trash can (MWD food)	5-gallon water can
MWD training records	Medical supplies	Engineer tape
MWD medical records	Carpenter kit/tools	Cots
MWD health certificate	Safe for narcotics aides	Fire extinguishers

Table E-1. Deployment Equipment List (Continued)

Additional Equipment		
MWD first aid kit	Explosives and narcotics training aides	Axes, shovels, rakes, brooms
Arm protector, sleeve (hidden)	Parasite preventative mediation	Cooling vest (MWD)
Soldier first aid kit	Light set	Paw pad protectors (2 sets)
Bite suit (optional)	Generator	Foot lockers (2) or blood boxes
Dog shampoo	Transformer	Equipment duffel bag
Heaters and fans	Warning signs (MWD)	Camouflage netting with poles

Appendix F
Force Protection Conditions

The FPCONs outlined below describe the progressive level of a terrorist threat to all US military facilities and personnel under *DOD Directive 2000.12*. As approved by the Joint Chiefs of Staff, the terminology and definitions in this appendix are recommended security measures designed to ease interservice coordination and to support US military antiterrorism activities. Each installation must localize these measures for its specific circumstances, accounting for service and commander requirements, local laws, and SOFAs as a minimum. See *AR 525-13* for further details concerning FPCONs. This appendix details what each condition is and the suggested actions of the MWD teams for each level and measure. Commanders are encouraged to employ MWD teams wherever an MWD team can enhance security and force protection measures that are already established.

FORCE PROTECTION CONDITION NORMAL

F-1. FPCON normal applies when a general global threat of possible terrorist activity exists. It warrants a routine security posture.

FORCE PROTECTION CONDITION ALPHA

F-2. FPCON Alpha applies when there is an increased general threat of possible terrorist activity against personnel or facilities, the nature and extent of which are unpredictable. Alpha measures must be capable of being maintained indefinitely. The following bullets explain the FPCON Alpha measures:

- **Alpha measure 1.** Inform personnel and family members of the general situation at regular intervals. Also, remind them to be alert for and report suspicious activities such as the presence of unfamiliar personnel and vehicles, suspicious/unattended parcels, and possible surveillance attempts.
- **Alpha measure 2.** Ensure that the duty officer or personnel with access to building plans as well as the plans for area evacuations are available at all times. Plans to execute access control procedures should be in place. Key personnel required to implement security plans should be on call and readily available.
- **Alpha measure 3.** Secure and randomly inspect buildings, rooms, and storage areas not in regular use.

Note: MWD should randomly inspect buildings after normal duty hours. The KM should ensure that no inspection time or location patterns are ever set by MWD teams.

- **Alpha measure 4.** Increase security spot checks of vehicles and persons entering installations under the jurisdiction of the United States.

Note: MWD should perform vehicle checks randomly at entry points.

- **Alpha measure 5.** Limit access points for vehicles and personnel commensurate with a reasonable flow of traffic.

- **Alpha measure 6.** Apply Alpha measures 4, 5, 7, or 8 either individually or in combination with each other as a deterrent.
- **Alpha measure 7.** Review intelligence, counterintelligence, and operations dissemination procedures.
- **Alpha measure 8.** Review and, if necessary, implement security measures for HRP.
- **Alpha measure 9.** Consult local authorities on the threat and mutual antiterrorism measures as appropriate.
- **Alpha measure 10.** Review all plans and be prepared to implement a higher FPCON.
- **Alpha measure 11 (spare).** Add additional measures as deemed appropriate by the commanders.

FORCE PROTECTION CONDITION BRAVO

F-3. FPCON Bravo applies when an increased or more predictable threat of terrorist activity exists. Sustaining Bravo measures for a prolonged period may affect operational capabilities and impact relations with local authorities. The following bullets explain the FPCON Bravo measures:

- **Bravo measure 1.** Continue, or introduce, all measures in the previous FPCON level.
- **Bravo measure 2.** Check plans for the implementation of the next FPCON level.
- **Bravo measure 3.** Identify critical and high-occupancy buildings. Move cars and objects (such as crates and trash containers) away from buildings to reduce vulnerability to bomb attacks. Consider centralized parking.
- **Bravo measure 4.** Secure and inspect all buildings, rooms, and storage areas not in regular use.
- **Bravo measure 5.** Inspect the interior and exterior of buildings in regular use for suspicious packages at the beginning and end of each workday, as well as at random intervals. Patrol teams search the exterior of high-occupancy buildings at the beginning and end of each day.

Note: Use PEDD to search buildings and areas.

- **Bravo measure 6.** Implement mail-screening procedures to identify suspicious letters and parcels.

Note: Search the mail distribution center with both PEDD and PNDD teams randomly.

- **Bravo measure 7.** Inspect commercial deliveries randomly. Advise family members to check home deliveries.
- **Bravo measure 8.** Increase presence around and surveillance of domestic accommodations, schools, messes, clubs, and other soft targets to improve deterrence and defense and build confidence among staff and dependents.

Note: Randomly patrol and perform walking patrols of populated areas with MWDs.

- **Bravo measure 9.** Implement plans to enhance off-installation security of DOD facilities. In areas with threat levels of moderate, significant, or high, coverage includes facilities (such as DOD schools and daycare centers) and transportation services and routes (such as bus routes) used by DOD employees and family members.
- **Bravo measure 10.** Inform local security committees of actions being taken.
- **Bravo measure 11.** Verify the identity of visitors and randomly inspect their suitcases, parcels, and other containers.
- **Bravo measure 12.** Conduct random patrols to check vehicles, people, and buildings.
- **Bravo measure 13.** Implement additional security measures for HRP.
- **Bravo measure 14.** Identify and brief personnel who may augment guard forces. Review specific rules of engagement including the use of deadly force.

- **Bravo measure 15.** Test the mass notification system.
- **Bravo measure 16.** Verify the identity of all personnel entering buildings as deemed appropriate.
- **Bravo measure 17.** Identify off-installation safe havens for use at the next FPCON.
- **Bravo measure 18.** Review and implement proper OPSEC, communications security (COMSEC), and information security (INFOSEC) status.
- **Bravo measure 19.** Consider using tactical deception techniques.
- **Bravo measure 20 (spare).** Add additional measures as deemed appropriate by the commanders.

FORCE PROTECTION CONDITION CHARLIE

F-4. FPCON Charlie applies when an incident occurs or intelligence is received indicating that some form of terrorist action or targeting against personnel or facilities is likely. Implementing FPCON Charlie measures creates hardship and affects the activities of the unit and its personnel. The following bullets explain the FPCON Charlie measures:

- **Charlie measure 1.** Continue, or introduce, all measures in the previous FPCON.
- **Charlie measure 2.** Ensure that duty personnel are cognizant of actions required to implement antiterrorism and force protection plans. Recall required personnel.
- **Charlie measure 3.** Reduce access control to a minimum and strictly enforce entry. Randomly search vehicles.

Note: Increase the PEDD and PNDD teams' search times at entry points.

- **Charlie measure 4.** Increase standoff from sensitive buildings based on the threat.
- **Charlie measure 5.** Issue weapons and appropriate communication equipment to guards according to SOFAs. Include specific orders on the issue of ammunition.
- **Charlie measure 6.** Increase patrols of the installation to include waterside perimeters, if appropriate.

Note: Use MWD assets to expand the perimeter patrol area and detect intruders.

- **Charlie measure 7.** Implement a barrier plan to prevent a vehicle-borne attack.
- **Charlie measure 8.** Protect all designated vulnerable points. Give special attention to vulnerable points outside the military establishment.
- **Charlie measure 9.** Consult local authorities about closing public (and military) roads and facilities to reduce the vulnerability to attack.
- **Charlie measure 10.** Consider searching suitcases, briefcases, and packages being brought onto the installation through ACPs and consider randomly searching suitcases, briefcases, and packages leaving.

Note: Use PEDD and PNDD teams to search the items above.

- **Charlie measure 11.** Review personnel policy procedures to determine a course of action for family members.
- **Charlie measure 12.** Review access procedures for all non-US personnel and adjust as appropriate.
- **Charlie measure 13.** Consider escorting children to and from DOD schools.
- **Charlie measure 14 (spare).** Add additional measures as deemed appropriate by the commanders.

FORCE PROTECTION CONDITION DELTA

F-5. FPCON Delta applies in the immediate area where a terrorist attack has occurred or when intelligence has been received that terrorist action against a specific location or person is imminent. Normally, FPCON Delta is declared as a localized condition. The following bullets explain the FPCON Delta measures:

- **Delta measure 1.** Continue, or introduce, all measures in the previous FPCON.
- **Delta measure 2**. Augment guards as necessary.
- **Delta measure 3.** Identify all vehicles within operational or mission support areas.
- **Delta measure 4.** Search all vehicles and contents before allowing entrance to the installation.

Note: Use PEDD and PNDD as needed to search vehicles and contents.

- **Delta measure 5.** Control access and implement positive identification of all personnel with no exceptions.
- **Delta measure 6.** Search all suitcases, briefcases, and packages brought into the installation.

Note: Use PEDD and PNDD as needed to search the items above.

- **Delta measure 7.** Escort children to and from DOD schools or close the schools.
- **Delta measure 8.** Make frequent checks of the exterior of buildings and of parking areas. Use MWD teams to make frequent checks of the exterior of buildings and parking areas.
- **Delta measure 9.** Restrict all nonessential movement.
- **Delta measure 10.** Close public and military roads and facilities if permitted.
- **Delta measure (spare).** Add additional measures as deemed appropriate by the commanders.

Appendix G
After-Action Reviews

The AAR process is used to document and discuss MWD operations and training. AARs are focused on performance standards that enable other MWD handlers, training NCOs, KMs, program managers, and concerned agencies to discover what happened, why it happened, and how to sustain strengths and improve on weaknesses. Leaders, planners, and developers can use AARs to get the maximum benefit from every mission or task conducted during predeployment, deployment, and redeployment.

AFTER-ACTION REVIEW PURPOSE

G-1. AARs must provide—

- Candid insights into specific soldier, leader, and MWD strengths and weaknesses from various perspectives.
- Feedback and insight critical to future successes.
- Details that evaluation reports alone often lack.

G-2. The USAMPS MWD Career Management NCO collects historical data, end-of-tour reports, and other organizational specific issues. With the submission of AARs, USAMPS can build a historical database to benefit all in the improvement of Army MWD team tactics, techniques, and procedures. Ultimately, information will be submitted to the Center for Army Lessons Learned (CALL) for distribution to the Army.

AFTER-ACTION REVIEW SUBMISSION

G-3. After the completion of TDY missions, special operations, deployments or other significant events, MWD handlers submit AARs to their respective MACOMs. MACOMs forward the AARs to HQDA Office of the Provost Marshal General (OPMG) and USAMPS.

POINT OF CONTACT

G-4. The point of contact for the MWD Career Management NCO is commercial: (573) 563-8040 and defense switched network (DSN): 676-8040/7946. Send hard-copy correspondence to: US Army Military Police School, ATTN: ATSJ-P, 401 MANSCEN Loop, Suite 1058, Fort Leonard Wood, MO 65473-8926.

EXAMPLE AFTER-ACTION REVIEW

G-5. *Figure G-1,* page G-2, should be used as the format for AARs submitted to the USAMPS MWD Career Management NCO. Refer to *TC 25-20* for comprehensive guidance on AARs.

DEPARTMENT OF THE ARMY
42^{D} Military Police Detachment
Ft. Bragg, NC 27380

OFFICE SYMBOL

DATE

MEMORANDUM FOR RECORD

SUBJECT: After-Action Review, Task Force Iron Dog, Sep 03 – Nov 03

1. State what was supposed to happen: This paragraph contains a brief introduction of the mission that was being performed by the MWD team or section. Include operation objectives, the overall mission, and the intent from the task force commander as well as any other relevant information.

2. State what happened: This paragraph contains pertinent information about what actually happened during the mission. Was the mission accomplished as stated above or were other objectives established inside of the area of operation?

3. Determine what went right and wrong during the mission:

   a. Topic: What was right and wrong with what happened during the mission.

   b. Discussion: Give the who, what, when, where, and how for the mission. Give a good summary of what the issue was.

   c. Recommendation: Give recommendations to improve the issues that were identified as needing improvement.

4. Point of contact for this memorandum is the undersigned at DSN: 123-4567 or jon.doe@us.army.mil.

Jon Doe
SSG, USA
Kennel Master / PEDD Handler

Note: If you have a positive topic just use Topic and Discussion sections.

Figure G-1. Sample AAR Format

Appendix H

Health and Welfare Inspections Briefing Guide

Commanders frequently request MWD support during health and welfare inspections. Commanders should be made aware that the conduct and responsibility of health and welfare inspections lies within their purview and that the MWD team is not the authority for the mission. This appendix identifies the actions that need to be addressed with the commander to ensure a successful and lawful health and welfare inspection. These actions are needed to establish and act on probable cause determinations. Questions concerning these matters should be directed to the local Staff Judge Advocate Office for clarification.

COORDINATION

H-1. The KM coordinates and approves each health and welfare inspection. After the KM receives the request for assistance, he assesses the needs of the mission and determines which detector dog team (if any) will assist the commander. The KM notifies the unit commander or 1SG and accepts or denies the mission. Once the mission is accepted, the KM briefs them on their roles and responsibilities and informs them that if his criteria are not met, the KM will terminate the search.

HANDLERS BRIEFING

H-2. Prior to the inspection, the KM, plans NCO, or handler ensures that the commander requesting support is aware that the building being inspected must be unoccupied and that all doors are unlocked (or keys available on site). On the arrival of the MWD team at the location of the health and welfare inspection, the handler performs a commander's briefing and demonstration. The handler—

- Explains the probable cause folder to the commander.
- Shows the commander the training aids used for detection.
- Demonstrates the dog's ability to detect.

H-3. A briefing document shows handlers how to record the details of each health and welfare inspection (*Figure H-1*, page H-3). The commander is provided with a copy of the briefing document containing the following information:

- The MWD teams used for the inspection.
- The canines' names and tattoo numbers.
- The scope of the proficiency training of the MWD teams. Discuss with the commander the standards and scope of the detector training required to maintain the certification of assigned MWD teams.
- The MWDs' accuracy rates. Explain to the commander that the accuracy percentage rate is based on quarterly training.
- The MWDs' detection experience. Inform the commander of the amount of time that the MWDs have actually spent conducting detection missions. Include any possible finds in prior inspections.
- The MWDs' capabilities. Explain to the commander that the MWDs are capable of detecting either drugs or explosives.

- The MWDs' alert response. Describe what the MWDs will do when they indicate the presence or odor of found drugs or explosives.
- The unit representatives. Notify the commander that one SSG or above is required to escort each MWD team through out the inspection. The commander may not be a unit representative.

UNIT REPRESENTATIVE DUTIES AND RESPONSIBILITIES

H-4. Ensure that the unit representatives are informed that they—

- Escort the MWD team at all times.
- Unlock doors or areas as needed.
- Carefully observe the MWD team.
- Follow the handler's instructions.
- Obey safety precautions.
- Spot areas that are missed by the MWD team.

DOCUMENTATION

H-5. It is important that the MWD handler documents the commander's briefing on every occasion. *Figure H-1*, page H-3, gives an example for documenting the briefing given to the commander before a health and welfare inspection. Advise the commander to maintain the handlers briefing and other documentation concerning each response of the MWD. Advise the commander that a positive response by a narcotic detector dog provides probable cause for a command-directed urinalysis. Ensure that the commander signs the briefing form and then provide a copy for filing in case of a judicial action.

INSPECTION

H-6. After the briefing has been conducted, the inspection may begin. Remember to advise the commander and unit representatives that this health and welfare inspection is the unit's mission and that the MWD team is assisting in the mission. The MWD team is not in charge of the health and welfare inspection but can be used as a tool by the commander.

ATZP-PM-K DATE: ____________

SUBJECT: Commander's Briefing for Health and Welfare Inspections Utilizing Detector Dogs

1. (commander's name) has requested narcotics detector dog support for a health and welfare inspection on (dd/mm/yyyy).

2. Prior to canine(s) being utilized, the building must be unoccupied and all doors must be unlocked (or keys available).

3. The following provides the commander with information essential to a successful inspection:

_____ A. Team(s) to be utilized: (number of PEDD and PNDD to be used).

_____ B. MWD name(s) and tattoo number(s): (name and tattoo number for each MWD)

_____ C. The scope of formal training of the canine team(s). Discuss detector training. Discuss the standards that must be successfully completed to maintain certification of detector dog team(s).

_____ D. The canine's accuracy rate(s) is/are (accuracy rates for each MWD). This is based on the previous 3 months training records.

_____ E. The extent of actual MWD detection field experience (experience for each MWD)

_____ F. Explain the types of drugs or explosives that the canine(s) is/are trained to detect.

_____ G. Describe the canine's specific behavior(s)/actions when it responds, indicating the presence or the odor of narcotics (detector response of each MWD).

_____ H. Ensure that the commander(s) is/are aware that one representatives (SSG or above) is needed to act as an escort for each team. The commander may not be one of the individuals accompanying canine team(s).

_____ I. Ensure that the escorts are briefed on their duties and responsibilities. Discuss the safety precautions that they must follow, and ensure the safety of others.

_____ J. Ensure that the commander(s) is/are aware that the unit is responsible for maintaining proper documentation (date, time, room number, common or personal area and occupant[s] name[s]), concerning the response of the canine(s). This information should be kept on file with the commander's copy of this briefing sheet, in case of a judicial action.

_____K. Ensure that the commander is aware that a positive response by a narcotic detector dog is probable cause for a command-directed urinalysis on the occupant or occupants of that room.

I (commander's name) have received a personal briefing covering all the information stated above and now understand the capabilities of the canine team(s). I am satisfied that the canine team(s) have the training and experience necessary to provide reliable information upon which to authorize a search and/or perform a command-directed urinalysis. I authorize the canine team(s) to inspect the following area(s): (areas to be inspected)

Figure H-1. Health and Welfare Briefing Document Sample

HANDLER(S) SIGNATURE

1. ______________________________ 2. ______________________________

3. ______________________________ 4. ______________________________

COMMANDER'S INFORMATION	**1SG'S INFORMATION**
NAME: ______________________	NAME: ______________________
RANK: ______________________	RANK: ______________________
UNIT: ______________________	UNIT: ______________________
SIGNATURE: ______________________	SIGNATURE: ______________________
UNIT REPRESENTATIVE	**UNIT REPRESENTATIVE**
NAME: ______________________	NAME: ______________________
RANK: ______________________	RANK: ______________________
UNIT: ______________________	UNIT: ______________________
SIGNATURE: ______________________	SIGNATURE: ______________________

Figure H-1. Health and Welfare Briefing Document Sample (Continued)

Appendix I

Terrorist Bomb Threat Stand-off Distances

Table I-1 contains information from the DOD technical support working group. Commanders, PMs, and MWD planners can use this information as a guide to determine the minimum distances necessary to evacuate personnel from suspected explosive threats.

Table I-1. Minimum Distances for Personnel Evacuation

Threat Description	***Explosives Capacity* (TNT Equivalent)***	***Building Evacuation Distance*****	***Outdoor Evacuation Distance******
Pipe bomb	5 lbs/2.3 kg	70 ft/21 m	850 ft/259 m
Briefcase/suitcase bomb	50 lbs/23 kg	150 ft/46 m	1,850 ft/564 m
Compact sedan	500 lbs/227 kg	320 ft/98 m	1,500 ft/457 m
Sedan	1,000 lbs/454 kg	400 ft/122 m	1,750 ft/534 m
Passenger/cargo van	4,000 lbs/1,814 kg	640 ft/195 m	2,750 ft/838 m
Small moving van/delivery truck	10,000 lbs/4,536 kg	860 ft/263 m	3,750 ft/1,143 m
Moving van/water truck	30,000 lbs/13,608 kg	1,240 ft/375 m	6,500 ft/1,982 m
Semitrailer	60,000 lbs/27,216 kg	1,570 ft/475 m	7,000 ft/2,134 m

* Based on maximum volume or weight of explosives (TNT equivalent).

** Governed by the ability of an unstrengthened building to withstand severe damage or collapse.

*** Governed by either the fragment throw distance or glass breakage/falling glass hazard distance (whichever throws fragments farther than vehicle bombs).

NOTES:

1. All personnel must either seek shelter inside a building away from windows and exterior walls (with some risk) or move beyond the outdoor evacuation distance.
2. The preferred area is beyond the outdoor evacuation distance for the evacuation of personnel in buildings. This is mandatory for personnel outdoors.

Glossary

Acronyms and Abbreviations

1SG	first sergeant
AAR	after-action review
ACP	access control point
AF	Air Force
agg	aggression
AR	Army regulation
ASG	area support group
ASI	additional skill identifier
ASP	ammunition supply point
ATTN	attention
BDU	battle dress uniform
C2	command and control
C4	composition 4
CA	corrective action
CALL	Center for Army Lessons Learned
CCTV	Closed-circuit television
cert	certification
CI	civilian internee
CID	Criminal Investigation Division
clg	cleaning
CMFO	centralized furnishings management office
comm	communications
COMSEC	communications security
CONEX	container express
cont	continuted
CP	command post
CPT	captain
CR	corrective response
DA	Department of the Army
d	deep
DARE	Drug Abuse Resistance Education
DCSPER	Dputy Chief of Stagg for Personnel
DC	detonating cord
DD	Department of Defense
DEA	Drug Enforcement Administration

Dec	December
det	detonating
DF	deficiency
DOD	Department of Defense
DODMWDVS	Department of Defense Military Working Dog Veterinary Services
DSN	defense switched network
DST	drug suppression team
EAC	echelon above corps
EOD	explosive ordnance disposal
EPW	enemy prisoner of war
eval	evaluaiton
FM	field manual
FMO	Furnishing managemetn office
FPCON	force protection condition
ft	foot; feet
GE	Germany
GOV	government owned vehicle
GSA	General Services Administration
h	high
HF	hole, variable
HRP	high-risk personnel
HRT	high-risk target
HVAC	heating, ventilation, and air conditioning
I/R	internment and resettlement
IDS	Intrusion Detection System
IED	improvised explosive device
INFOSEC	information security
IPB	intelligence preparation of the battlefield
iso	isolation
kg	kilogram
KM	kennel master
LAFB	Lackland Air Force base
LAN	local area network
LBS	pounds
LBV	load-bearing vest
LCE	load-carrying equipment
LED	light-emitting diode
LIN	line item number
LTC	lieutenent colonel
m	meter

MACOM	major Army command
MANSCEN	Maneuver Support Center
MCI	military customs inspector
MCIPM	military customs inspector program manager
MCM	Manual for Courts-Martial
mech	mechanical
METL	mission-essential task list
METT-TC	mission, enemy, terrain, troops, time available, and civilian considerations
MEVA	mission-essential vulnerable areas
MILSTRIP	military standard transportation and issue procedures
min	minimum
MMSO	maneuver mobility support operations
MOPP	mission-oriented protective posture
MPR	military police report
MSR	main supply route
MTOE	modification table of organization and equipment
MWD	military working dog
MWDMS	Military Working Dog Management System
NBC	nuclear, biological, and chemical
NCO	noncommissioned officer
NCOIC	noncommissioned officer in charge
no	number
Nov	November
NSN	national stock number
OB	obstacle
Oct	October
ODCSOPS	Office of the Deputy Chief of Staff for Operaitons and Plans
off/L	off leash
on/L	on leash
OIF	Operation Iraqi Freedom
OPMG	Office of the Provost Marshal General
OPSEC	operational security
pam	pamphlet
PC	potassium chlorate
PD	patrol dog
PEDD	patrol explosive detector dog
PIO	police intelligence operations
PM	provost marshal
PMO	provost marshal office

PNDD	patrol narcotic detector dog
POV	privately owned vehicle
PT	physical training
prep	preparation
RAMP	random antiterrorism measures program
rec	recreation
S2	intelligence section
SC	sodium cholorate
sec	seconds
Sep	September
SF	standard form
SFC	sergeant first class
SGT	sergeant
SOFA	Status of Forces Agreement
SOP	standard operating procedures
SP	smokeless powder
SRT	special-reaction team
SSG	staff sergeant
STX	Semtex
TDA	table of distribution and allowancees
TDY	temporary duty
TISA	troop issue subsistence activity
TMP	transportation motor pool
TNT	trinitrotoluene
TOE	table of organization and equipment
TRADOC	United States Army Training and Doctrine Command
Tues	Tuesday
US	United States
USA	United States of America
USC	United States Code
USCS	United States Customs Service
USSS	United States Secret Service
v	version
VCO	veterinary corps officer
vet	veterinarian
VIP	very important person
VTF	Veterinary Treatment Facility
VTR	Veterinary Treatment Record
WT	weight

References

DA Forms are available on the Army Publishing Directorate web site <http://www.apd.army.mil>. DD forms are available from the Office of the Secretary of Defense web site <http://www.dtic.mil/whs/directives/infomgt/forms/formsprogram.htm>.

SOURCES USED

These are the sources quoted or paraphrased in this publication.

AR 190-12. *Military Working Dogs.* 30 September 1993.

AR 190-14. *Carrying of Firearms and Use of Force for Law Enforcement and Security Duties.* 12 March 1993.

DA Pam 190-12. *Military Working Dog Program.* 30 September 1993.

DA Pam 611-21. *Military Occupational Classification and Structure.* 31 March 1999.

DOCUMENTS NEEDED

These documents must be available to the intended users of this publication.

AR 700-81. *DOD Dog Program.* 5 May 1971.

DA Form 2807-R. *Military Working Dog Training and Utilization Record.*

DA Form 3992–R. *Narcotics or Explosives Detector Dog Training and Utilization Record.*

FM 3-19.15. *Civil Disturbance Operations.* 18 April 2005.

READINGS RECOMMENDED

These sources contain relevant supplemental information.

AF Form 1256. *Certificate of Training.*

AR 190-47. *The Army Corrections System.* 5 April 2004.

AR 190-8. *Enemy Prisoners of War, Retained Personnel, Civilian Internees and Other Detainees.* 1 October 1997.

AR 350-1. *Army Training and Education.* 9 April 2003.

AR 40-1. *Composition, Mission, and Functions of the Army Medical Department.* 1 July 1983.

AR 40-3. *Medical, Dental, and Veterinary Care.* 12 November 2002.

AR 40-905. *Veterinary Health Services.* 16 August 1994.

AR 525-13. *Antiterrorism.* 4 January 2002.

AR 600-8-101. *Personnel Processing (In-, Out-, Soldier Readiness, Mobilization, and Deployment Processing).* 15 July 2003.

AR 600-9. *The Army Weight Control Program.* 10 June 1987.

AR 614-200. *Enlisted Assignments and Utilization Management.* 30 September 2004.

DA Form 2028. *Recommended Changes to Publications and Blank Forms.*

DA Form 3161. *Request for Issue or Turn-In.*

DA Form 5513-R. *Key Control Register and Inventory.*

DA Form 6. *Duty Roster*

DA Form 7281-R. *Command Oriented Arms, Ammunition, & Explosives Security Screening and Evaluation Record.*

DA Pam 710-2-1. *Using Unit Supply System (Manual Procedures).* 31 December 1997.

DD Form 1834. *Military Working Dog Service Record.*

DEA Form 223. *Controlled Substance Registration Certificate.*

DOD Directive 2000.12. *DoD Antiterrorism (AT) Program.* 18 August 2003.

DOD Directive 5030.49. *DoD Customs and Border Clearance Program.* 4 May 2004.

DOD Directive 6400.4. *DoD Veterinary Services Program.* 22 August 2003.

DOD Instruction 3025.12. *Military Assistance for Civil Disturbances (MACDIS).* 4 February 1994

FM 21-20. *Physical Fitness Training.* 30 September 1992.

FM 27-10. *The Law of Land Warfare.* 18 July 1956.

FM 3-19.40. *Military Policy Internment/Resettlement Operations.* 1 August 2001.

Lackland AFB Form 375. *Patrol Dog Certification.*

Lackland AFB Form 375a. *Detector Dog Certification.*

MCM, Part III Military Rules of Evidence; Rule 311, *Evidence obtained from unlawful searches and seizures.* 2000.

MCM, Part III Military Rules of Evidence; Rule 313, *Inspections and inventories in the armed forces.* 2000.

MCM, Part III Military Rules of Evidence; Rule 314, *Searches not requiring probable cause.* 2000.

MCM, Part III Military Rules of Evidence; Rule 315, *Probable cause searches.* 2000.

MCM, Part III Military Rules of Evidence; Rule 315(f)(2), *Probable cause determination.* 2000.

MCM, Part III Military Rules of Evidence; Rule 316, *Seizures.* 2000.

SF 702. *Security Container Check Sheet.*

TC 19-210. *Access Control Handbook.* 4 October 2004.

TC 25-20. *A Leader's Guide to After Action Reviews.* 30 September 1993.

USC Title 18, Crimes and Criminal Procedure; Part I, Crimes; Section 1385, *Use of Army and Air Force as posse comitatus.*

Index

Made in the USA
Monee, IL
08 November 2026

68028159R00243